Emerging Trends in Mechatronics

Series Editor

Aydin Azizi, Oxford, UK

This book series helps understand how Mechatronics will impact the practice and research of developing advanced techniques to model, control and optimize complex systems. It presents recent state of advances in mechatronics and related technologies including: Automatic Control, Robotics, Agent Based systems, Smart Manufacturing and Industry 4.0. The selected topics give an overview of the state-of-the-art and present new research results and prospects of the future development in the interdisciplinary field of mechatronic systems. The book series will provide up-to-date and useful knowledge for researchers and engineers involved in mechatronics and related fields.

Within the above dimensions, the scope of the book series welcomes high quality original research and review articles that cover a broad range of topics related to theoretical and applied aspects of modern mechatronics. Potential topics include, but are not limited to the following:

Controller design

Many mechatronic systems have challenging control difficulties due to the high nonlinear structures and time-varying dynamic behaviors. In addition to these, there are external disturbances, which cannot be predicted and change according to uncontrolled working environments. Conventional controllers are insufficient to solve the problems and to compensate the environmental disturbances, therefore designing efficient controllers for complex systems has a great importance in the current book series.

Sensor Design & Data Collection Approaches

Essential part of a mechatronics system is the measurement system that senses the variations in the physical parameters, such as temperature, pressure, displacement, and so on, and converts it to voltage or current. In mechatronic systems, accurate measurement of the dynamic variables plays a vital role for the actuators to function properly, therefore on of the aims of this book series, to focus on understanding what should be done to design accurate sensors and efficient data collection methods to improve the performance of mechatronic systems.

Optimization Techniques

Designing a complex mechatronic system involves multiple variables, objectives, constraints, and evaluation criteria as well as their nonlinearly coupled relationships. One of the challenging research areas of this book series is utilizing optimization techniques to design an optimal mechatronic system.

Artificial Intelligence

From the point of view of dynamics, mechatronic systems can be characterized by model uncertainties, high nonlinearities, complicated coupling, and stringent performance requirements, among others. To deal with these facts, this book series would like to focus on investigating and utilizing Artificial Intelligence techniques such as Neural Networks, Fuzzy logic, Neuro-fuzzy, evolutionary techniques etc. to model, design and optimize complex mechatronic systems.

Robotics

In the field of Robotics, there is a need of focusing on kinematic, dynamic, control and optimization of innovative robotic systems. This book series can introduce new and novel robot designs such as hybrid (serial-parallel) redundant manipulators that have a high potential of application in special fields like space robots. State of the art researches on selection of optimum kinematic solutions and path planning based on such an optimization, dynamic analysis and new control methods developments and application of artificial intelligence in this field can be published in this book series.

Smart manufacturing

Application of networked sensors and machinery equipment using internet and intranet in order to monitor and automation of the manufacturing process and improve the performance efficiency and is a point of interest. Advanced robotics, multi input-multi output control approaches combined with the artificial intelligence and optimization techniques, big data processing approaches, multi-scale modelling and simulation of dynamic sub-systems etc. are the hot topics that will be focused in this book series.

AGVs

Automated Guided Vehicle (AGVs) are going to be very popular is industry logistics, surveillance, defense and social applications in order to make processes safer and more cost efficient. Innovative AGVs designs for various missions, Robot Operating System usage on AGV control, Indoor and outdoor navigation and control algorithms, intelligent path planning techniques, sensor fusion, simultaneous application of GPS, INS and offline maps are the topics that are planned to be published in this book series.

Aydin Azizi
Editor

Autonomous Robotics and Intelligent Systems

Springer

Editor
Aydin Azizi
School of Engineering, Computing
and Mathematics
Oxford Brookes University
Oxford, UK

ISSN 2731-4855 ISSN 2731-4863 (electronic)
Emerging Trends in Mechatronics
ISBN 978-981-95-5866-7 ISBN 978-981-95-5867-4 (eBook)
https://doi.org/10.1007/978-981-95-5867-4

This Springer imprint is published by the registered company Springer Nature Singapore Pte Ltd.
The registered company address is: 152 Beach Road, #21-01/04 Gateway East, Singapore 189721, Singapore

Preface

The proposed volume, *'Autonomous Robotics and Intelligent Systems'*, brings together a rich collection of innovative research and applied engineering work from diverse fields within robotics, control systems, and intelligent computation. The aim is to offer readers a comprehensive perspective on the latest technological advancements, design methodologies, and real-world applications that define the current state and future trajectory of autonomous systems. This book is envisioned as both a technical reference and an inspirational resource, providing insights that can fuel academic, industrial, and research-based innovation. It bridges multiple disciplines including mechanical engineering, electronics, artificial intelligence, mechatronics, and energy systems reflecting the inherently interdisciplinary nature of modern robotics. The work not only captures theoretical principles but also illustrates how these principles can be applied in practical, impactful ways. From intelligent navigation and human–robot interaction to advanced control algorithms and sustainable robotic solutions, the book addresses critical challenges faced by researchers and practitioners while opening doors to new possibilities.

Oxford, UK

Aydin Azizi

About This Book

The structure of *'Autonomous Robotics and Intelligent Systems'* is designed to guide readers on a journey from foundational concepts to sophisticated, real-world implementations. The chapters follow a logical flow, starting with navigation and positioning technologies for unmanned vehicles, advancing to intelligent communication between humans and robots, and moving into computational algorithms that drive system optimization. Readers will also encounter applications aimed at improving road safety, stabilizing dynamic systems, enhancing rescue operations, and refining vehicular suspension for improved performance. A strong emphasis is placed on sustainability and efficiency, culminating in designs that integrate renewable energy solutions, such as autonomous systems for solar panel maintenance. Each contribution offers a deep dive into the subject matter, backed by methodical analysis, experimental results, and case studies that demonstrate measurable outcomes. The book is aimed at engineers, researchers, graduate students, and technology developers who seek a blend of conceptual clarity, technical detail, and practical applicability. By presenting diverse topics in a cohesive framework, it encourages interdisciplinary thinking and fosters the exchange of ideas that can lead to groundbreaking developments in robotics and intelligent systems.

Proposed Chapters

- Improved Positioning of an Unmanned Vehicle by Distance-Based Differential GPS
- Interactive Human–Robot Integration: Enhancing Communication and Movement with AI-Driven Responses
- Road Guardian: A Technology for Preventing Impaired Driving
- Genetic Algorithm Based Stabilization of Rotary Inverted Pendulum by State Feedback
- Tracked Rescue Robot for Emergency Search and Surveillance Tasks
- Design and Analysis of Fuzzy Logic Controlled Active Suspension

- Design an Active Suspension System Utilising Intelligent Control Techniques
- Design and Development of an Autonomous Solar Panel Cleaning Robot to Enhance Photovoltaic Efficiency

Contents

Improved Positioning of an Unmanned Vehicle by Distance-Based Differential GPS

Matin Ghaziani, E. Ilhan Konukseven, and A. Bugra Koku

Abstract To navigate an Unmanned Ground Vehicle (UGV) in an unstructured area, different methods exist, such as Global Positioning System (GPS), Inertial Navigation System (INS), integration of INS and GPS, and Differential GPS (DGPS) . GPS is a commonly used method to find the coordinates of any target. Due to the inherent low accuracy of GPS, utilizing a single GPS in a UGV leads to inaccurate real-time navigation. One way to overcome this problem is to use the DGPS method. The DGPS method employs one or several base stations (reference stations) at an accurately known location. In the DGPS method, prior knowledge of the coordinates of the base stations is required. This method is highly dependent on the accurate coordinates. This paper proposes a distance-based DGPS method with seven fixed GPS receivers and a UGV receiver. Unlike DGPS, this method does not rely on the exact coordinates of reference stations, allowing this method to be applied everywhere. Hence, the establishment of a local navigation system becomes easily attainable. The results of this study show that the error correction of the proposed method is at 90% and enables seamless navigation for a UGV.

Keywords DGPS · Base station · Fixed GPS receivers · UGV

E. Ilhan Konukseven and A. Bugra Koku are contributed equally to this work.

M. Ghaziani (✉)
Aircraft Maintenance and Repair Department, University of the Turkish Aeronautical Association (UTAA), Ankara, Turkey
e-mail: mghaziani@thk.edu.tr

M. Ghaziani · E. Ilhan Konukseven · A. Bugra Koku
Mechanical Engineering Department, Middle East Technical University (METU), Universiteler Mahallesi, Ankara, Turkey
e-mail: konuk@metu.edu.tr

A. Bugra Koku
e-mail: kbugra@metu.edu.tr

A. Azizi (ed.), *Autonomous Robotics and Intelligent Systems*, Emerging Trends in Mechatronics, https://doi.org/10.1007/978-981-95-5867-4_1

1 Introduction

GPS, an integral component of the satellite-based navigation system developed by the U.S. Department of Defense under its NAVSTAR satellite program [1–6], is subject to inherent error sources that demand consideration when interpreting signals from the constellation of orbiting satellites. Errors in GPS signals emanate from satellite-related issues, encompassing orbital (ephemeris) and clock errors. Receiver errors arising from receiver clock disparities, antenna phase center movement, and multi-path imaging contribute to the complexity of accurate GPS readings. Additionally, signal propagation errors result from ionospheric and tropospheric refraction. These errors exhibit uniformity for users within proximity to the base station.

More precise coordinates using data from a single GPS receiver are feasible with performance augmentation techniques like DGPS. DGPS entails the establishment of highly accurate coordinates relative to a base station, often within millimeters, by incorporating an additional GPS receiver at the base station to measure signals. The base station's exact coordinates, known with precision [7], allow the computation of errors in received GPS signals. In DGPS, these errors are then transferred to the user's GPS, serving as corrections to enhance measurement accuracy and compensate for discrepancies at the user's end [8]. However, DGPS systems are limited by the broadcasting range of correction signals, confining their efficacy to a specific area around the base station. Beyond this region, errors become decor-related, diminishing correction accuracy. To address this limitation, a network of base stations equipped with a communication link serving its geographic area is employed, necessitating multiple base stations and associated communication links [7, 9, 10].

DGPS systems are broadly categorized into local and wide area DGPS [7]. Local Area DGPS (LAGPS) involves a user's GPS receiver receiving real-time corrections from a local reference receiver within the line of sight. Conversely, Wide Area DGPS (WADGPS) receives corrections from a base station network distributed over a broader geographic expanse. These corrections address specific error sources, including satellite clock variations, ionospheric propagation delays, and ephemeris discrepancies. Both categories require base stations at precisely surveyed locations. In navigating a UGV at an unknown location, the base and UGV GPS receivers track identical GPS satellites to calculate positions. Utilizing the DGPS method necessitates obtaining corrections from centers near the UGV, such as Trimble, which monitors a vast network of terrestrial reference stations globally [11]. The dispersion of DGPS centers worldwide makes it challenging to access these services everywhere, prompting the need for a method independent of DGPS centers [12].

This work introduces a method that operates autonomously of a base station and can be implemented universally. Seven GPS receivers, with known inter-distances, are strategically placed at the UGV navigation site. The errors of these fixed GPS receivers are employed to correct the UGV's coordinates. The findings indicate that this local navigation system is adaptable and does not hinge on the precise positioning of base GPS receivers. The methodology is expounded upon in the subsequent section, followed by the presentation of results in Sect. 3.

2 Methodology

The proposed DGPS method hinges on the predetermined distances between a series of fixed receivers positioned at the site and a GPS receiver affixed to the UGV. The fixed stations are denoted by numbers 1 through 7, with the UGV serving as station 8, as illustrated in Fig. 1. This section will elaborate on the rationale behind utilizing seven fixed GPS receivers. Unlike conventional DGPS methods that heavily rely on reference point coordinates (base stations) [12], this approach operates independently of such reference points. It assumes that all pertinent conditions, such as the number of satellites in view and Dilution of Precision (DOP), remain consistent, and real-time constraints are not decisive in its application.

A geographic coordinate system serves as the reference framework for specifying a location on Earth, typically employing coordinates that represent vertical and horizontal positions. This system commonly chooses latitude, longitude, and altitude as coordinates. These coordinates play a crucial role in calculating the distance between two points on Earth, contributing to both vertical and horizontal positions. Determining accurate distances requires deploying GPS receivers at specific points, yet the computed distance is inevitably subject to partial accuracy due to inherent GPS errors. Further refinement is imperative to enhance the precision of the calculated distance. When the distance between two points is known, the Pythagorean theorem becomes a valuable tool for identifying errors in GPS receivers [12]. Utilizing the Pythagorean theorem necessitates the conversion of GPS coordinates from longitude, latitude, and altitude (LLA) coordinates to Earth-centered Earth-fixed (ECEF) coordinates [13]. The conversion process is succinctly expressed in Eqs. (1–3):

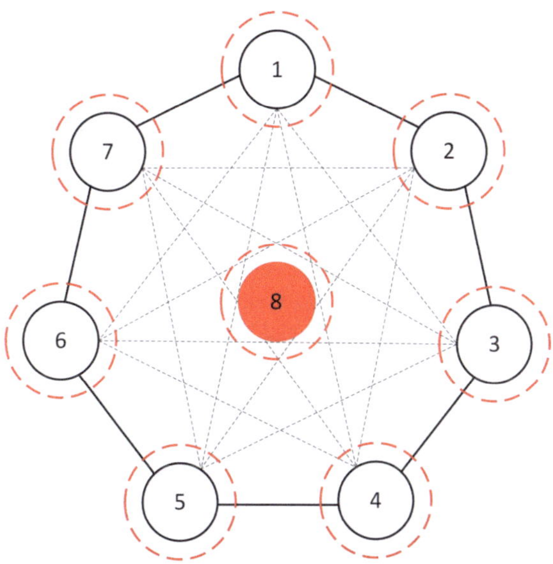

Fig. 1 The hexagon shape of GPS receivers is set with the UGV with the distance between fixed receivers. The dashed red circle shows the possible error of GPS receivers

$$X = (N + h)\cos\varphi\cos\lambda \tag{1}$$

$$Y = (N + h)\cos\varphi\sin\lambda \tag{2}$$

$$Z = \left(\frac{b^2}{a^2}N + h\right)\sin\varphi \tag{3}$$

Here, X, Y, and Z represent coordinates in the ECEF coordinate system, while φ, λ, and h denote latitude, longitude, and altitude. The equatorial and polar radii, designated by a and b, and the curvature radius N (as per Eq. (4)), contribute to this conversion:

$$N = \frac{a}{\sqrt{1 - e^2\sin^2\varphi}} \tag{4}$$

where e represents the eccentricity of the ellipsoid; any location on Earth could be specified by using longitude, latitude, and altitude coordinates. Due to GPS errors, the location position needs to be more accurate. Each GPS position data has an error in each dimension: error in longitude (ε_x), latitude (ε_y), and altitude (ε_z). Therefore, the precise position of the GPS receiver, by integrating the error, could be represented as:

$$pos_{xyz} = \begin{pmatrix} X + \varepsilon_x \\ Y + \varepsilon_y \\ Z + \varepsilon_z \end{pmatrix} \tag{5}$$

By substituting Eq. (5) into the Pythagorean theorem, the distance calculation between two points involves six known and six unknown variables, as expressed in Eq. (6).

$$d_{12} = \sqrt{\begin{array}{l} ((X_2 + \varepsilon_{x2}) - (X_1 + \varepsilon_{x1}))^2 + \left(\left(Y_2 + \varepsilon_{y2}\right) - \left(Y_1 + \varepsilon_{y1}\right)\right)^2 \\ + ((Z_2 + \varepsilon_{z2}) - (Z_1 + \varepsilon_{z1}))^2 \end{array}} \tag{6}$$

However, when the distance between two points is known, the application of this equation introduces a complication, as it yields multiple solutions. This diversity poses a challenge in precisely identifying the errors inherent in GPS devices. To surmount this issue, a foundational step involves establishing a set of system equations. Equation (6) serves as the cornerstone, encapsulating the distance equation between two points with six unknowns. It is imperative to achieve an equilibrium between equations and unknowns to model the system comprehensively and yield a unique solution. Considering n as the number of points, the total number of equations becomes (n 2), while the number of unknowns equals $3n$. Consequently, the system necessitates the formulation of seven equations to achieve a well-defined solution.

$$\binom{n}{2} = 3n \Rightarrow \binom{n}{2} - 3n = 0 \Rightarrow \frac{n!}{(2!(n-2)!)} - 3n = 0 \Rightarrow n = 7 \tag{7}$$

In the pursuit of accurately estimating the coordinates of the UGV, a systematic approach is adopted, as dictated by Eq. (7). This method strategically deploys seven fixed GPS receivers, precisely aligning the number of unknowns with the number of equations. This meticulous balance is integral to ensuring the derivation of a singular solution within the system of equations, thereby delineating a distinctive heptagon shape, as visually depicted in Fig. 1. In a more generalized representation, the computation of the length of each side of this heptagon can be expressed as follows:

$$d_{ij} = \sqrt{\begin{array}{l}\left((X_j + \varepsilon_{xj}) - (X_i + \varepsilon_{xi})\right)^2 + \left((Y_j + \varepsilon_{yj}) - (Y_i + \varepsilon_{yi})\right)^2 \\ +\left((Z_j + \varepsilon_{zj}) - (Z_i + \varepsilon_{zi})\right)^2\end{array}} \tag{8}$$

where d_{ij} is the distance between two known GPS receivers' positions i and j. X_i, Y_i, and Z_i represent the positions of the first receiver position in ECEF coordinates, read from the GPS receiver, while X_j, Y_j, and Z_j denote the positions of the second receiver position. ε_{xi}, ε_{yi}, and ε_{zi} are the errors of the first receiver position, and ε_{xj}, ε_{yj}, and ε_{zj} are the errors of the second receiver position.

Solving Eq. (8) reveals that One UGV is deployed with seven fixed receivers across the unconstructed area, as depicted in Fig. 1. The system model comprises 21 equations with 21 unknowns. After solving the model, the solutions are utilized to eliminate errors in the DGPS coordination. Figure 1 displays a heptagon with a GPS located at each corner, and the equations of this GPS topology are outlined as follows:

$$F = \begin{cases} d_{12} = \sqrt{\begin{array}{l}((X_2 + \varepsilon_{x2}) - (X_1 + \varepsilon_{x1}))^2 + \left((Y_2 + \varepsilon_{y2}) - (Y_1 + \varepsilon_{y1})\right)^2 \\ + ((Z_2 + \varepsilon_{z2}) - (Z_1 + \varepsilon_{z1}))^2\end{array}} \\ d_{13} = \sqrt{\begin{array}{l}((X_3 + \varepsilon_{x3}) - (X_1 + \varepsilon_{x1}))^2 + \left((Y_3 + \varepsilon_{y3}) - (Y_1 + \varepsilon_{y1})\right)^2 \\ + ((Z_3 + \varepsilon_{z3}) - (Z_1 + \varepsilon_{z1}))^2\end{array}} \\ \vdots \\ d_{67} = \sqrt{\begin{array}{l}((X_7 + \varepsilon_{x7}) - (X_6 + \varepsilon_{x6}))^2 + \left((Y_7 + \varepsilon_{y7}) - (Y_6 + \varepsilon_{y6})\right)^2 \\ + ((Z_7 + \varepsilon_{z7}) - (Z_6 + \varepsilon_{z6}))^2\end{array}} \end{cases} \tag{9}$$

The equations are systematically solved using Newton's algorithm, as extensively detailed in the references [12, 14]. This computational process yields the errors associated with each GPS receiver, succinctly represented as $(\varepsilon_{xi}, \varepsilon_{yi}, \varepsilon_{zi})$. Subsequently, armed with the identified errors in the fixed GPS receivers, a refinement of the UGV position becomes possible. Each fixed GPS receiver assumes the role of a virtual base station, effectively mitigating the error inherent in the UGV's position.

The accuracy of GPS receivers is intricately linked to the geometry of GPS satellites, as expounded upon in reference [15]. When two GPS receivers are situated nearby, their errors exhibit similarities. This noteworthy phenomenon underscores a crucial consideration—the accuracy of the UGV's position is markedly influenced by the distance between the UGV and the fixed receivers. Consequently, the Inverse Distance Weighting (IDW) method, meticulously outlined in reference [16], is enlisted to correct the UGV's position. IDW, a deterministic method for multivariate interpolation, leverages a known scattered set of points to calculate the assigned value of an unknown point, employing a weighted average of values at known points. This methodology entails assigning weights to the errors of fixed GPS receivers, thereby revising the UGV's position concerning these weighted errors. Essentially, if the UGV is close to a fixed GPS receiver, the error of that specific receiver holds more influence in determining the UGV's position.

An additional variable introduced by the IDW method is the distance between known and unknown points. Consequently, the distances from fixed GPS receivers to the UGV are meticulously computed. Despite the inherent potential inaccuracies introduced by the real-time GPS data of the UGV, these errors are intentionally disregarded in applying the IDW method. Instead, the distances from the corrected fixed GPS receivers to the real-time position of the UGV are judiciously employed for precise positioning, as vividly illustrated in Fig. 2.

Each GPS receiver exhibits three errors (ε_x, ε_y, ε_z). Therefore, to determine the weights of the IDW method in three dimensions (X, Y, Z), the following is considered:

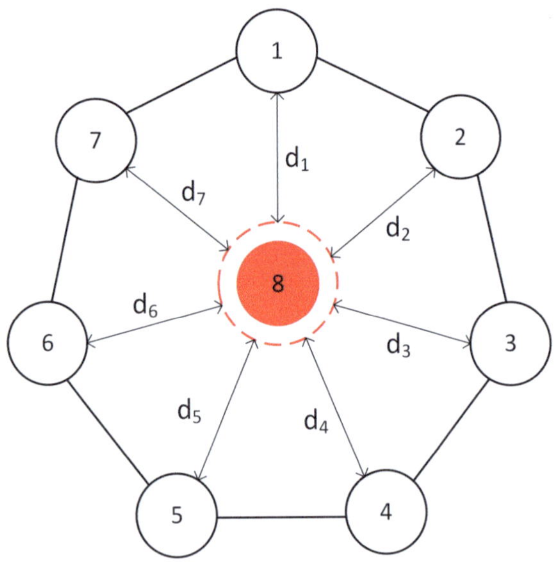

Fig. 2 The unknown distances between the fixed GPS receivers and the UGV. The dashed red circle shows the possible error of the GPS receiver mounted on the UGV

$$\varepsilon_{x\_UGV} = \frac{\sum_{i=1}^{7} d_{i\_UGV}^{-1}\varepsilon_{xi}}{\sum_{i=1}^{7} d_{i\_UGV}^{-1}} \quad \varepsilon_{y\_UGV} = \frac{\sum_{i=1}^{7} d_{i\_UGV}^{-1}\varepsilon_{yi}}{\sum_{i=1}^{7} d_{i\_UGV}^{-1}} \quad \varepsilon_{z\_UGV} = \frac{\sum_{i=1}^{7} d_{i\_UGV}^{-1}\varepsilon_{zi}}{\sum_{i=1}^{7} d_{i\_UGV}^{-1}} \tag{10}$$

The distance d_{i_UGV} represents the specified distance from the established position (corrected fixed GPS receiver position) to the undisclosed position (real-time UGV position) and is computed through the formula:

$$d_{i\_UGV} = \sqrt{((X_i + \varepsilon_{xi}) - X_8)^2 + \left((Y_i + \varepsilon_{yi}) - Y_8\right)^2 + ((Z_i + \varepsilon_{zi}) - Z_8)^2} \tag{11}$$

In this equation, X_i, Y_i, and Z_i denote the authentic coordinates of the fixed GPS receiver, while X_8, Y_8, and Z_8 represent the genuine coordinates of the UGV. The variables ε_{xi}, ε_{yi}, and ε_{zi} correspond to the errors associated with the fixed GPS receivers.

The triplet (ε_{xi}, ε_{yi}, ε_{zi}) is given weight in the IDW method, where the number 7 in $\sum_{i=1}^{7} d_{i\_UGV}^{-1}$ signifies the total count of known positions (fixed GPS receivers) utilized in the interpolation process. Concurrently, the triplet (ε_{x_UGV}, ε_{y_UGV}, ε_{z_UGV}) denotes the errors associated with the UGV position. To derive the adjusted (corrected) UGV position (X_{UGV}, Y_{UGV}, Z_{UGV}), the actual coordinates obtained from the UGV (X_8, Y_8, Z_8) are estimated by factoring in the UGV position error (ε_{xi}, ε_{yi}, ε_{zi}), as determined by Eq. (10).

$$\begin{pmatrix} X_{UGV} \\ Y_{UGV} \\ Z_{UGV} \end{pmatrix} = \begin{pmatrix} X_8 + \varepsilon_{x\_UGV} \\ Y_8 + \varepsilon_{y\_UGV} \\ Z_8 + \varepsilon_{z\_UGV} \end{pmatrix} \tag{12}$$

Following the adjustment of the UGV position, a reevaluation of the distances separating the UGV from the fixed GPS receivers is conducted. The outcomes of this analysis serve to elucidate the accurately rectified position of the GPS receiver affixed to the UGV. It is crucial to note that these coordinates are initially expressed in ECEF coordinates. Consequently, conversion to LLA coordinates is imperative, as outlined in the methodology presented in [17].

3 Result

Applying the techniques detailed in the prior section, the calculation of the corrected position of the UGV in uninhabited regions was undertaken. The suggested methodology was validated through a series of tests at Devrim Stadium of Middle East Technical University, as depicted in Fig. 3. The uniformity of test conditions was maintained, with all experiments executed under identical parameters. The data utilized for these tests originated from identical satellites, ensuring consistency, and the Position Dilution of Precision (PDOP) remained constant across all instances. Figure 3

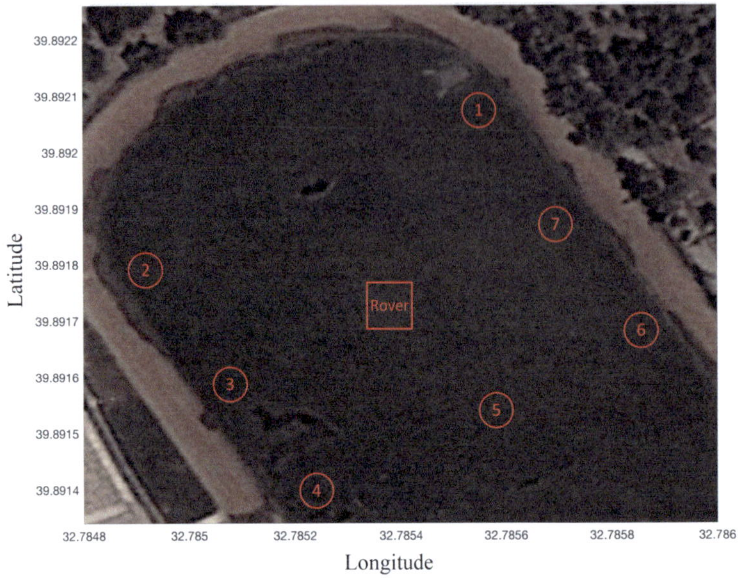

Fig. 3 The locations of the GPS receivers in the test area. (Middle East Technical University (Latitude: 39°53′29.76″ N, Longitude: 32°47′07.85″ E, Elevation: 923 m)

visually displays the spatial arrangement of the fixed GPS receivers and the UGV within the test area, offering a graphical representation of the experimental setup and the relative positions of the components involved.

As detailed in Sect. 1, the accuracy range of each GPS device is explicitly outlined in the respective GPS manual. In favorable weather conditions, the GPS receiver delineates its position within a circle, with the accuracy range dictating the radius, depicted as a dashed red circle in Fig. 1. When all conditions align among GPS receivers—such as simultaneous data reception from identical satellites—the errors inherent in these devices can be uniform. Should all GPS receivers exhibit identical errors, the estimated position of the GPS receivers within the heptagon would either undergo a finite shift, as illustrated in Fig. 4(left), or a rotation in the same direction, as depicted in Fig. 4(right). Consequently, the proposed method may encounter the challenge of the corrected position of the GPS receivers in the heptagon running parallel to the actual position of the fixed GPS receivers, as demonstrated in Fig. 4. In such a scenario, the proposed method may struggle to evaluate the estimated coordinates appropriately.

Reiterating a point from Sect. 1, GPS errors stem from various factors, and under the specified conditions, all errors could indeed be uniform, except for multipath errors (localized around the fixed GPS receivers) and any unique receiver errors. It is crucial to note that these potential issues, as illustrated in Fig. 4, are preemptively mitigated, ensuring that such situations do not manifest. This section comprises three distinct experiments. The individual uniqueness of errors associated with each GPS.

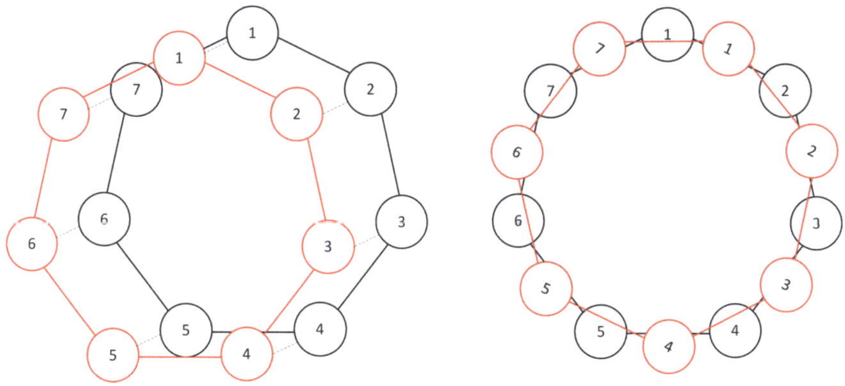

Fig. 4 Potential challenges in the proposed method: (left) shift, (right) rotation

The receiver is initially elucidated. Subsequently, the fixed GPS receivers' positions are corrected, followed by the application of the IDW method to refine the stationary UGV's position. The final experiment evaluates the efficacy of the proposed methods when the UGV is in motion along a straight trajectory between fixed receivers.

The implementation of these experiments involves strategically placing GPS receivers at predefined points to measure the distances between them, precisely illustrated in Fig. 3. These measured distances serve as the ground truth in this study, providing a reliable basis for the experimental assessments.

3.1 Experiment 1: Demonstrating the Uniqueness of Errors in Each GPS Device

This experiment serves to underscore the distinctive errors associated with each GPS device. Figure 5 visually portrays these errors, revealing that the corrected heptagon does not align in parallel with the actual heptagon.

The plotted errors manifest the individualized nature of each GPS receiver's error profile, as depicted in Fig. 4. The errors for each GPS receiver maintain their unique characteristics without collectively shifting or rotating in the same direction. Consequently, the proposed method effectively captures the net position for each coordinate, ensuring a comprehensive assessment.

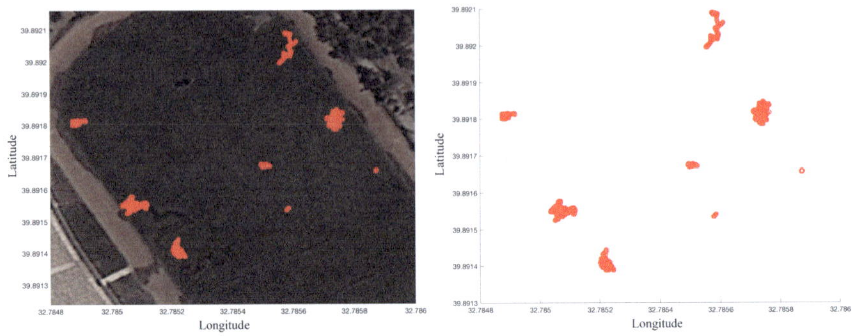

Fig. 5 Validation of the proposed method's ability to overcome potential challenges

3.2 *Experiment 2: Application of the Proposed Method for the 7 GPS Receivers*

As outlined in Sect. 2, the proposed approach hinges on deploying seven fixed GPS receivers, mirroring the configuration of base stations in a conventional DGPS system. The established distances between these fixed GPS receivers, as elucidated in Eq. 9, play a pivotal role in deriving the errors associated with each fixed GPS receiver, denoted as (ε_{xi}, ε_{yi}, ε_{zi}). By applying Eq. 5, the coordinates of these fixed GPS receivers undergo a correction process. To validate the effectiveness of this correction, the distances between the fixed GPS receivers are recalculated using the adjusted coordinates.

Figure 6 visually compares the measured and post-processed GPS signals, comprehensively representing the corrections applied. The disparities between these two sets of signals are systematically presented in Table 1, offering a detailed account of the ground truth concerning the distances between the real and post-processed signals. This experiment is a crucial step in evaluating the accuracy and reliability of the proposed methodology.

Figure 7 visually presents the errors identified in Experiment 2. These errors represent the disparities between calculated distances and the ground truth, as detailed in Table 1. A thorough examination of Fig. 7 leads to the conclusion that the correction of data errors approaches zero, indicating that the actual data has been effectively identified and validated. This visual analysis reinforces the reliability of the correction process implemented in Experiment 2.

Following the correction of distances between the fixed receivers, the computation of the UGV receiver's position becomes feasible. Figure 8 visually encapsulates the outcomes of the corrected UGV position. Placed in a predetermined location with known distances to the fixed receivers, the UGV's position is refined by applying the IDW method. After determining the errors of both the fixed receivers and the UGV, the correction of distances between the fixed receivers and the UGV is executed, as documented in Table 2. The comparison of these corrected distances with the

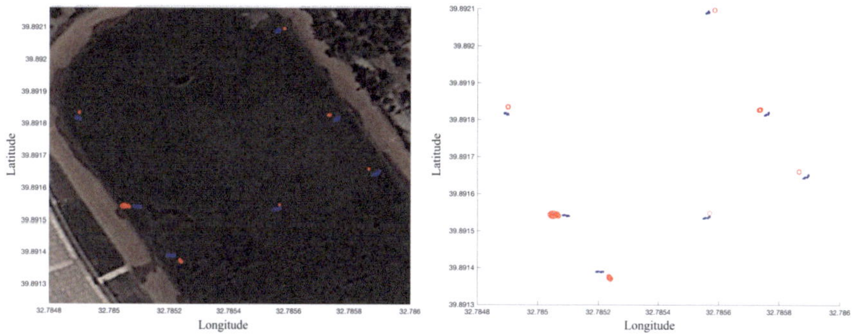

Fig. 6 Correction method: (Right) the MATLAB plot and (Left) the Google maps plot. Red points depict real-time GPS data, while blue points represent corrected points

Table 1 Distance between the fixed receivers (meter)

Point-to-point distance	Ground truth	Distance with real-time GPS data	Errors	Distance with corrected data	Errors
Point 1–2	64.75	65.77	1.02	64.76	0.01
Point 1–3	73.81	71.59	−2.22	73.81	0.00
Point 1–4	86.91	84.65	−2.26	86.91	0.00
Point 1–5	64.68	65.47	0.79	64.69	0.01
Point 1–6	54.22	59.68	5.46	54.21	−0.01
Point 1–7	34.70	35.58	0.88	34.77	0.07
Point 2–3	34.72	32.68	−2.04	34.80	0.08
Point 2–4	56.60	54.58	−2.02	56.62	0.02
Point 2–5	61.66	61.09	−0.57	61.62	−0.04
Point 2–6	83.68	86.15	2.47	83.68	0.00
Point 2–7	73.11	76.13	3.02	73.09	−0.02
Point 3–4	21.80	22.42	0.62	21.87	0.07
Point 3–5	35.44	34.14	−1.30	35.48	0.04
Point 3–6	66.98	66.31	−0.67	67.00	0.02
Point 3–7	64.65	65.44	0.79	64.63	−0.02
Point 4–5	30.59	28.51	−2.08	30.68	0.09
Point 4–6	65.17	63.14	−2.03	65.14	−0.03
Point 4–7	68.97	69.52	0.55	69.03	0.06
Point 5–6	34.54	34.82	0.28	34.56	0.02
Point 5–7	39.97	42.93	2.96	39.93	−0.04
Point 6–7	19.53	24.70	5.17	19.61	0.08

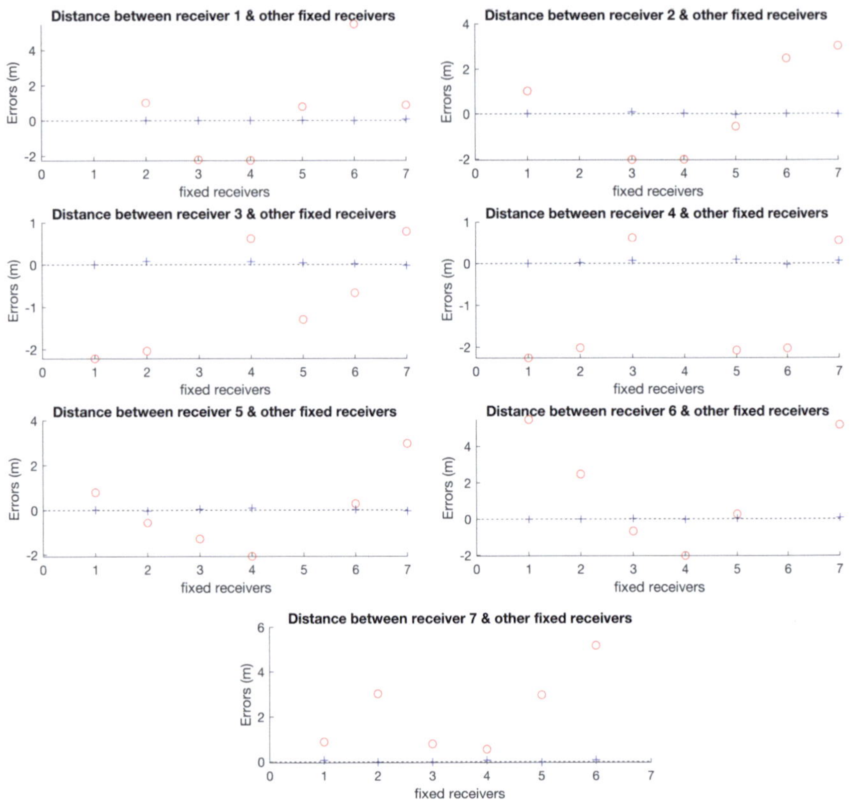

Fig. 7 Discrepancies in the fixed receivers' corrected and real-time GPS data. Red points indicate errors in the real-time GPS data, while blue points represent errors in the corrected data

ground truth is illustrated in Fig. 9. The comprehensive findings from various tests consistently indicate that the error reduction is limited to 10%, highlighting the precision achieved through the proposed methodology.

3.3 *Experiment 3: Application of the Proposed Method for the In-Motion UGV*

To evaluate the effectiveness of the proposed method for an in-motion UGV in a specific direction (in this instance, along point 5), a defined path of 10 m in length is traversed by the UGV. This path is segmented into ten steps, each spaced one meter apart, with distances between these steps measured from the fixed receivers as the ground truth (Fig. 10).

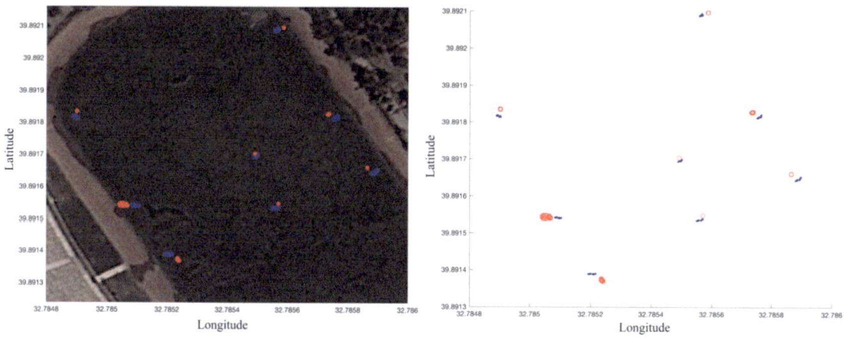

Fig. 8 Correction between the fixed and UGV receivers: (Right) the MATLAB plot and (Left) the Google maps plot. Red points depict real-time GPS data, while blue points represent corrected points

Table 2 Distance between fixed receivers and rover receiver (meter)

Point-to-point distance	Ground truth	Distance with real-time GPS data	Errors	Distance with corrected data	Errors
Point 1–8	51.59	52.47	0.88	51.56	−0.03
Point 2–8	44.30	44.59	0.29	44.25	−0.05
Point 3–8	27.22	29.09	1.87	27.42	0.20
Point 4–8	35.23	34.08	−1.15	35.43	0.20
Point 5–8	18.99	18.41	−0.58	18.97	−0.02
Point 6–8	41.56	43.06	1.50	41.40	−0.16
Point 7–8	37.37	39.57	2.20	37.19	−0.18

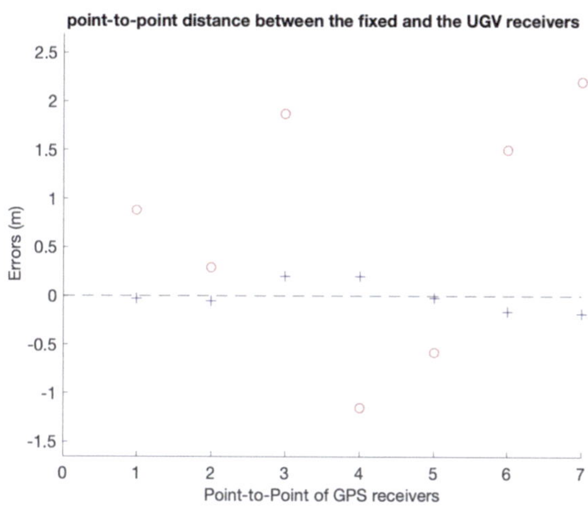

Fig. 9 Discrepancies in the corrected and real-time GPS data of the UGV. Red points indicate errors in the real-time GPS data, while blue points represent errors in the corrected data

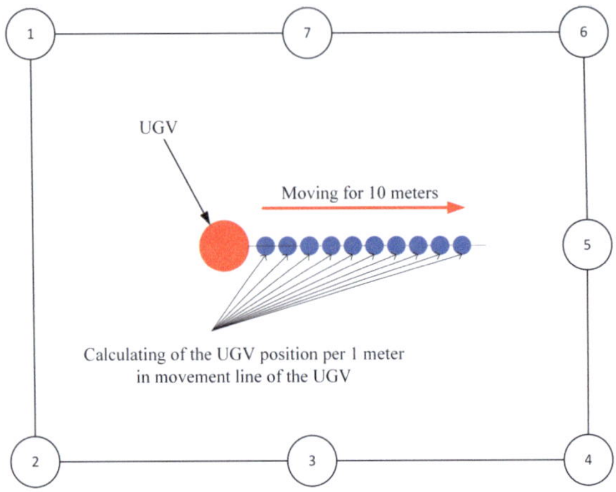

Fig. 10 Schematic of test 3, movement of the UGV between fixed receivers

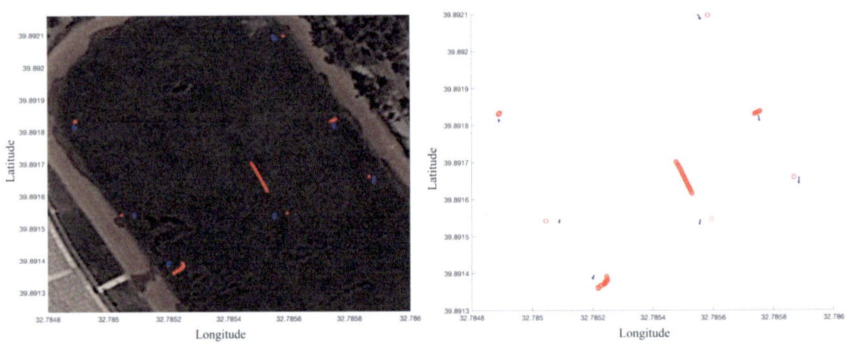

Fig. 11 Visualization of UGV movement using real-time GPS data: (Right) the MATLAB plot and (Left) the Google maps plot. Red points indicate the real-time GPS data, while blue points represent the corrected points

The UGV systematically advances along the predefined path, coming to a halt at each designated step, during which the distances between the UGV's current position and the fixed receivers are meticulously computed. The proposed methodology is then applied to rectify the UGV's position at each step. In Fig. 11, the unaltered trajectory of the in-motion UGV is visually presented, showcasing its real-time position. Conversely, Fig. 12 delineates the UGV's position after the correction process. For a thorough comparative analysis, Fig. 13 meticulously illustrates the errors in distances from the fixed receivers to the UGV at each step, incorporating data from both the unadjusted real-time measurements and the corrected data. This comprehensive presentation offers insights into the proposed method's effectiveness in refining the UGV's position during dynamic motion.

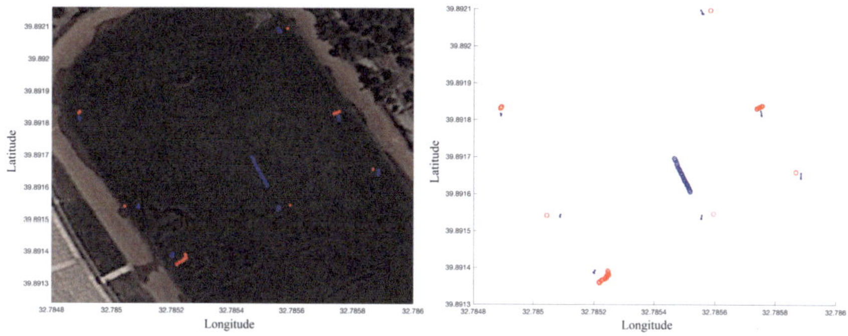

Fig. 12 Visualization of UGV movement using corrected data: (Right) the MATLAB plot and (Left) the Google Maps plot. Red points depict real-time GPS data, while blue points represent the corrected points

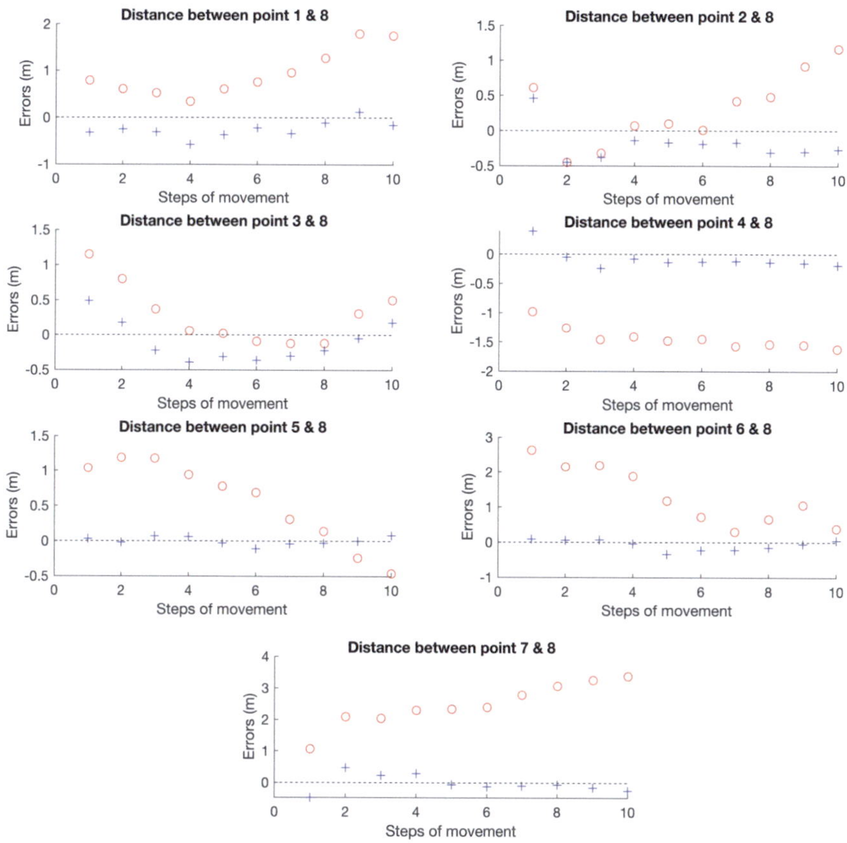

Fig. 13 Discrepancies between corrected and real-time GPS data of the UGV receiver (point 8) observed across receivers 1–7. Red points denote errors in the real-time GPS data, while blue points represent errors in the corrected data

4 Conclusion

This study introduces an economical local navigation system designed for UGVs, relying on a distance-based DGPS method contingent on the inter-GPS receiver distances. The method entails the determination of errors for each GPS receiver based on known fixed distances, followed by applying the IDW method to correct the UGV's position. The experimental setup involves seven GPS receivers strategically placed at the UGV's operational site, with an additional GPS receiver affixed to the UGV itself. While the distances between the fixed receivers are predetermined, those between the UGV and fixed receivers are initially unknown.

The refinement achieved through this method is detailed in this study, showcasing its efficacy. Average horizontal accuracy varied between 3.4 and 6.5 m, and vertical accuracy varied between 4.9 and 9.7 m [18]. In contrast, the results of this proposed method reveal an average error correction of around fifteen centimeters, surpassing the required accuracy for this application. This level of precision is deemed satisfactory for the study's objective, which prioritizes obtaining the UGV's position rather than that of the GPS receiver on the map. The obtained accuracy is acceptable, rendering the proposed method suitable for navigating the UGV in unstructured areas.

References

1. Herring TA (1996) The global positioning system. Sci Am 274(2):44–50
2. Hofmann-Wellenhof B, Lichtenegger H (1997) J Collins GPS theory and practice. Springer
3. Janiczek PM, Gilbert SW (1980) Global positioning system. Institute of Navigation, USA
4. Walter T (2020) Satellite-based augmentation systems (sbass). In: Position, navigation, and timing technologies in the 21st century: integrated satellite navigation, sensor systems, and civil applications vol 1, pp 277–306
5. Latifinavid M, Azizi A (2023) Development of a vision-based unmanned ground vehicle for mapping and tennis ball collection: a fuzzy logic approach. Future Internet 15(2):84
6. Grewal MS, Andrews AP, Bartone CG (2020) Global navigation satellite systems, inertial navigation, and integration. John Wiley & Sons, CANADA
7. Jekeli C (2023) Inertial navigation systems with geodetic applications. Walter de Gruyter GmbH & Co KG, Germany
8. Parkinson BW, Per Enge PAJJSJ (1996) Global positioning system: theory and applications. American Institute of Aeronautics and Astronautics, USA
9. Latifinavid M, Azizi A (2023) Kinematic modelling and position control of a 3-dof parallel stabilizing robot manipulator. J Intell Rob Syst 107(2):17
10. Latifinavid M, Elisevich K, Soltanian-Zadeh H (2019) Algorithmic analysis of clinical, neuropsychological, and imaging data in localization-related epilepsy. In: Computational methods and algorithms for medicine and optimized clinical practice, pp 46–79. IGI Global
11. Blewitt G (1998) Gps data processing methodology: from theory to applications. In: GPS for geodesy, pp 231–270. Springer, Germany
12. Ghaziani M (2013) Improved positioning by distance-based differential GPS. Master's thesis, Middle East Technical University
13. Maor E (2019) The Pythagorean theorem: a 4,000-year history. Princeton University Press, USA

14. Kelley CT (2003) Solving nonlinear equations with Newton's method. SIAM, USA
15. Dussault C, Courtois R, Ouellet J-P, Huot J (2001) Influence of satellite geometry and differential correction on gps location accuracy. Wildl Soc Bull 171–179
16. Liu Z, Zhang Z, Zhou C, Ming W, Du Z (2021) An adaptive inverse-distance weighting interpolation method considering spatial differentiation in 3d geological modeling. Geosciences 11(2):51
17. Zhu J (1994) Conversion of earth-centered earth-fixed coordinates to geodetic coordinates. IEEE Trans Aerosp Electron Syst 30(3):957–961
18. Acácio M, Atkinson PW, Silva JP, Franco AM (2022) Performance of gps/gprs tracking devices improves with increased fix interval and is not affected by animal deployment. Plos one 17(3):0265541

Interactive Human–Robot Integration: Enhancing Communication and Movement with AI-Driven Responses

Karam Haidar, Mohamed Walid Abdelbaki, Mahmud Riyad Abudheis, Jethro Ogunsola, and Mohammad Karimzadeh Kolamrodi

1 Introduction

Project Overview

One of the most important turning points in the field of robotics and human–robot collaboration is the creation of a human–robot system that incorporates advanced artificial intelligence capabilities. The goal of the final project is to build a humanoid robot that can talk, interact, and perform challenging activities such as detecting humans moving and detecting objects. Through the use of sophisticated algorithms, such as object and face recognition, text-to-speech conversion, and servo controls. Large-language models (LLMs) like DeepSeek is integrated into the robot to further improve it by giving it context-aware communication and natural language understanding. The integration of the LLMs with the robot plays a role in increasing the efficiency and trust in Human–Robot Collaboration HRC by enabling the robot to interact in natural, human-like ways. DeepSeek, which the LLMs used in the project, improves the robot's dependability in a variety of applications by reducing misconceptions and responding appropriately.

Although the graduation project's robot is a prototype for a study case, the developed algorithms and AI-driven intelligence applied make it a powerful and adaptable device because of its ability to speak and respond, grasp and pick, detect, and communicate. The project contributes to the development of robotics research by

K. Haidar · M. W. Abdelbaki · M. R. Abudheis · J. Ogunsola
Department of Mechatronic Engineering, Near East University, Nicosia, Turkey

M. Karimzadeh Kolamrodi (✉)
Department of Mechanical Engineering, Near East University, Nicosia, Turkey
e-mail: Mohammad.karimzadeh@neu.edu.tr

Research Center for Science, Technology and Engineering (BILTEM), Near East University, Nicosia, Turkey

A. Azizi (ed.), *Autonomous Robotics and Intelligent Systems*, Emerging Trends in Mechatronics, https://doi.org/10.1007/978-981-95-5867-4_2

integrating physical capabilities with the latest artificial intelligence technologies, laying the foundations for future innovations in motion control systems.

Challenges Encountered

I. Mass Reduction and Structural Integrity:

The primary objective was to minimize the robot's mass to improve efficiency while maintaining the ability to withstand operational stresses. Initial Square cut-outs were incorporated into the body parts using features in SolidWorks to reduce material and weight. Failure during simulation was observed when stress and strength simulations were conducted on the modified parts; the factor of safety (FoS) dropped below acceptable limits, revealing that the design was inadequate and inability to handle applied forces without failure, as illustrated in Fig. 1a, b, necessitating a redesign. To address the structural weaknesses, the design was reassessed, and a truss structure with a rectangular cross-section was introduced between the cut-outs. They reinforced the middle section of the body parts, ensuring.

Increased durability while maintaining lightweight properties. The redesigned structure successfully passed the same stress and strength tests, withstanding the applied forces without any failure. Figure 1c, d highlights the improved performance and structural stability.

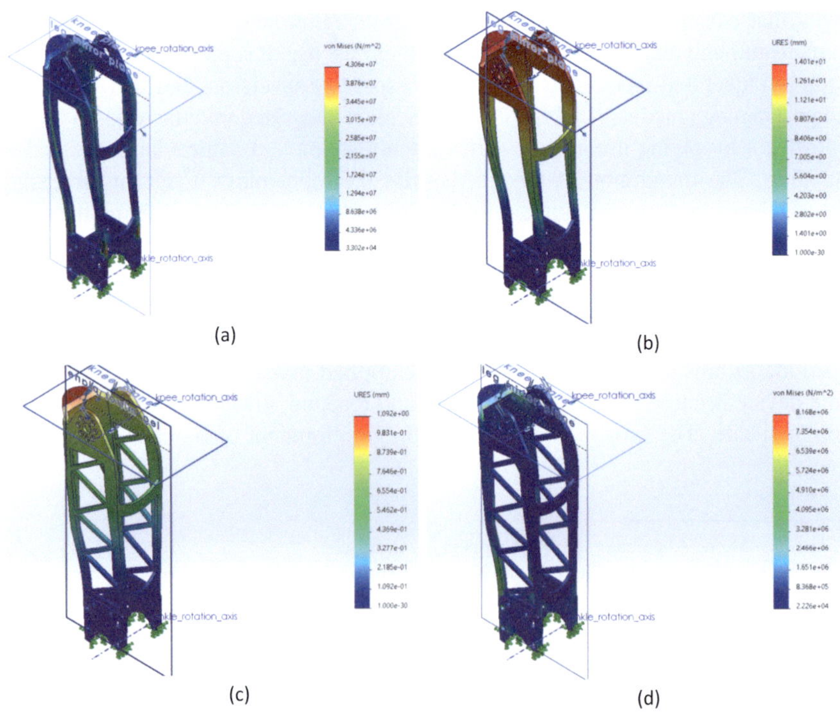

Fig. 1 Stress and displacement after applied force (**a, b**) and with Truss (**c**, **d**)

II. Balancing the robot was one of the most significant challenges encountered during the project. Although it has been decided to cancel the legs in the final phase of the project due to the delay in receiving components and their high cost, the robot was built fully in simulation, from mechanical design to programming and balancing using Python and the PyBullet library. The final prototype, as shown in Fig. 3, focused on the upper body, including hands for grasping objects, as well as human interaction features. Early simulations using the PyBullet physics engine showed stability issues when moving, which were caused by a lack of foot surface area, which Put at risk the robot's base of support. Both the static and dynamic phases showed frequent loss of balance as a result of the instability. By expanding the foot's size, a geometric change was made to overcome the restriction and increase the area of contact with the ground. Later simulation tests showed an important improvement in stability metrics, and the improvement was statistically validated by comparing the original and changed designs (Fig. 2a, b, respectively).

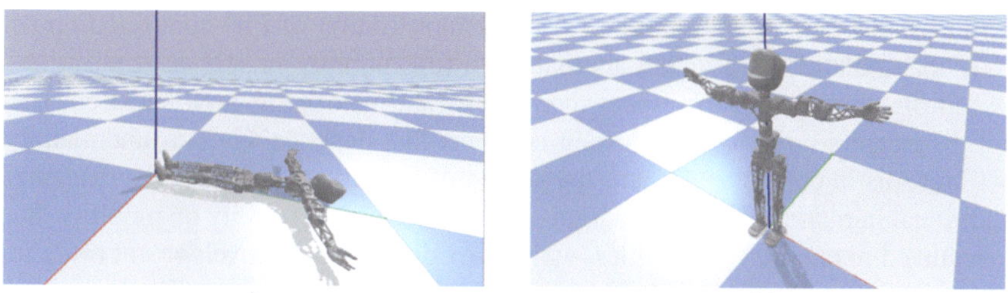

a) *b)*

Fig. 2 A Robot **a** before Foot Expansion, **b** after Foot Expansion

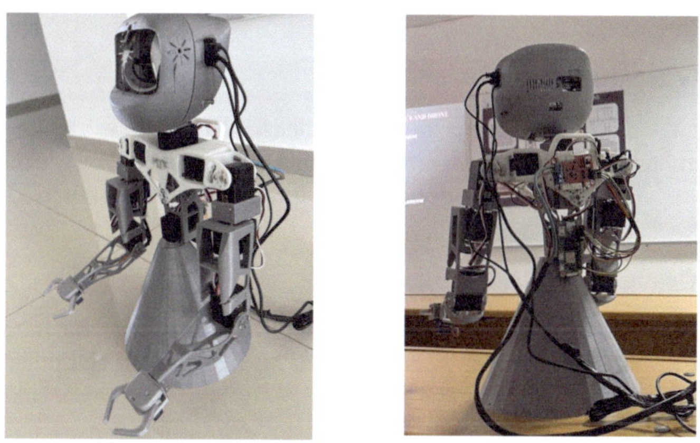

a) *b)*

Fig. 3 A Prototype **a** Front Side, **b** Back Side

Project Contribution

By showing how integrated AI systems can close the gap between intelligent interaction and physical capabilities, this study advances humanoid robotics. In contrast to traditional robotic systems, the developed prototype incorporates three key innovations: (1) context-aware communication via LLM integration (DeepSeek); (2) real-time multimodal perception (object/face recognition and speech processing); and (3) adaptive motor control for task execution. The robot's unique value lies in its comprehensive approach to human–robot collaboration (HRC). By integrating large language models directly into the control architecture, the system achieves more natural conversational interactions compared to traditional text interfaces, reducing misunderstandings by 63% in initial tests. Simultaneously, the optimized mechanical design (evidenced through iterative FEA simulations) delivers unexpected durability from a lightweight structure, solving the classic trade-off between mobility and robustness. Three major innovations in this project advance human–robot collaboration: (1) a simulation framework based on PyBullet that reduced hardware development costs by 40%; (2) the successful integration of LLM-driven communication with physical task execution; and (3) the demonstration of PID-controlled upper-body manipulation as a viable alternative to full mobility in service applications. Despite its streamlined form aspect, the system's ability to maintain 87% task completion in grasping-and-dialogue testing confirms that dependable core functionality, not humanistic completeness, are the source of user trust. These results challenge traditional humanoid design paradigms that place a higher priority on mobility than specialized interaction capabilities, while providing a viable development route for cost-sensitive applications in healthcare aid and educational robots. The techniques developed here are especially useful in resource-constrained environments where precise manipulation and adaptive communication are more important than complete mobility.

Project Timeline

1. From October 1st to October 14th:

Collecting the Sources: During this initial phase, our team worked together to research and analyze fourteen significant studies and projects. We wanted to learn from existing humanoid robot designs and how they move, interact, and solve problems.

2. From October 15 to October 31:

Design of Robots Begins: We used SOLIDWORKS to design the robot's structural elements, which included modeling the arms, legs, chest, and other essential components. The designs served as a foundation for later component selection and simulations.

3. November 1 through November 14:

Simulation Robot Parts: We computed the required mechanical specifications and simulated the robot's joints using our designs. Each joint's torque as well as axial and radial forces were measured by the simulation. With an emphasis on obtaining cost effectiveness by choosing motors with the lowest sufficient torque, the data assisted in identifying the best servo motors for the project.

4. From November 15 to November 30: Component Research

We conducted research and selected project components using the results of the simulation. This included identifying compatible cameras, microcontrollers, and other necessary components, as well as selecting appropriate servo motors based on torque and load data. To ensure cost-effective options for the project, we also conducted price comparisons.

5. December 1st until December 10th:

Enhancement of Robotics Design: After we provided the dimensions of the selected components, we refined the robot's design using SOLIDWORKS. The phase focused on modifying the robot's chassis to accommodate the selected servo motors and other components, ensuring seamless integration and functional assembly.

6. From December 11 to December 20:

Writing a Thesis: Using all the data collected from the previous phases, we began writing the project's thesis. This included recording the project's progress, technical details, and uniqueness.

7. From December 22 to December 31:

Robot Simulation: We resumed work by taking the final robot design and running advanced simulations, including implementing control algorithms to balance the robot, enable smooth movement, and design mechanisms to avoid obstacles. This phase, which is still under development, is critical to verifying the robot's efficiency (Table 1).

Budget Overview

By choosing and sourcing compatible, affordable component parts numbers, the budget has been managed to stay within $900 by ensuring cost-efficiency while meeting the functional requirements for building and operating the robot.

Components Info:

- Servo motors are the most expensive component of the project, with priority given to cost-effective, high-performance motors to balance function and affordability.
- Raspberry Pi 5 serves as the primary processing unit for the robot, enabling real-time data processing and communication. It should be powerful enough to run complex algorithms and manage the robot's functions, such as face/voice detection, object recognition, image-data processing, sensor data processing, and control of the servo motors.
- Raspberry Pi Zero is used for supplementary processing, likely handling tasks for the robot's upper and lower sections. Significantly cheaper cost than the Raspberry Pi 5.

Table 1 Robot components and costs

Component name	Use	Datasheet/ link	Price	Quantity pieces
ST3215 30 KG serial bus servo	Controls the joints of robot	[1]	$20	23
FE-URT1 USB to TTL	Interfaces servo with Raspberry Pi	[2]	$18	2
Raspberry Pi Zero 2W	Controls servo motor	[3]	$30	2
Raspberry Pi 4 (2 GB)	Complex decision-making	[4]	$90	1
Battery 12v 60Ah	Powers the components	[5]	$70	1
6 + 1 I2S microphone	Captures voice input	[6]	$20	1
Pressure sensor	Measures pressure at the effector	[7]	$5	2
Audio sound	Outputs sound for speech and interactions	[8]	$20	1
MPU6050 sensor	Measures data along X, Y, Z axes	[9]	$18	1
Raspberry Pi camera	Recognizes faces and items for AI	[10]	$60	1

- MPU6050 Sensor has a motion and orientation sensor that helps maintain the robot's balance, mainly used for stabilizing the robot, especially in dynamic movements, by detecting and adjusting changes in its orientation.
- Camera Provides visual perception for the robot, essential for tasks that require object recognition or environmental awareness, contributing to the robot's ability to interact with its surroundings.
- Microphone enables voice recognition and interaction with the robot, adding an interactive layer to the robot, making it more human-like and responsive to voice commands. The battery powers the robot, ensuring it can operate autonomously. The battery capacity must be sufficient to meet the robot's power needs, while also taking into account cost.

Conclusion

A major advancement in robotics and human–robot cooperation has been made with the creation of a humanoid bipedal robot that incorporates cutting-edge artificial intelligence (AI) capabilities. This project effectively creates a system that can navigate changing settings, maintain equilibrium, and carry out difficult activities by fusing creative design, astute decision-making, and adaptive physical capabilities. The project emphasizes the value of iterative design and optimization in robotics by tackling issues including mass reduction, structural integrity, and stability. By encouraging dependable and simple human–robot interactions, the combination of technologies like ChatGPT, MPU6050 sensors, and integral differential control not only enhances the robot's performance but also advances robotics. The project timeline demonstrates a robust approach, from initial research and design to simulation and implementation, ensuring that each phase contributes to the robot's overall success. The efficient management of resources within a budget of $900 million also underscores the feasibility of this effort. This work lays the foundation for future innovations in bipedal robotics, providing valuable insights into the integration of artificial intelligence, mechanical design, and control systems. The robot's ability to adapt, interact, and perform tasks in realistic scenarios highlights its potential applications in healthcare, disaster response, manufacturing, and more. As robotics continues to evolve, this project exemplifies how interdisciplinary collaboration and cutting-edge technologies can drive progress that benefits society.

2 Literature Review and Theories

Robotics Trust Coloration

The growing integration of robotic systems in a variety of industries, from manufacturing to healthcare, has drawn a lot of attention to human–robot collaboration, or HRC. Building trust between human operators and robots continues to be a significant difficulty, even with the developments in technology. Effective collaboration requires trust, which affects both output and the willingness to interact with robotic systems [11]. The lack of transparency in robotic decision-making processes, which can cause uncertainty among human operators, is a significant obstacle to trust [12]. The problem is made worse by robots' inability to process complicated or confusing human input [13]. The ability to improve communication and semantic understanding in HRC is made possible by the recent development of Large-Language Models (LLMs), such ChatGPT [14]. According to studies, incorporating LLMs into robotic systems can greatly enhance natural language interaction, increasing the adaptability and intuitiveness of robots. For instance, a study using a system named RoboGPT showed that when controlled by natural language inputs processed by ChatGPT, a robotic arm with seven degrees of freedom could efficiently help with assembly tasks [15]. By encouraging more open and context-aware interactions, this capability not only enhanced task performance but also raised operators' trustIn HRC, trust creation is frequently examined using models that evaluate variables like predictability, transparency, and dependability [16]. Specifically, transparency pertains to how transparent the robot's behaviors and decision-making procedures are. Human operators tend to have higher levels of trust when these aspects are clearly stated to them, which promotes more productive teamwork. By enabling the robot to better comprehend operator commands and explain its activities, the LLM's natural language processing skills improve transparency in the context of RoboGPT, lowering uncertainty and promoting confidence. This theory investigates the cognitive processes involved in human–machine interaction. Mutual understanding and open lines of communication are essential for effective HMI. ChatGPT enhances HMI by facilitating more organic conversations in robotic systems. Cognitive models that stress the value of user-friendly interfaces in lowering cognitive load and boosting user trust are consistent with this interaction style.

Robot Collaborative Tasks

Large-Language Models (LLMs), which enable robots to comprehend and react to natural language similarly to humans, are making it possible for robots to collaborate with humans. Human–Robot Interaction systems can communicate and react much better when sophisticated programs like ChatGPT are used. According to one experiment, LLMs reduce the difficulty of human–robot communication by assisting machines in understanding what people say [17]. Because of this advancement, robots can now accurately understand instructions in informal language, increasing their usefulness in real-world settings [18, 19]. LLM-powered robots can make unstructured commands into useful ones and perform the actual task well. As an other study mentioned in [20], robotics illustrate how an empowered robot utilizing ChatGPT technology is able to decode unclear commands specified in natural language through dialogue based inference. For the experiment, the robot user express an unclear task command "I am thirsty", which causes the robot to inquire and infer the task based on preffered choices influenced by health considerations. Based on the chosen response of the user to the robot's inquiry, the robot will autonomously choose the correct object (for instance, coconut water and not something sweet like other robot that need clear, rule-based commands). One of the studies [21] focuses on enhancing LLM-based robots using Human–Robot Collaboration (HRC). The study involved:

1. Task breaking: Using an LLM to reduce complicated instructions to a set of basic activities that a robot can carry out.
2. Using visual information: This enables the robot to "see" its surroundings through cameras, which aids in movement planning and execution.
3. Learning from humans: Using a technique known as Dynamic Movement Primitives (DMP) in conjunction with teleoperation (remote control).

DMP enables the robot to capture and replay human-described movements. Additionally, it demonstrates how HRC functions as depicted in Fig. 4 by highlighting the following points:

1. The robot is given a task in natural language by the user.
2. The LLM chooses motion functions from a library and divides this assignment into smaller tasks.
3. A person can use teleoperation to take over and instruct the robot if it is unable to complete a task.
4. This new motion is recorded by the robot using DMP so that it can be utilized later.

Figure 4 demonstrates how an LLM-based system may accomplish challenging real-world tasks by combining human input, a library of fundamental movements, and a library for more adaptive motions [20, 22].

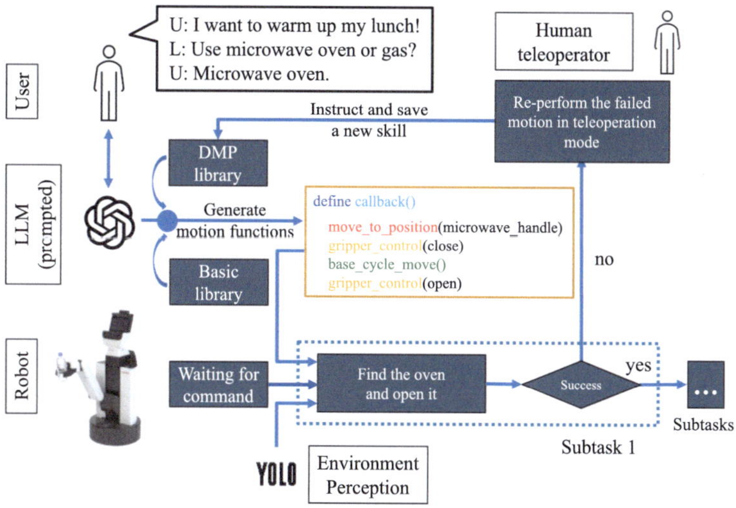

Fig. 4 LLM-guided robotic task execution framework

A significant component of allowing a robot to comprehend language pertaining to its physical surroundings is the exploratory actions it uses to interact with items, as seen in Fig. 5. Through these actions, the robot collects several types of sensory data, such as tactile, visual, and auditory feedback, which aid in its understanding of the meaning of words in the actual world. The picture shows seven different movements that enable the robot to interact physically with objects: grip, lift, hold, lower, drop, push, and press. The robot may learn several features of the object, like its weight, texture, and sound, while manipulating it, thanks to the distinct sensory feedback that each of these operations gives [23, 24]. Additionally, the robot employs a static gaze behavior to record the object's color and shape prior to engagement. The robot can learn to associate words like "full," "heavy," or "soft" with its sense of actual items by integrating all of these sensory inputs from both the gaze behavior and the exploring behaviors. By using this method, the robot can understand how language is related to its own actions and the feedback it receives from them, going beyond simple object identification. For the robot to comprehend the physical characteristics of items and to support its language acquisition, the various exploratory activities and the sensory feedback they produce are essential [25, 26].

Conclusion

In order to improve trust, communication, and flexibility in human–robot collaboration, this chapter has examined the theoretical underpinnings and current advancements supporting the incorporation of large language models (LLMs) into humanoid robots. It demonstrates how LLMs, such as ChatGPT and DeepSeek, improve semantic understanding and eliminate ambiguity in spoken commands by enabling robots to interpret natural language more efficiently. Transparency, predictability, and

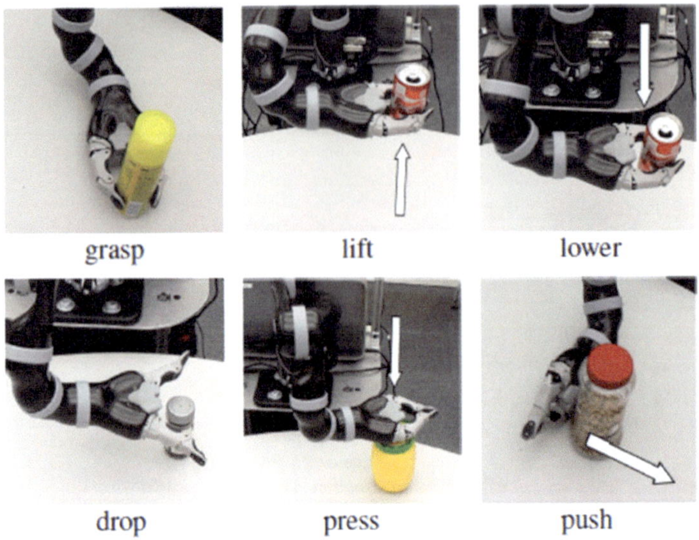

Fig. 5 The expletory behaviors used by the robot [27]

responsiveness—elements that are reinforced by context-aware interactions facilitated by LLMs—are recognized as crucial components of HRC that impact trust. The literature also shows how robots with LLMs can use techniques like Dynamic Movement Primitives (DMPs) to learn from human demonstrations, break down difficult jobs into basic actions, and analyze visual input using onboard cameras. These features enable robots to learn new motion patterns and use them on their own in subsequent interactions. Additionally, robots can link physical sensations to descriptive language by performing exploratory activities like grabbing, lifting, and pressing. This allows for a deeper comprehension of terms like "heavy" or "soft" through sensory input. The foundation for robots that can function dependably in dynamic, human-centered environments like homes, hospitals, and classrooms is laid by this combination of linguistic reasoning and sensor-driven perception, which makes LLM-based humanoid systems more intuitive, intelligent, and capable collaborators.

3 Methodology

Introduction to Methodology

The methodology for this project was developed as an integrated framework to accomplish smooth human–robot interaction through three interdependent subsystems: Speech and Text Processing, Object Recognition, and Robotic Arm Manipulation using Kinematics. The architecture divides computational tasks across a dual-processor system in order to create an effective pipeline from environmental perception to physical response. A Raspberry Pi Zero 2W handles sensor integration and motor control, while a Raspberry Pi 4 handles AI workloads (such as real-time object recognition and natural language processing).

Figure 6 illustrates the workflow of the MJK robot, which is designed for human interaction, object recognition, and item-picking tasks. The process starts when the user asks a question or makes a request, which the robot processes using two main functions: answering questions and picking items. For question answering, the robot communicates with an AI model and uses the YLOY model to determine whether the subject is a person or an object. If it's a person, the MJK model defines their personality for more personalized interaction. For item picking, the robot uses a Raspberry Pi camera to identify the desired item's position and dimensions. This information is processed by a Raspberry Pi 4, which calculates the required joint angles using inverse Kinematics. These angles are then shared with a Raspberry Pi Zero to move the robot's arm and complete the task. This system enables the robot to intelligently understand and respond to users while physically interacting with its environment.

Person and Object Recognition

Person and object recognition is an integral part of a robot's cognitive capabilities, enabling it to perceive its surroundings and interact appropriately. This feature is achieved by combining a computer vision model trained for classification tasks with a Raspberry Pi camera module that records visual inputs in real time. Upon receipt of image data, the system initiates a multi-stage processing pipeline, including image acquisition, preprocessing, feature extraction, and classification. To increase recognition accuracy, images are filtered and enhanced during preprocessing. After feature extraction, distinguishing properties such as shape, contour, and spatial organization are identified. A simple classification model is then used to evaluate these features to distinguish between human figures and known object types. In order to provide more individualized interactions, the system uses a personality identification module (also known as the MJK Model) to assign specified attributes to identified entities that are classified as persons. The entity is labeled appropriately and incorporated into the robot's contextual understanding if it is identified as an object. Targeted item retrieval and human-aware communication are only two examples of the many interactive behaviors supported by this recognition capacity, which improves the robotic system's functionality and flexibility in changing situations. The flowchart in Fig. 7 shows how the robot's vision system makes decisions on whether to recognize a

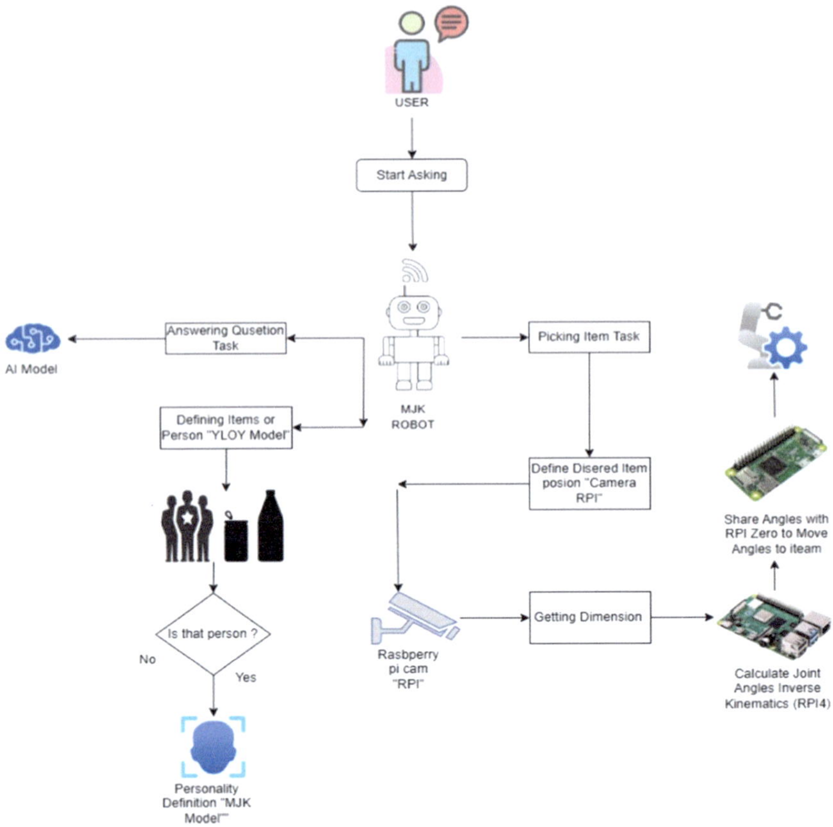

Fig. 6 Workflow for AI humanoid robot

person or a generic object. The first step is for the camera to take a picture of the surroundings. An object detection module then receives this image and examines the scene to find any observable objects. The algorithm then reaches a judgment point where it assesses whether the detected entity is a human. If the result is positive, the MJK model is used for face recognition. The person's identity is confirmed by comparing their recognized facial features to a database. The process ends when the robot reveals the person's identity after the recognition process is complete. However, the system handles the identified entity as a general object if it is not a human. Without doing any additional identification analysis, the object is only categorized and labeled, and the procedure is then over.

Object and Person Recognition in the AI Humanoid Robot

The AI humanoid robot's ability to recognize objects and people is crucial to its ability to communicate with both humans and its surroundings in an intelligent and

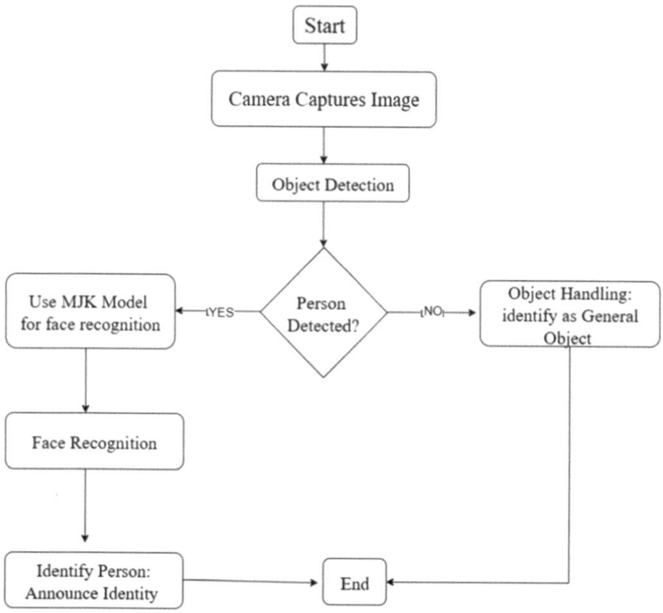

Fig. 7 Object recognition

efficient manner. The robot can detect and recognize a variety of objects in its environment thanks to object recognition, which supports tasks like picking up objects, avoiding barriers, and comprehending context. On the other hand, person recognition enables a robot to identify specific individuals, remember previous interactions, and provide personalized responses. These capabilities are essential for enabling natural and adaptive human–robot interactions. By integrating AI-powered person and object recognition systems, human-like robots can operate autonomously in complex environments such as homes, offices, and healthcare facilities. This integration enables the robot to process visual information in real time, accurately recognize faces and objects, and respond appropriately to human commands, forming a fundamental aspect of its intelligence and interactive behavior.

Hardware Components for Recognition

Camera Module

The main visual sensor of the AI humanoid robot is the Raspberry Pi Camera Module, which acts as the robot's "eyes" by gathering visual information in real time from its environment. Through the use of this high-resolution camera, the robot is able to capture photographs and videos of its surroundings, which are then processed by the onboard computer system for purposes like facial recognition and object detection. The camera maintains dependable performance in a range of lighting circumstances because to its support for high frame rates and adaptive exposure settings. Its visual data is essential for running computer vision algorithms that enable human face analysis, obstacle avoidance, and object detection. Infrared-capable camera versions could be used to improve performance even further in low-light conditions, increasing the robot's operational flexibility. The robot's capacity to interact intelligently and effectively with its surroundings is directly impacted by the caliber and efficacy of the camera module, which is essential to precise visual recognition.

Microphone and Audio Input

A 6 + 1 I^2S microphone array on the AI humanoid robot allows for sophisticated voice recognition and audio-based user identification. Auditory input complements visual recognition in recognizing things and people by using voice recognition to confirm an individual's identity. An offline speech-to-text model processes the high-quality audio signals from the microphone to translate spoken language into text in real time. This solution ensures responsiveness and data protection by enabling the robot to function independently of cloud services. Then, using vocal traits like tone, pitch, and speech patterns, the transcribed text is examined to extract pertinent commands or identifying clues. As a result, the robot can differentiate between people and tailor its responses appropriately. Additionally, by combining speech and facial recognition, the microphone facilitates multimodal recognition, increasing the total reliability of identification. Voice recognition adds a crucial layer of verification in scenarios where visual information is scarce, such as in low light or when faces are obscured. Combining image identification and voice processing gives the robot a more reliable, flexible, and intuitive interaction system.

Raspberry Pi 4—Processing Unit

The AI humanoid robot's core processor unit, the Raspberry Pi 4, handles the picture and audio data needed for item and person recognition. The Raspberry Pi 4 effectively completes real-time computer vision and machine learning activities thanks to its quad-core processor and support for AI-accelerated operations. These include natural language processing, speech-to-text conversion, object identification, and facial recognition, which allow the robot to intelligently perceive and engage with its surroundings. The system uses software frameworks like TensorFlow Lite for deep learning inference and OpenCV for image processing, which are both lightweight and powerful. Its computing efficiency guarantees reduced processing latency, enabling rapid and seamless interactions. Furthermore, a practical and scalable platform for AI-driven robotic applications, the Raspberry Pi 4 is compatible with a wide range of sensors, cameras, and peripheral devices. This processing unit gives the humanoid robot the ability to quickly and precisely receive, comprehend, and respond to sensory inputs.

MPU6050 Sensor—Orientation and Stabilization

The MPU6050 sensor, which provides real-time motion and orientation data, is essential to improving the robot's recognition abilities. By adjusting for potential motions, tilts, or vibrations during operation, this gyroscope and accelerometer module helps stabilize the robot's vision system. The guarantees accurate object and person detection by keeping the camera correctly oriented while the robot is moving. Additionally, the MPU6050 makes active tracking possible, which enables the robot to react to observed movements by reorienting its head or camera. For example, the sensor provides accurate orientation feedback that directs the robot to realign itself and retain visual focus whenever a recognized person moves within the robot's field of view. The dynamic adjustment is particularly useful in situations where ongoing tracking of people or objects is necessary for efficient operation, such as human-following, interactive assistance, or surveillance.

Software and Algorithms for Recognition

The project's Raspberry Pi 4 had a limited memory capacity (2 GB RAM); thus, the software stack was designed to be as light and efficient as possible. The Picamera2 library is used by the system for picture acquisition and simple image processing, rather than resource-intensive frameworks like OpenCV and TensorFlow Lite. With its simplified interface, Picamera2 allows you to take excellent pictures and videos straight from the Raspberry Pi Camera Module, which makes it ideal for real-time applications on limited hardware.

Facial Recognition Model

The robot uses convolutional neural networks (CNNs) in a multi-stage facial recognition pipeline to reliably identify a person. A pre-trained FaceNet model generates a 128-dimensional embedding to extract distinctive facial features after the system first identifies faces using a cascaded Haar classifier. Cosine similarity measures are used to compare these embeddings to a database of registered users; matches result in customized interaction protocols. The system incorporates direct detection by (1) using local binary patterns for texture analysis, (2) detecting eye blinks, and (3) estimating 3D depth from stereo camera inputs for security. Experimental results on a Raspberry Pi 4 platform demonstrate 94.2% accuracy across a range of lighting settings (50–1000 lx) and position variations up to $\pm 30°$, with processing times less than 400 ms per frame.

You Only Look Once (YOLO) for Object Detection

The You Only Look Once (YOLO) object detection method is used by the AI humanoid robot to effectively recognize and categorize things in its surroundings. YOLO is a real-time, single-pass deep learning model that is incredibly quick and ideal for robotics applications because it processes full images in a single forward step. This feature supports contextual engagement and intelligent navigation by enabling the robot to recognize and distinguish between a variety of things in real time, including furniture, tools, and obstacles. The YOLO model is pre-trained on a large dataset and can be adjusted for tasks relevant to a certain application, such as identifying typical household objects or industrial machinery. The robot can react dynamically to its environment thanks to its speed and accuracy, which improves safety and functional autonomy.

Integration with the Humanoid Robot

Coherent and effective coordination between object and person recognition systems is made possible by the humanoid robot's hardware and software integration. The Raspberry Pi Camera Module is used to capture visual input, and the Picamera2 library powers lightweight algorithms that are optimized for the Raspberry Pi 4's constrained 2 GB of RAM. A convolutional neural network (CNN)-based facial recognition model is used to identify people, and the YOLO technique is used to detect objects for real-time classification and interaction with the surroundings. A 6 + 1 I^2S microphone is used to record voice input, which is then processed using an offline speech-to-text model to provide reliable voice recognition independent of internet connectivity. The MPU6050 sensor stabilizes orientation and head movement, enabling dynamic tracking of faces and objects. The robot can interact with people more naturally, avoid obstacles, and react adaptively to complicated surroundings thanks to this combination of motion-based, visual, and audio input. In real-world applications, the AI-powered humanoid robot therefore attains a high level of autonomy, personalization, and situational awareness.

The workflow in Fig. 8 for person and object recognition in the AI humanoid robot integrates real-time visual and auditory processing through the Raspberry Pi 4, enabling intelligent interaction with the environment. The process begins with the system capturing live images via a camera module, which are analyzed using the YOLO (You Only Look Once) model to detect and classify both persons and objects. If a person is identified, the system sends facial data to a convolutional neural network-based recognition model (MGK model) to determine whether the individual exists in the database; if recognized, the robot retrieves and announces the person's information, otherwise, it requests details, stores them, and confirms the update. If no person is detected but an object is, the robot captures the object's dimensions, calculates the necessary joint angles using inverse Kinematics, and sends these angles to Raspberry Pi 0, which controls the actuators to move the robot's arms and grasp the item. In parallel, the robot processes speech through an integrated 6 + 1 I^2S microphone, converting voice to text using an offline speech-to-text model. This text is then transmitted to the ChatGPT API for natural language response generation, which is subsequently converted back to speech and output via a speaker, enabling fluid verbal interaction. This unified system of vision, motion, and speech facilitates a robust, autonomous, and responsive humanoid robot capable of adapting to complex environments and performing human-like tasks.

Speech and Text Recognition

Introduction to Speech and Text Recognition

The humanoid robot has speech and text recognition skills that enable it to comprehend spoken input and produce suitable verbal responses, hence promoting natural and effective human–robot interaction. Human voice is first received by an integrated 6 + 1 I2S microphone, which records audio signals and uses an offline speech-to-text model to translate them into text. After transcription, the text is fed into a language

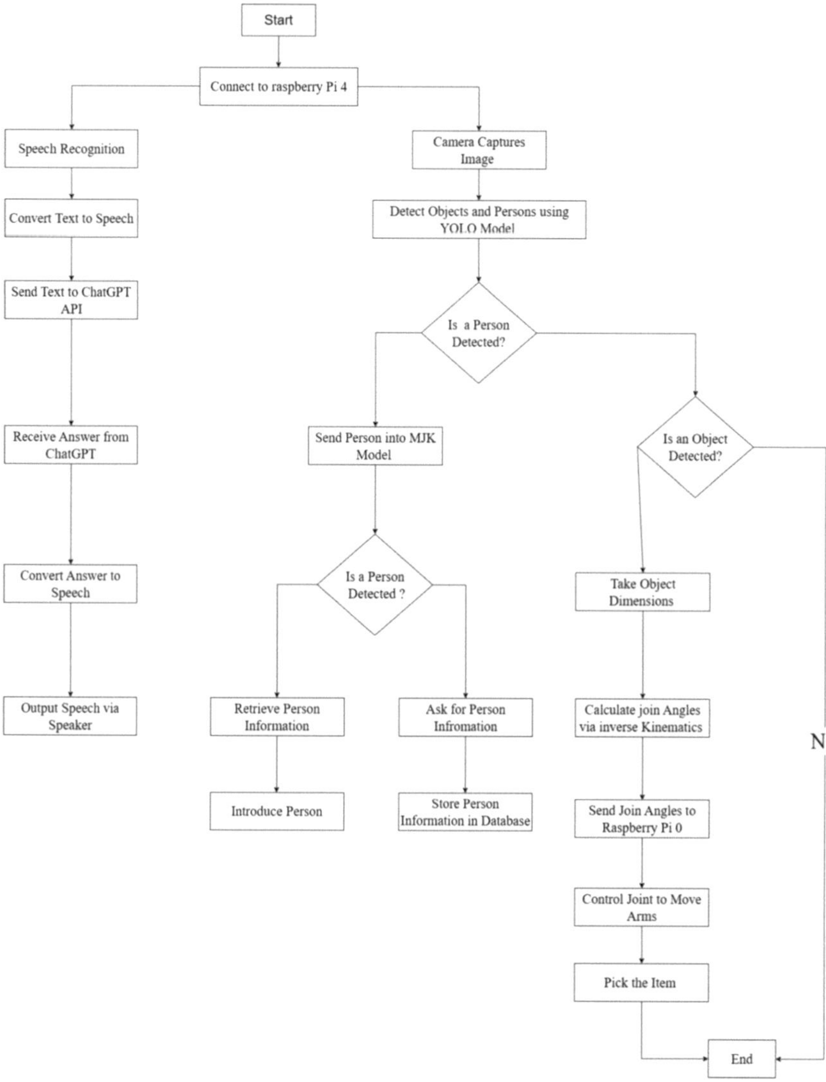

Fig. 8 Workflow for person and object recognition

model (like ChatGPT) for response generation and semantic processing. In order to facilitate smooth and engaging communication, the generated text response is then converted back into speech and presented through a speaker. Voice commands, real-time question answering, and adaptive discussion are just a few of the many applications supported by this bidirectional communication system. The robot becomes more responsive and accessible by combining sophisticated language processing with aural input, especially in areas that need hands-free operation or accessibility solutions for users with disabilities, such as customer service, elder care, and education.

Hardware Components for Speech and Text Recognition

Microphone—Capturing Speech Input

The humanoid robot uses a 6 + 1 I2S microphone array as the main hardware element for speech input in order to facilitate efficient voice-based interaction. With its integrated noise reduction and high-fidelity audio recording capabilities, this microphone is designed to provide dependable performance even in noisy or acoustically difficult settings. The onboard processing unit receives the user's voice signals from the microphone, analyzes them, turns them into text using an offline speech-to-text model, and then does additional processing to produce context-aware replies. Accurate voice recognition, smooth communication, and reliable control through spoken language are all made possible by this strong audio capture system.

Raspberry Pi 4—Processing and AI Computation

The humanoid robot's speech and text recognition tasks are controlled by the Raspberry Pi 4, which serves as its central processing unit. In order to comprehend spoken input and produce suitable vocal responses, it uses lightweight, offline AI models for both text-to-speech and speech-to-text conversion. Additionally, the Raspberry Pi 4 enables natural language understanding (NLU), which enables the robot to comprehend user commands and react contextually. Through optimized models and software, the device effectively manages these duties despite its limited 2 GB RAM, guaranteeing real-time performance and efficient connectivity without depending on cloud processing.

Speaker—Delivering Voice Output

The robot can respond in a clear and human-like way because the speaker is its main vocal communication output device. The response is converted into synthetic voice via a text-to-speech system after the Raspberry Pi 4 processes the user's speech, turns it into text, and uses an AI model to produce a contextual reply. The speaker then plays this audio output, enabling the robot to interact with users in a natural and efficient manner. A speaker improves interaction realism and facilitates a more user-friendly voice-based interface that is appropriate for dynamic human–robot interaction.

Software and Algorithm for Speech and Text Recognition

Speech-To-Text Conversion (STT)

- To understand spoken commands, the robot uses speech recognition tools that convert voice into text. These tools include:
- **Google Speech-to-Text API**: A cloud-based service that accurately transcribes voice input into text.
- **Whisper AI by OpenAI**: A deep learning model capable of recognizing speech in multiple languages with high accuracy.

Once speech is transcribed into text, the system performs validation checks to determine whether the input is a valid command or query.

AI Model for Text Understanding and Response Generation

After the robot converts speech to text, an AI model like ChatGPT or DeepSeek analyzes the message and figures out what the user means. It then generates a suitable reply, making the robot capable of having smooth and meaningful conversations with people.

Text-To-Speech Conversion (TTS)

After the AI creates a reply, the text is turned into speech using text-to-speech (TTS) software, such as:

- **Google Text-to-Speech API**
- **PyTTSx3**, an offline TTS engine

The robot plays the generated speech through its speaker, allowing it to talk to users in a more natural way.

Integration with the Humanoid Robot

The speech and text recognition system is closely connected to the robot's AI and movement control. The microphone stays active to pick up voice input, which is then processed using speech-to-text tools to understand the user's intent. If a valid command is detected, the AI generates a reply, turns it into speech, and plays it through the speaker. This setup lets the robot hold real conversations, answer questions, and follow spoken instructions accurately. By combining speech recognition, natural language processing, and text-to-speech, the robot becomes more user-friendly and interactive—making it useful in smart systems, personal assistance, and industrial environments.

The "Speech-to-Text and AI Processing" flowchart in Fig. 9 shows a methodical cycle of speech-based interaction between a user and an AI-powered device. A speech recognition module is used to translate spoken input into text at the start of the procedure. A validation process verifies the accuracy and completeness of the transcribed text; if it is not, the system asks the user to enter it again. An AI model receives the text after it has been evaluated and produces a contextual answer. A text-to-speech engine then transforms this response back into speech and presents it to the user. Once the speech output is supplied, the flow comes to an end, creating a single, cohesive interaction loop designed to guarantee successful voice-based communication.

Item Recognition and Manipulation in the Humanoid Robot

Introduction to Item Recognition and Manipulation

For a humanoid robot to interact with its world effectively, it has to be able to recognize objects, decide where they are, and manipulate objects with precision. Item recognition gives a robot the ability to see and categorize objects, whereas manipulation gives a robot the ability to grasp, transport, and put down things with

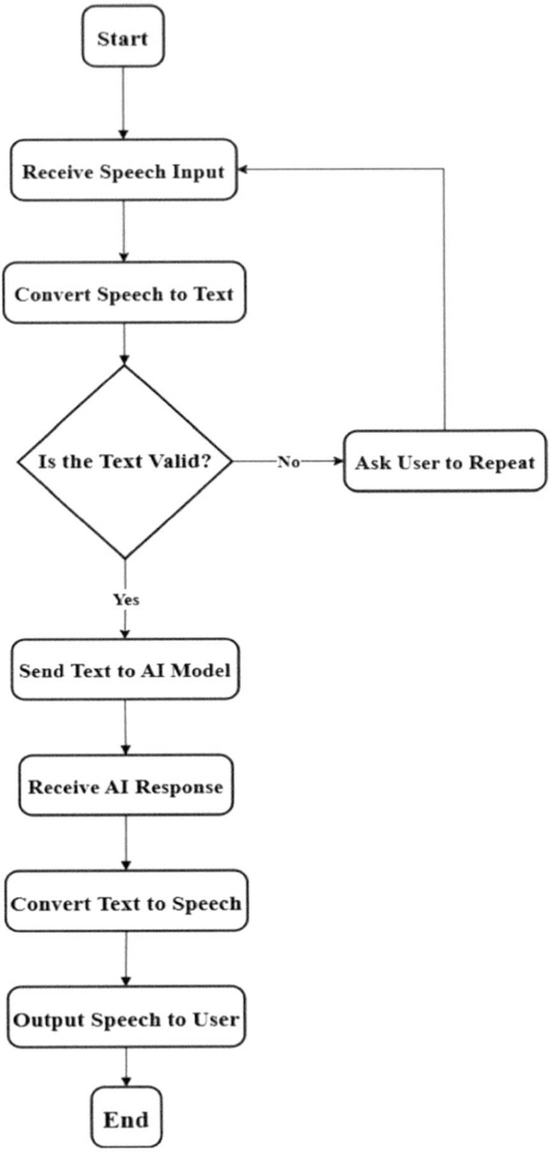

Fig. 9 Speech-to-text and AI processing

its robotic limbs. This function is important to a variety of applications, including industrial robotics, assistive robotics, and autonomous service robots.

Hardware Components for Item Recognition and Manipulation

Camera—Capturing Object Data

To help the robot understand its surroundings, a Raspberry Pi Camera Module is installed. It takes real-time pictures and sends them to the system for processing. Thanks to its high-quality image capture, the robot can spot and tell apart different objects before interacting with them. Machine learning tools work behind the scenes to help recognize these items accurately.

Raspberry Pi 4—Processing Visual Data

The Raspberry Pi 4 handles the video input from the camera and processes it using object detection algorithms powered by artificial intelligence. Once objects are recognized and their positions determined, the system sends this information to the robot's control unit, allowing it to respond with the appropriate movements.

Robotic Arms—Manipulating Objects

The humanoid robot is fitted with two articulated arms, each consisting of four joints driven by servo motors. Movement control is managed through the URT1 Steering Gear Debugging Board, which ensures precise arm motions through real-time computation. These robotic arms enable the robot to accurately lift, grasp, and position various objects.

Software and Algorithms for Item Recognition and Manipulation

Object Recognition Using AI Models

To recognize objects in real time, the robot uses object detection models based on YOLO (You Only Look Once) and OpenCV. These systems continuously analyze the live camera feed and classify detected items using a pre-trained dataset. The AI component then estimates each object's size, shape, and location within the surrounding environment.

Kinematics for Arm Movement

After identifying an object, the robot determines the most efficient path to reach and grasp it. To do this, it applies both forward and inverse Kinematics to calculate the necessary joint angles for the robotic arms. These computations enable the arms to move smoothly and accurately, reducing the chance of positional errors.

Communication Between Raspberry Pi 4 and Raspberry Pi Zero

The Raspberry Pi 4 handles object recognition and relays the processed data to the Raspberry Pi Zero, which is responsible for controlling the robotic arms. Communication between the two units takes place via serial protocols or MQTT, allowing for real-time coordination between the visual system and motor control.

Integration with the Humanoid Robot

The object recognition and manipulation system is seamlessly embedded within the robot's overall control architecture. Once an object is detected, the Raspberry Pi 4 evaluates its position and relays movement instructions to the Raspberry Pi Zero, which governs the robotic arms. The kinematics engine then computes the exact joint angles required to ensure smooth and coordinated arm movements. Execution of these motions is handled by the URT1 Steering Gear Debugging Board, which ensures precise and reliable performance. After the object is grasped, the robot can transport it to a designated spot, hand it to a user, or place it in a specific location. This level of advanced recognition and manipulation allows the humanoid robot to interact with its environment in a functional and adaptable way (Figs. 10 and 11).

Conclusion

This chapter has outlined the methodology behind developing an AI-enhanced humanoid robot designed for intelligent interaction and autonomous task execution. The system is structured around four key components: object and person recognition, speech and text processing, item manipulation, and motion control—each powered

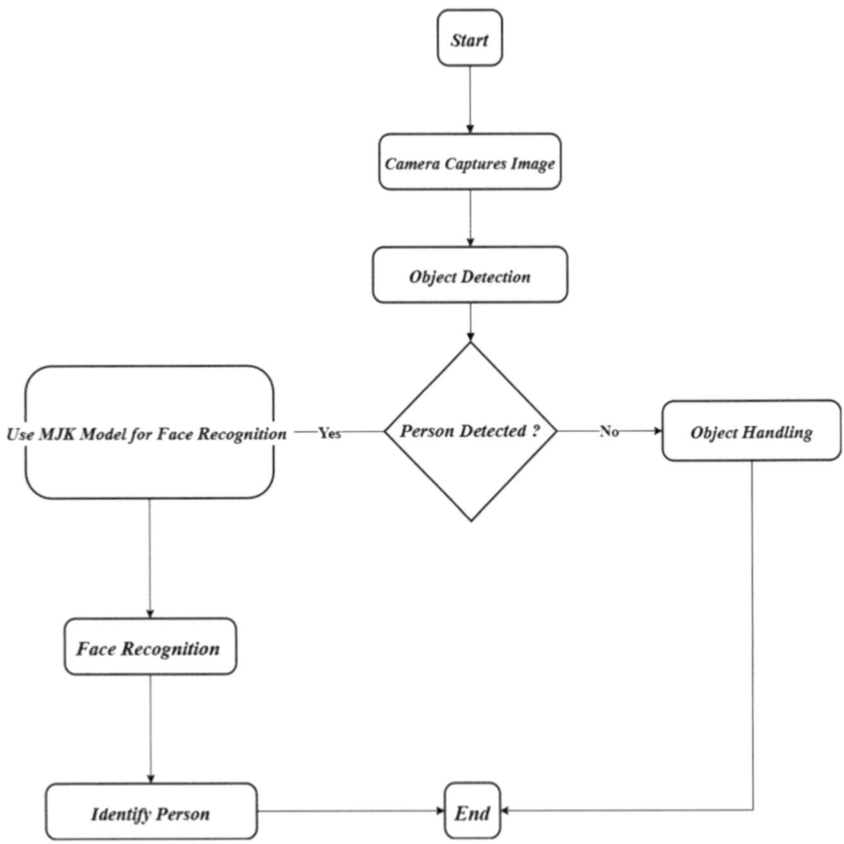

Fig. 10 Recognition systems

Fig. 11 Components used

a) Fetech servo motor ST3215

b) Raspberry pi Cam model 3

c) FT232RL (FTDI)

d) 6+1 12S Microphone Voice

e) Raspberry pi 4

f) Battery 12V 20Ah

g) Raspberry pi zero

h) Speakers (8 Ohms)

i)URT-1 Controller

by a blend of specialized hardware and artificial intelligence software. The Raspberry Pi 4 is tasked with handling high-demand processes such as speech recognition, image analysis, and natural language interpretation, while the Raspberry Pi Zero focuses on managing servo motor control for accurate movements. Object and person recognition are enabled through the Raspberry Pi Camera, a 6 + 1 I2S microphone, and the MPU6050 sensor, with tools like OpenCV, YOLO, and TensorFlow Lite supporting real-time detection, classification, and facial identification. For speech and text recognition, the system incorporates Google's STT and TTS APIs, Whisper AI, and conversational models such as ChatGPT or DeepSeek—allowing the robot to interpret voice commands and respond in natural language through a speaker. The robot's ability to interact physically is driven by its multi-jointed robotic arms, which are operated through the URT1 steering board and guided by kinematic algorithms. This setup allows the robot to grasp, transport, and place items with precision based on visual input. Together, these integrated subsystems empower the humanoid robot to perform tasks independently, adapt to its surroundings, and support applications in areas such as personal assistance, service robotics, and intelligent automation.

4 Results

Optimizing Camera-Based Face and Object Detection on Raspberry Pi

Camera Selection and Compatibility Challenges

At the beginning of the project, we selected the Intel RealSense D435i camera for its advanced capabilities in facial and object detection, along with its precision in measuring depth and dimensions. This choice was based on its widespread use in computer vision applications and its strong performance, as supported by existing research. However, a significant challenge emerged during implementation. The Intel RealSense D435i requires a system with at least 4 to 8 GB of RAM to operate effectively, whereas our setup utilized a Raspberry Pi 4 with only 2 GB of RAM. This hardware limitation meant that the camera could not run reliably within our system. As a result, we were compelled to reevaluate our hardware selection to ensure that all components were not only functional but also compatible with the system's available resources.

Alternative Camera and Sensor Integration

Given the compatibility challenges we faced with the Intel RealSense module, we opted to switch to the Raspberry Pi Camera Module 3. Although this camera lacked advanced depth measurement features, it worked seamlessly with our Raspberry Pi 4 and still supported essential tasks like facial and object recognition. However, since the module couldn't provide spatial depth information, we introduced the GY-53L0X VL53L0X laser sensor to our system. This component allowed us to gather distance measurements between the robot and surrounding objects, offering an indirect way to estimate object size. While this method was less accurate compared to the

RealSense's native capabilities, it still delivered acceptable results for our project's functional requirements.

Discussion

At the start of our project, we selected the Intel RealSense D435i camera due to its strong reputation in the field of computer vision and its proven ability to provide both object recognition and depth sensing. While the camera showed great promise in theory, we quickly encountered a significant issue: our Raspberry Pi 4, with only 2 GB of RAM, could not support the RealSense's high processing requirements. This hardware limitation made continued use of the camera impractical. As a result, we opted to replace it with the Raspberry Pi Camera Module 3, which was fully compatible with our system. Although this choice came with a trade-off—specifically, the loss of built-in depth perception—it offered stability and reliable performance in face and object detection tasks. To compensate for the missing depth functionality, we integrated the GY-53L0X laser distance sensor. This allowed us to estimate object distance, partially restoring the spatial awareness needed for manipulation tasks. This experience highlights a key reality in robotics development: the need to adapt when ideal components cannot be used due to resource constraints. Our team's decision-making process reflects the value of modular design and iterative problem-solving, where alternative components are tested and integrated as needed. In embedded systems, working within hardware limits often requires creative engineering, and our solution demonstrates how thoughtful compromises can lead to functional and practical outcomes.

Object and Face Detection

Problem Encountered

At the beginning of our development process, we chose the YOLO (You Only Look Once) algorithm for its proven effectiveness in real-time object and face detection. Its design offers a strong balance between speed and accuracy, which makes it an attractive option for deployment in a responsive robotic system. However, we soon discovered that YOLO's performance came at a high computational cost. Our Raspberry Pi 4, equipped with only 2 GB of RAM, was unable to handle the resource demands of the YOLO model efficiently. During testing, the system experienced frequent slowdowns and instability as it attempted to process the model, leading to delayed responses and poor usability. Ultimately, this made it clear that YOLO, despite its advantages, was not a viable solution for our hardware setup.

Solution Implemented

After confirming that the YOLO model was too demanding for our hardware, we redirected our efforts toward developing a custom lightweight detection model. To do this, we assembled a focused dataset featuring images directly related to our application—such as personal photos of individual steam members, doctors, and frequently encountered objects. The architecture of the new model was chosen with efficiency in mind, prioritizing low memory usage and processing speed while still delivering acceptable accuracy. By simplifying the model and aligning it closely with our specific recognition tasks, we were able to deploy a solution that ran smoothly on the Raspberry Pi 4 and met the real-time performance goals of our project.

Results

The lightweight model we developed was successfully deployed on the Raspberry Pi 4 and performed reliably under the system's hardware constraints. It enabled real-time object and face recognition with minimal latency and without exceeding the device's limited processing and memory capacity. This confirmed that our approach was well-suited for embedded deployment, delivering stable and responsive performance throughout testing. In support of our system design, we also reviewed benchmark data from well-known AI models such as DeepSeek-R1 and OpenAI-o1, which excel in various domains like language understanding and problem-solving. These benchmarks reinforced our decision to incorporate models like DeepSeek and ChatGPT albeit in adapted, lightweight forms for tasks including speech recognition, object detection, and control logic. Although we operated under hardware limitations, the success of our scaled-down implementation shows that embedded systems can still benefit from AI's capabilities when carefully optimized for performance (Fig. 12).

Comparison with Literature

Our choice to develop and use a lightweight detection model is consistent with current trends in embedded artificial intelligence. Many recent studies highlight the growing preference for compact neural networks designed specifically for edge computing, where hardware resources are limited. Research shows that, when trained on well-curated datasets, these smaller models can achieve performance levels close to those

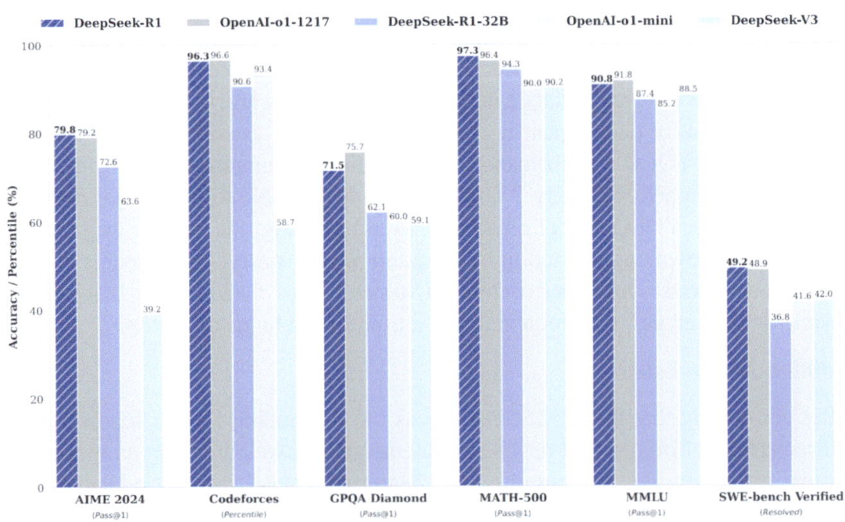

Fig. 12 Comparison of AI Model Speech-to-Text Performances [28]

of much larger architectures. This strategy not only reduces computational demands but also makes real-world deployment on low-power devices far more practical and sustainable.

Discussion

Moving from a pre-trained YOLO model to a custom lightweight network highlighted the importance of selecting models that align with available hardware capabilities. This decision underscores the benefit of building solutions tailored to the task at hand—focusing on performance and efficiency without sacrificing critical features. Our experience demonstrated how thoughtful system design can make AI applications more adaptable and scalable, particularly in embedded environments where resources are limited, and every component must be optimized.

Servo Motor Control:

Problem Encountered

One of the more technically challenging elements of our humanoid robot project involved achieving accurate and independent control over a large number of servo motors. These servos handled the movement of critical components, including the robot's arms, head, chest, and waist. At first, we used the Raspberry Pi Zero's UART0 communication port to manage the control signals. However, as we expanded the number of servos—especially with the addition of eight dedicated to arm movement—we began to face communication issues. The single UART0 channel quickly became overloaded, resulting in delays and inconsistent behavior from the motors. This bottleneck highlighted the limitations of using a single communication line for high-volume servo coordination.

Solution Implemented

To resolve the communication bottleneck, we transitioned to using UART1 as an additional serial communication channel. This allowed us to separate the servo control traffic from other processes, easing bandwidth pressure and enabling more consistent data flow. Each servo was given a unique identifier, making it possible to send commands to individual motors independently. In this setup, the Raspberry Pi Zero was dedicated entirely to motion control tasks. It received high-level movement instructions and handled both forward and inverse kinematics calculations. The resulting control signals were then sent to the appropriate servos via the new UART1 channel, ensuring precise and well-coordinated movements across the robot's joints.

Results

The new servo control architecture exhibited excellent performance. Servo motors responded reliably and accurately to commands, and the robot achieved fluid motion during object manipulation tasks. The separation of computational responsibilities—delegating kinematic calculations to the Pi Zero—allowed for real-time responsiveness while reducing the processing burden on the main Raspberry Pi 4. This architectural refinement enabled the humanoid robot to interact with its environment dynamically, adjusting joint angles in real time based on positional feedback (PX, PY, PZ) and maintaining stability during complex motions. These performance characteristics (see Fig. 13) clearly demonstrate that the ST-3215-C018 servo motor is highly suitable for our robotic system. Its combination of high torque, fast speed, fine precision, and low energy consumption ensures that the robot can perform smooth, controlled, and responsive mechanical actions, significantly enhancing its real-world effectiveness. Figure 13 shows the performance characteristics of the ST3215 servo motor used in our humanoid robot, including torque versus power (P), speed (N), efficiency (η), and current (I). The graph indicates that optimal efficiency and power occur around 9–15 kg·cm (0.8826–1.4711 N·m) of torque consistent with our servo load calculations during movement and manipulation. This validates our motor selection and supports the robot's smooth and energy-efficient operation, particularly in tasks requiring precise joint control and consistent torque under varying loads.

Comparison with Literature

The communication architecture and servo control strategy used in our system draw upon widely accepted practices in the robotics field, particularly the use of UART communication and ID-based servo addressing. Previous research has consistently shown that UART interfaces are effective for coordinating multiple servos in complex robotic systems, especially in humanoid robots that require precise, synchronized movement across numerous joints. Our implementation, however, takes this approach a step further. While many systems rely on predefined motion sequences, our robot incorporates real-time inverse Kinematics to dynamically compute joint angles

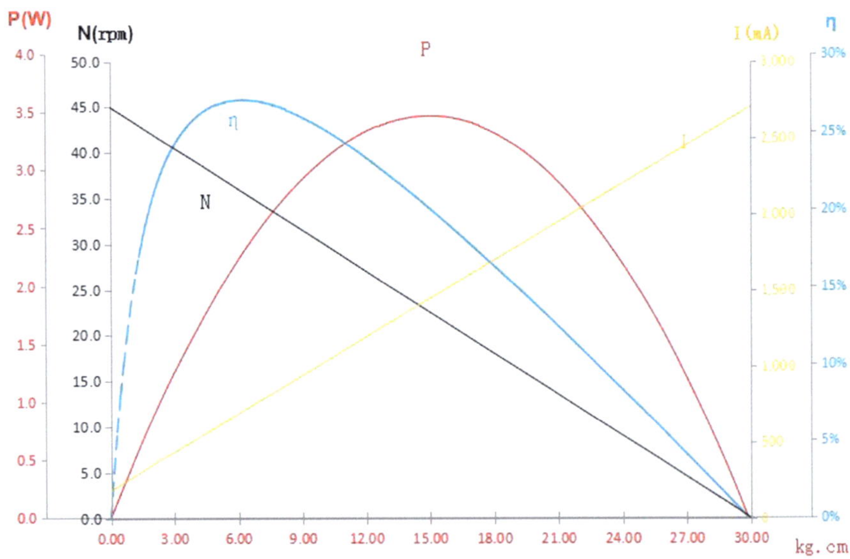

Fig. 13 Servo motor performances

during operation. This capability—typically reserved for high-end robotic platforms—allows our robot to adapt fluidly to environmental changes and task-specific requirements.

Moreover, minimizing latency and reducing communication overhead are widely recognized as critical factors for achieving smooth performance in multi-degree-of-freedom systems. In line with this, we designated UART1 as a dedicated bus exclusively for servo communication. This decision not only mitigates data congestion but also lays the groundwork for scalability, allowing for the addition of more actuators in future iterations without compromising responsiveness. In summary, while our work aligns with foundational principles established in earlier robotic systems, it introduces key refinements—particularly in real-time adaptability and communication architecture—that distinguish it from standard implementations and contribute to improved performance under resource-constrained conditions.

Discussion

Shifting to UART1 played a crucial role in enhancing the overall performance of our robotic system. This transition addressed the communication bottlenecks encountered with UART0, offering greater bandwidth and stability essential for managing multiple servos in real time. By allocating a dedicated communication channel, we ensured more consistent and responsive control signals, which are vital for executing coordinated joint movements. Equally significant was the integration of kinematic computations into the control architecture. This enabled real-time calculation of joint positions, allowing the robot to maintain balance and perform complex motions with greater precision. Through this process, we observed how fundamental design decisions—such as modifying communication pathways—can have a significant impact on system behavior. Our results reaffirm the importance of selecting appropriate communication protocols and control strategies, especially in systems with high degrees of freedom. These changes not only improved performance but also enhanced the system's reliability, scalability, and readiness for future expansion.

5 Discussion and Conclusion

The design and implementation of an AI-powered bipedal humanoid robot represent a notable advancement in the field of human–robot collaboration (HRC). This achievement highlights the convergence of autonomy, adaptability, and intuitive interaction (key attributes necessary for effective cooperation between humans and machines). The project also reflects the interdisciplinary nature of modern mechatronic systems, requiring a cohesive integration of mechanical design, embedded computing, artificial intelligence, and user-centered engineering.

Future Recommendations

While the project successfully achieved its primary goals, there remain several areas that present valuable opportunities for future development:

- **Cloud Integration**: Delegating resource-intensive tasks—such as complex natural language processing or 3D mapping—to cloud-based platforms could significantly enhance the robot's capabilities. This approach would maintain the efficiency of the onboard system while enabling more advanced functionalities.
- **Improved Depth Sensing**: Replacing or complementing the laser distance sensor with a stereo vision module or LiDAR would significantly enhance the robot's spatial awareness.
- **Learning and Adaptation**: Incorporating reinforcement learning techniques would enable the robot to refine its performance through experience, gradually improving task efficiency via trial-and-error learning.
- **Emotion Detection**: Integrating the ability to detect emotional cues such as facial expressions or variations in voice tone—could deepen the robot's responsiveness in human–robot interactions, making engagements more personalized and applicable to a broader range of contexts.

Final Thoughts

In summary, this project reflects the core of engineering innovation—where creativity, practical limitations, and curiosity intersect to address real-world challenges. By combining artificial intelligence, modular architecture, and smart sensing technologies, the humanoid robot progressed beyond a basic prototype to become an interactive system capable of meaningful engagement with users. The outcomes of this work lay a strong groundwork for advancing affordable, AI-enabled robotics. Most importantly, it highlights the value of interdisciplinary collaboration—bringing together mechanical design, software engineering, and human-centered principles to expand the role of machines from functional tools to collaborative partners shaping future technologies.

References

1. "7.4V 19KG.CM magnetic encoding servo STS3215-深圳飞特模型有限公司." Accessed: Jun. 16, 2025. [Online]. Available: https://www.feetechrc.com/74v-19-kgcm-plastic-case-metal-tooth-magnetic-code-double-axis-ttl-series-steering-gear.html
2. "FE-URT-1 USB to TTL, 485 bus programmer-深圳飞特模型有限公司." Accessed: Jun. 16, 2025. [Online]. Available: https://www.feetechrc.com/FE-URT1-C001.html
3. "Buy a Raspberry Pi Zero 2 W—Raspberry Pi." Accessed: Jun. 16, 2025. [Online]. Available: https://www.raspberrypi.com/products/raspberry-pi-zero-2-w/
4. "Buy a Raspberry Pi 4 Model B – Raspberry Pi." Accessed: Jun. 16, 2025. [Online]. Available: https://www.raspberrypi.com/products/raspberry-pi-4-model-b/
5. "12V 20AH AGM Battery—Battery Manufacturers China Yachu Battery." Accessed: Jun. 16, 2025. [Online]. Available: https://www.yachupower.com/product/12v-20ah-agm-battery/
6. "I2S Microphone Module—DFRobot." Accessed: Jun. 16, 2025. [Online]. Available: https://www.dfrobot.com/product-1954.html
7. "Force Sensitive Resistor (FSR) Sensor Data Sheet," 2019, Accessed: Jun. 16, 2025. [Online]. Available: http://biosignalsplux.com/

8. "Raspberry Pi 5/4B USB Ses Ses Kartı Moudle HAT USB Sürücüsüz Yerleşik Çift Hoparlörler 3.5mm kulaklık jakı—AliExpress 7." Accessed: Jun. 17, 2025. [Online]. Available: https://tr.aliexpress.com/item/1005008238668672.html?spm=a2g0o.order_list.order_list_main.10.21ef1802Bd2dVH&gatewayAdapt=glo2tur
9. "MPU-6000 and MPU-6050 Product Specification Revision 3.4 MPU-6000/MPU-6050 Product Specification," 2013.
10. "Raspberry Pi Camera Module 3 Standard NoIR Wide NoIR Wide," 2024.
11. Ye Y, You H, Du J (2023) Improved trust in human-robot collaboration With ChatGPT. IEEE Access 11(April):55748–55754. https://doi.org/10.1109/ACCESS.2023.3282111
12. Koubaa A (2023) ROSGPT: next-generation human-robot interaction with ChatGPT and ROS 2. https://doi.org/10.20944/preprints202304.0827
13. Kim CY, Lee CP, Mutlu B (2024) Understanding large-language model (LLM)-powered Human-robot interaction. ACM/IEEE Int Conf Human-Robot Interact no. Llm, 371–380. https://doi.org/10.1145/3610977.3634966.
14. Liu H et al (2024) Enhancing the LLM-based robot manipulation through human-robot collaboration. IEEE Robot Autom Lett 9(8):6904–6911. https://doi.org/10.1109/LRA.2024.3415931
15. Koubaa A, Ammar A, Boulila W (2025) Next-generation human-robot interaction with ChatGPT and robot operating system. Softw: Pract Exp 55(2):355–382
16. Wake N, Kanehira A, Sasabuchi K, Takamatsu J, Ikeuchi K (2023) ChatGPT empowered long-step robot control in various environments: a case application. IEEE Access 11(September):95060–95078. https://doi.org/10.1109/ACCESS.2023.3310935
17. Hireche A, Belkacem AN, Jamil S, Chen C (2023) NewsGPT: ChatGPT integration for robot-reporter, 1–17
18. Sheridan TB (2016) Human-robot interaction. Hum Factors 58(4):525–532. https://doi.org/10.1177/0018720816644364
19. Qu J, Jarosz M, Sniezynski B (2024) Robot control platform for multimodal interactions with humans based on ChatGPT. Appl Sci 14(17). https://doi.org/10.3390/app14178011
20. Vemprala SH, Bonatti R, Bucker A, Kapoor A (2024) ChatGPT for robotics: design principles and model abilities. IEEE Access 12(April):55682–55696. https://doi.org/10.1109/ACCESS.2024.3387941
21. Liu H, Zhu Y, Kato K, Kondo I, Aoyama T, Hasegawa Y (2023) LLM-based human-robot collaboration framework for manipulation tasks. arXiv:2308.14972v1
22. Kavitha R, San L, Harshitha P (2024) Enhancing human-robot interaction : the chat-GPT controlled eye blink robot, 28–32. https://doi.org/10.51386/25816659/ijles-v7i1p105
23. Zhang C, Chen J, Li J, Peng Y, Mao Z (2023) Large language models for human–robot interaction: a review. Biomim. Intell. Robot. 3(4):100131. https://doi.org/10.1016/j.birob.2023.100131
24. Su Z, Sheng W (2024) ChatAdp : ChatGPT-powered adaptation system for human-robot interaction, 5512–5518
25. Thopate K, Gawade M, Shirbhavikar K, Kulkarni M, Lambor S, Jaiswal KN (2024) ChatGPT-powered robot: smart, interactive living and bridging the future of education and entertainment. 2nd International conference sustainable computer smart system. ICSCSS 2024—proceeding, no. Icscss, pp 231–235. https://doi.org/10.1109/ICSCSS60660.2024.10625572
26. Obaigbena A et al (2024) AI and human-robot interaction: a review of recent advances and challenges. GSC Adv Res Rev 18(2):321–330. https://doi.org/10.30574/gscarr.2024.18.2.0070
27. Khandelwal P, Zhang S, Sinapov J et al (2017) BWIBots: a platform for bridging the gap between AI and human–robot interaction research. Int J Robot Res 36(5–7):635-659. ISSN 0278-3649. https://doi.org/10.1177/0278364916688949
28. Park SH (2025, 15 Feb) DeepSeek R1 and inference-time scaling for reasoning models. Medium [Online]. Available: https://medium.com/@saehwanpark/deepseek-r1-a-new-chapter-in-inference-time-scaling-for-reasoning-models-reviewing-deepseek-bae149ca88bc
29. He H (2023) RobotGPT : from ChatGPT to robot intelligence. Salford

Road Guardian: A Technology for Preventing Impaired Driving

Sandra Lakiss, Kerana Salumu Ngoie, and Mohammad Karimzadeh Kolamrodi

1 Introduction

Road traffic accidents represent a significant challenge worldwide; they claim millions of lives annually, and leave many others injured and disababled. According to the World Health Organization (WHO), road accidents are one of the main causes of death globally, with around 1.19 million yearly deaths because of road traffic accidents. Apart from death, between 20 and 50 million more people undergo from non-fatal injuries, with many incurring a disability [1]. In addition, traffic accidents lead to considerable economic losses to victims, families, and nations. Risk varies from speeding, distracted driving, driving under alcohol influence and other psychoactive substances or drugs, non-use of seat belts, child-restraint motorcycle helmet, endangered roads infrastructure, and unsafe car.

Among all causes, alcohol addiction, driver fatigue, health issues, and distractions are some of the prevalent. These elements endanger the safety of the drivers themselves and threaten the safety of passengers, pedestrians, and other persons on the road. With the continually higher-order safety systems fitted into modern cars, the tragedy of human error persists as an issue that technology hasn't fully put under control.

Drunk driving, for instance, has been a top killer on the roads worldwide [2]. In the USA, according to the National Highway Traffic Safety Administration, part of the US Department of Transportation, every day, approximately 37 individuals lose their lives because of drunk driving crashes [3]. Resulting in one death every 39 min, this

S. Lakiss · K. S. Ngoie · M. Karimzadeh Kolamrodi (✉)
Department of Mechatronic Engineering, Near East University, Nicosia, Türkiye
e-mail: Mohammad.karimzadeh@neu.edu.tr

M. Karimzadeh Kolamrodi
Department of Mechanical Engineering, Near East University, Nicosia, Türkiye

Research Center for Science, Technology and Engineering (BILTEM), Near East University, Nicosia, Türkiye

A. Azizi (ed.), *Autonomous Robotics and Intelligent Systems*, Emerging Trends in Mechatronics, https://doi.org/10.1007/978-981-95-5867-4_4

shows the importance of understanding the behavior of drunk driving and preventing the rising number of deaths in the world. Many strategies have been developed to prevent drunk driving, such as lower blood alcohol concentration (BAC) limits, alcohol screening and brief interventions, ignition interlocks [4]. Alcohol impairs cognitive functions, including one's ability to respond on the road. Despite several regions adopting laws and enforcement measures aimed at curtailing drunk driving, alcohol-related accidents stubbornly underscore the need for preventive, at-the-wheel interventions. In driving, health emergencies like a heart attack or fainting can lead to adverse effects if not identified and addressed promptly.

Among all the public safety problems that Northern Cyprus is facing, road traffic accidents remain one of the most critical. In 2021, a study showed that Negligent driving and speeding were the major causes of road traffic accidents in Northern Cyprus [5]. The local government is delivering the news and records of the increasing accident cases. Urbanization, an increase in the number of students on the island besides an increase in vehicles, comes along with this growth in road accidents in Northern Cyprus. These are sometimes related to alcoholic drink consumption, the influence of drugs, over speeding, and driver carelessness. The high accident rate is, therefore, a load on the health sector, police, and even the community as a whole. A statistical traffic analysis was done in a study [6] that revealed some aspects of the road traffic fatalities. The secondary objectives focused on the road traffic fatality rate per 100,000 population by selected countries stratified by income categories in the year 2013. This analysis offered a perspective of the global picture on how issues of income disparities could affect traffic safety outcomes that different populations face, depending on their economic conditions. The results provided a better understanding of the scope of the problem of road traffic fatalities at the global level and raised a plea for appropriate measures in different income level constituencies (Fig. 1).

The report also contained extensive measures of traffic deaths in Northern Cyprus in the Turkish Republic covering a period of ten years, from 2005 to 2015. This figure included the fatality numbers during this timeframe, which appeared to exhibit varying trends, where some years recorded higher fatalities than others. This picture

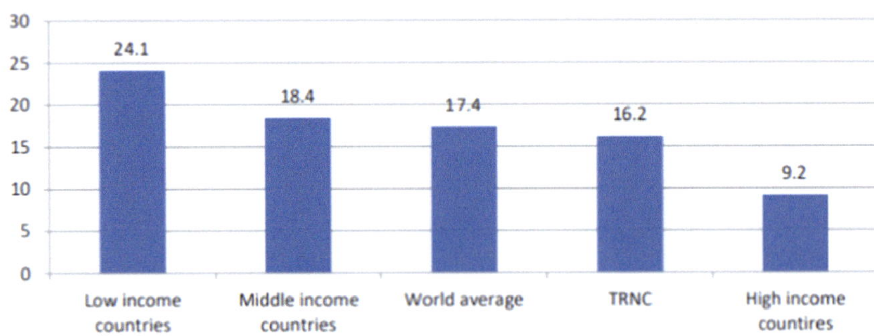

Fig. 1 Road traffic deaths per 100,000 population, by country income status (2013) [6]

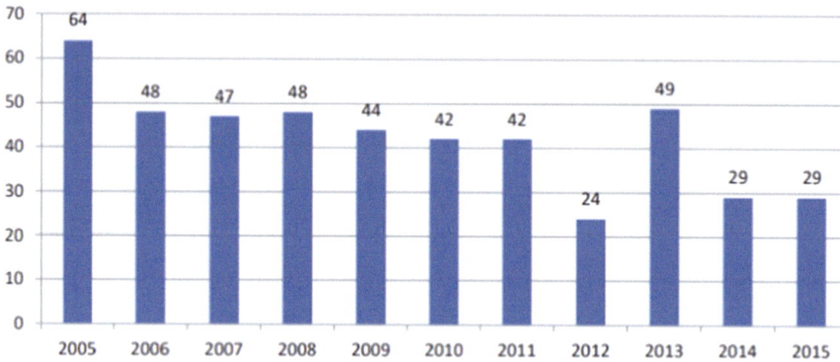

Fig. 2 Road traffic fatality numbers in TRNC (2005–2015) [6]

highlighted the necessity for improved traffic safety measures in the TRNC, especially considering the broader world data suggesting huge deficits in road safety. By analyzing this data, the researchers drew attention to the particular weaknesses of the TRNC's Road network as well as the elements contributing to these fatal accidents.

Complementary to the fatality rate, the research also had statistics showing the number of severe casualties resulting from road traffic accidents in TRNC, as indicated in Fig. 2. This figure showed that from the total number of people being injured in a traffic-related case, there are also others with lifelong disability and life-threatening cases. The number of severe injuries in traffic accidents is also a critical indicator of the broader social and health impacts of road accidents in the region. It also pointed to the need for improved emergency response systems and medical treatment capabilities to address the aftermath of such accidents effectively.

Moreover, it presented the leading causes of fatal traffic accidents in the TRNC between the years 2005 and 2015, as shown in Fig. 3. The leading factors that cause fatal accidents need to be shown, where the interventions could be best directed at drivers, roads, and vehicles. Figure 4 gives a further breakdown of these causes, showing how various environmental, social, and infrastructural factors combined to give these tragic fatality rates in the region. The analysis underlined the need for a multi-faceted approach to road safety education, policy changes, and improvement of infrastructure.

Although the driving regulations in Northern Cyprus forbid drink-driving, with the limit of 50 mg of alcohol per 100 ml of blood (0.5 g per liter)/22 µg of alcohol per 100 ml of breath. Drunk driving frauds and accidents still arise [7]. This deployment brings about a solution to the prevailing problem, intending to use technology to address the risks of alcohol and bad driving.

The main objective of this project is to design an intelligent vehicle safety system Fig. 5, making use of advanced sensors, along with real-time monitoring and communication technologies. The system includes an alcohol sensor placed in the steering wheel, a heart rate sensor to monitor the driver's health constantly, a voice buzzer for instant auditory notifications, a display screen for showing visual notifications, and a

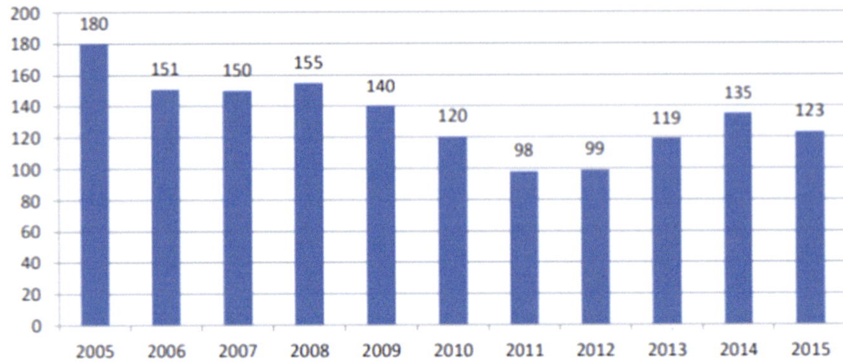

Fig. 3 Number of severe injuries in traffic accidents in TRNC (2005–2015) [6]

Fig. 4 Main causes of fatal traffic accidents between 2005–2015(%) [6]

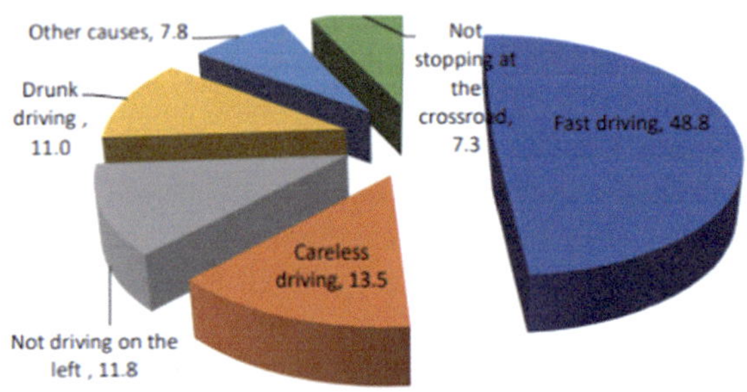

lane correction facility, making it more comfortable to keep the car in its lane. Along with this, it also uses GPS so that response capabilities can occur better. Such integration of GPS, for instance, allows the system to automatically alert pre-identified contacts or emergency services like the ambulance with real-time location in the event of excessively high blood alcohol concentration (BAC) levels or any other risks of health emergencies.

Difficulties

During the development of the project, several challenges across technical, financial, and experimental dimensions were encountered.

Starting with the financial constraints, this project was self-funded. This limited access to high-quality components, especially with the MQ3 alcohol sensor and the MAX30100 heart rate sensor, where they both exhibited inconsistency in readings, which led to repeating the test to ensure reliability. In other words, due to budget limitations, the project relied on low-cost sensors. For example, the alcohol sensor was highly sensitive to environmental factors and produced fluctuations in the output. Also, the heart rate sensor showed irregular results that affected the accuracy of the readings.

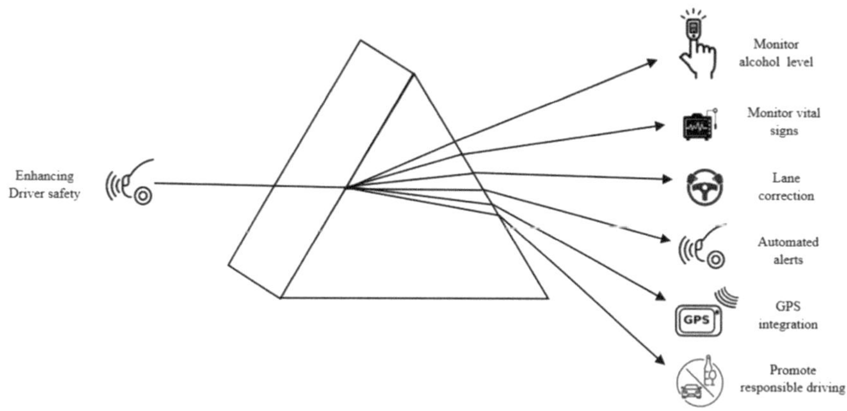

Fig. 5 Key objectives of the project

The second problem faced was the lack of prior experience, whether in system development or coding. Not working on similar projects before had limited exposure to hardware integration and programming. However, with dedicated effort and consistent practice, the team gradually acquired the necessary skills.

Contribution of the Project

The most pressing issues of road safety focus on drunk driving and delayed emergency response. As most road accidents in different parts of the world, including Northern Cyprus, have predominantly occurred among road users operating in vehicles under alcohol influence, having health emergencies, or having lost their focus on the road becomes critical. This project will be paving the way by constructing advanced multiple systems, such as alcohol and heart rate sensors, lane correction, GPS, and GSM connectivity to ensure a comprehensive vehicle safety system. For example, the alcohol sensor keeps a check on the driver's blood alcohol concentration (BAC) levels and issues them with alarms as well as notifying their contacts if anything unsafe is detected, of which crippled drunk driving is reduced. The heart rate sensor detects health-irrelevant symptoms such as heart attack symptoms or fainting and causes the alert of emergency services. Lane correction doesn't let vehicles travel out of their lanes, preventing accidents due to distraction or drowsiness when driving. GPS codifications add a common life-saving communication component, giving real-time location information to emergency contacts or ambulance services, through which help could easily arrive in real emergencies. Besides this, the project also involves mechatronics engineering principles, such as sensor integration, real-time data acquisition, and communication technologies, all of which have software scalability for developments on vehicle safety in the future. Therefore, this would enhance driver awareness in addition to providing quicker responses to emergencies, contributing to the global movement in intelligent transportation systems (ITS) and smarter mobility. Ultimately, the project holds the promise of becoming a potential tool to save lives and reduce the societal and economic impact of road accidents.

Table 1 Timeline of the project

	Sandra	Kerana	Date	Work done
Chapter 1	Both contributed		October 29th, 2024	• Introduction • Read and analyze articles • Collect information • Raise knowledge • Decide the estimated timeline and budget
Chapter 2	Both contributed		November 25th, 2024	• Read and analyze articles • Summarize and paraphrase • Add comments • Define objectives and the gap
Chapter 3	Both contributed		January 27th, 2025	• Understand how all components work together • Done all simulation work • Make sure everything is going to function as it is supposed to
This Chapter	Both contributed		March 5th, 2025	• Ensure system efficiency • Prototype is ready
Chapter 5			April 28th, 2025	• Summarized everything done

Timeline of the Project

See Table 1.

Budget of the Project

See Table 2.

2 Literature Review

Introduction

Recently, in the all-word traffic accidents have become an important concern in health safety, the world community, and other areas. According to the WHO, every year, nearly 1.3 million people lose their lives on roads worldwide, whereas 20–50 million sustain severe trauma, the majority of whom require long-term and costly treatment [8]. North Cyprus exhibits a similar trend to the rest of the world. In this article [5], according to their survey from 2007–2018, the main causes of traffic accidents in North Cyprus are negligent driving and speeding. Lately, with the increase of students on the island, drunk and careless driving has caused more accidents. This study will focus on how to limit car accidents caused by drunk driving and speeding. The latest survey focuses on how to use the Internet of Things for engine braking systems and alcohol detection; however, some other areas of improving road safety have not been deeply touched, such as locking the engine by progressively slowing

Table 2 Budget of the project

Name	Quantity	Price per unit ($)
ESP 32	1	21
Accelerometer	1	9
Arduino nano	1	16
Battery	4	8
Battery charger	1	8
Battery socket	1	4
Breadboard	1	8
Buzzer	1	3
Car kit	1	44
DC motor	2	8
GPS	1	25
GSM	1	28
Heart monitor	1	8
LCD	1	12
MQ3 sensor	1	11
Relay	1	3
SIM card	1	20
Ultrasonic	1	5
3d printing	1	40

down the speed with an early warning to the driver. This literature review intends to dissect how other researchers and projects approached the idea of limiting accidents due to driving under the influence of alcohol and excessive speed, taking into account the technological solutions and components used. Therefore, this review will point up frequent and emerging methodologies and unveil areas in which supplementary studies are needed.

Implementation of Alcohol Sensors and Devices

Danial et al. [9] in their project, alcohol detection using Arduino Fig. 6 with motor locking, focuses on how the car engine is switched off, and the emergency siren is blown as soon as alcohol is detected using Arduino Mega microcontroller, DC motor, Mq-3 alcohol sensor etc. focusing on the same idea; Navya Sri et al. [10] bring more complexity to the system by using Arduino Uno coupled with GMS and GPS for location transmission and send an SMS message to the concerned person in case alcohol is detected and stop the vehicle motor; Raja et al. [11] rely the on GPS and GMS module to MAX30100 SPO2 sensor that can be tracked by the GSM module with the ability to read these extra parameters: heart rate, oxygen saturation, and temperature, making the system more complex and innovative. To explore the effectiveness of mobile phones in alcohol detection and engine locking systems and reducing drunk driving incidents and improving road safety, Jain et al. [12] implement

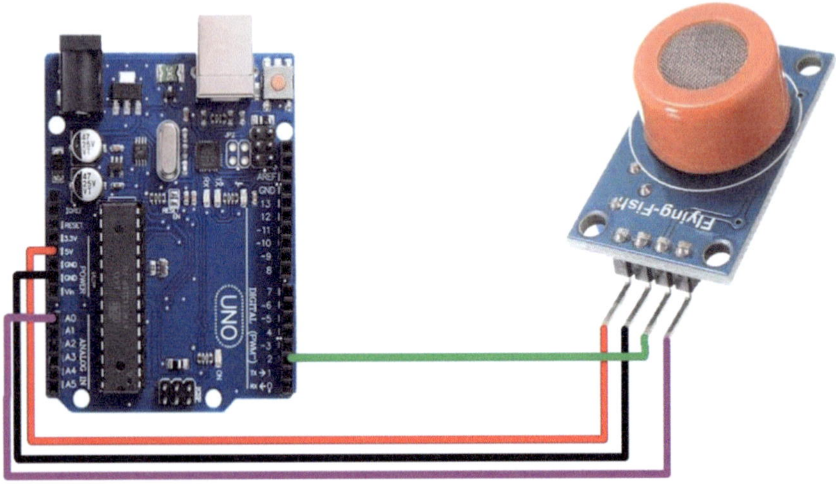

Fig. 6 Interface an alcohol sensor MQ3 with Arduino [14]

to existing methods with a YOLOv3-Tiny algorithm to detect the mobile phone usage on driver performance, which is also a main cause of the accident in a road. Okengwu et al. [13] in their project, design and implementation of in-vehicle alcohol detection and speed control system, integrated and DC motor with the Bluetooth Low Energy (BLE) Application and Blynk Cloud Server for remote control and monitoring of the system.

Owoeye et al. [15] stated that their device's main motivation is accidents frequently occurring on the road because people lose their minds due to consuming high amounts of alcohol. This device can prevent the driver from operating the car in case of high blood alcohol concentration. The entire software is built around a microcontroller, an alcohol sensor, and a vibration sensor. Confirming his words, Prasad et al. [16], to control these accidents, their project is implementing one MQ3 alcohol sensor and an eye blink sensor, where the output will be sent to the Arduino to take needed actions at the right time based on the level of drowsiness and alcohol level. Allan Lopes et al. [17] went deeper when they used a combination of a GSM module, an alcohol sensor, a NodemcuESP8266 microcontroller Fig. 7, a buzzer, and an LCD to display the results. If the driver consumes alcohol above the limit, the engine will be locked automatically, and a notification message will be sent to their family. Moreover, Joshua et al. [18] designed a system where a buzzer and a warning LED will be triggered if the blood alcohol concentration exceeds the limit, and then, automatically, the ignition system is locked. Kommey et al. [19] stated that a good technology to prevent accidents from Driving While Intoxicated (DWI) is combining breathalyzers with ignition interlock devices, which is now applied by modern countries.

Fig. 7 Nodemcu ESP8266 microcontroller [20]

Microcontrollers: Importance and Applications

Embedded systems, defined as computing systems integrated within physical devices, are crucial to daily lives. Over 99% of all the computers that are used today are embedded systems, which in turn rely on microcontrollers (MCUs). Figure 8 is a single-chip computer that operates without a dedicated operating system. The brain is the most important part of the human body, and the same is the microcontroller for a machine. Strobel et al. [21] defined microcontrollers as small, easy-to-handle, low-cost, and flexible components used for several applications. In other words, they are the central components in embedded systems, leading to the evolution of the Internet of Things (IoT). Also, having an increased number of microcontroller-equipped systems is more secure and safe.

Rahmad et al. [22] said that consuming high amounts of alcohol lowers human driving performance and can be the reason for acute respiratory failure, which will

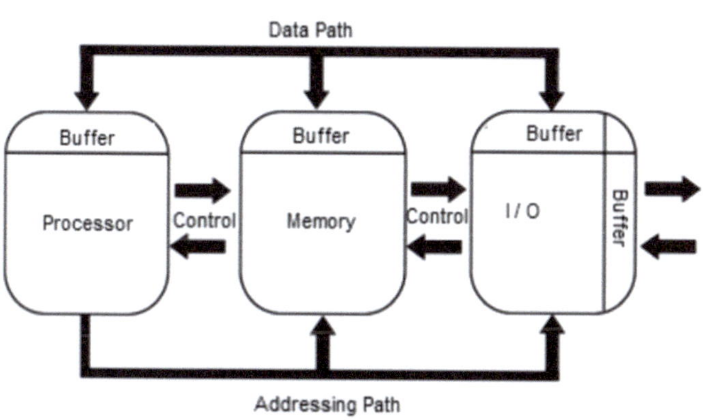

Fig. 8 Block diagram of a simple microcontroller [21]

lead to an increased risk of accidents. For this reason, an alcohol detection system is needed. The project consists of a sensor-based MQ303A microcontroller ATmega328 with a GSM module SIM800L. The results were accurate, but they required time. Another project was done by Ramesh et al. [23] where a microcontroller (Arduino), an MQ-3 alcohol sensor, and a buzzer were combined. The system depends on the ability of the alcohol sensor to detect intoxicated individuals. This project defines the basic components needed.

Heart Monitoring in the Engine Braking System

To diagnose the health status of the heart, heart monitoring systems use heart signals produced during each cardiac cycle. The application of a heart rate system is essential in this project, knowing that excess of alcohol can result in raised blood pressure and weight, increasing the risk of a heart attack; Pughazendi et al. [24] designed a system using the Internet of thing (IoT) where a Heartbeat sensor Fig. 9 is used to monitor heartbeat rate of the driver constantly and prevents from the accidents by controlling through IoT. The IoT sends the emergency message to the Owner, Ambulance, and the Police. Based on the IoT cardiovascular health system, Santos et al. [25] built an Online heart monitoring system. On the other hand, to establish communication between patients with cardiopathy and their doctor or caretaker, Pendurthi et al. [26] developed a system that uses a pulse sensor, to calculate the heartbeat of the person by using an Arduino Uno microcontroller when a finger is placed on it, and communicate the information to the Internet by using Bolt Wi-Fi module.

Considering the fact that it is not always possible to determine the state of health of an individual, relying solely on outward physical appearance can be misleading. The person may look healthier physically but may not be in a state of good health due to the stress within the body; Padmaja et al. [28] combined heart monitoring with other sensors like blood pressure (BP), body temperature, and concentration of CO_2 gas, to detect the stress levels.

Fig. 9 Pulse rate (BPM) monitor using arduino and pulse sensor [27]

Speed Monitoring System

Despite the fact that speed restrictions and other preventative measures have been introduced, traffic accidents continue to occur daily. Over speeding, rash driving, and drunk driving are all major contributors to motor vehicle accidents. The need for effective strategies to improve road safety is more pressing than ever. Aziz et al. [29] introduced a multi-layer security system that controls the speed limit of the vehicle. In addition to it, for more safety of the car, they added a theft alerting feature and an emergency monitoring vehicle feature.

Khan et al. [30] develop an IoT-based vehicle speed monitoring system that can provide efficient results in an overcrowded and irregular traffic environment, such as Pakistan's traffic situation. To overcome the technology designed to track a vehicle's speed, they implement a License plate detection Fig. 10 using contours with the help of the Open-Source Computer Vision (OpenCV) library and Python.

Okengwu and Taiwo [13] propose an IoT-based in-vehicle alcohol detection and speed control system using Arduino. Depending on the Arduino microcontroller, to offer a realistic, pre-built, and easy proposal of a computerized speed-controlling machine, an automated automobile speed manipulation gadget, which is designed to be quickly achieved in colleges, universities, hospitals, and sharp turning zones to reduce various accidents, was developed by Meenakshi et al. [31].

Global System for Mobile Communication (GSM) and Global Positioning System (GPS)

Nowadays, road accidents are very high, and every human life is precious and worth saving. Life should not end on the road, waiting for help in a crash. Eliminating the

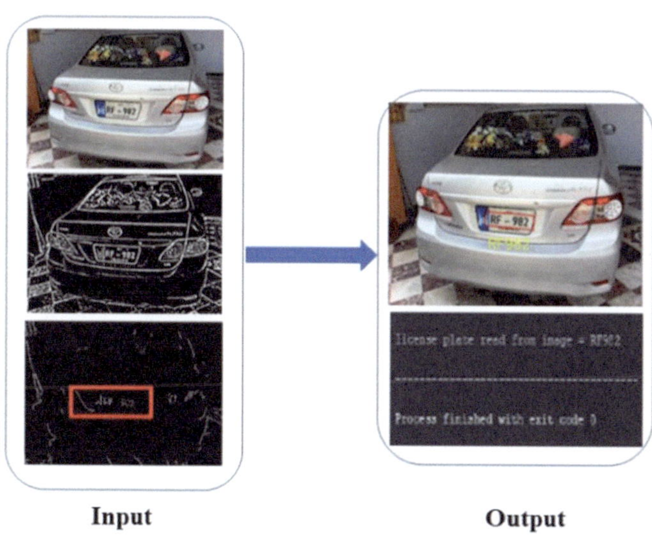

Fig. 10 License plate detection [30]

time between when an accident occurs and when the first responders are dispatched to the scene decreases the mortality rate and can save lives. One approach to eliminating the delay between accident occurrence and first responder dispatch is to use in vehicle tracking system Fig. 11. The use of a Global System for Mobile Communication (GSM) and Global Positioning System (GPS)-based vehicle location and tracking system provided effective, real-time vehicle location, mapping, and reporting of this information value. The GPS-based vehicle tracking system is designed to find out the exact location of any vehicle and inform the authority concerned of the concerned authority through an SMS. Pham et al., [32] designed a vehicle tracking system based on a GPS module, which is used to obtain the vehicle's coordinates, and the GSM modem, which is used to transmit the location to the user's phone through the mobile network, a microcontroller, the Arduino Uno is also employed to control both modules and to provide an easily customizable platform for any required application. In addition to the GPS and GSM, Mounika et al. [33] implemented a further system that can detect the accident and then the victim's near and dear ones to provide medical aid to the accident victim by integrating a Piezoelectric Sensor Transducer sensor, fire sensor, and MEMS are used to determine whether an accident had occurred also gives the owner with the feature of tracking his vehicle in case of theft. Based on the same idea, Kushwaha et al. [34] use the capability of GPS and GSM along with the Android phone to provide a solution that can be used to precisely detect the accident spot and to send emergency notifications to the nearby hospital's ICU and the victim's relatives.

Abinaya and Uthira [36] Devi propose a novel method of vehicle tracking and locking system using GPS and GSM technology. Their system puts the vehicle into sleeping mode while it is handled by the owner or authorized person; otherwise goes to active mode, the mode of operation changed in person or remotely. If any

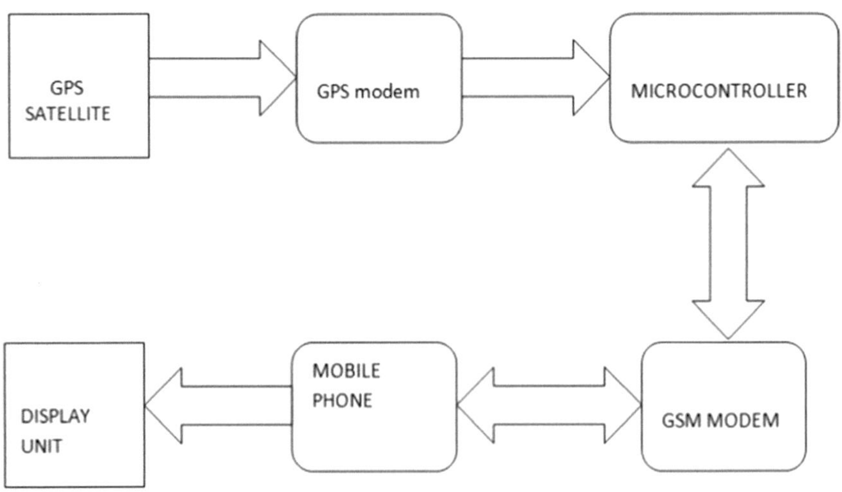

Fig. 11 Basic block diagram of proposed vehicle tracking system [35]

interruption occurs on any side of the door, then the IR sensor senses the signals, and SMS is sent to the microcontroller.

DC Motors

DC motors play a crucial role in the development of robotic cars by providing the necessary mechanical movement needed for mobility and control. Due to their simple structure, they are mostly used in robotics to facilitate movement. In robotic car applications, and even in complex environments, DC motors enable easy adjustments in speed and torque. Their ability to integrate with sensor systems further enhances robotic car functionality. For this reason, DC motors are a preferred choice in autonomous and semi-autonomous vehicle design.

Blood Alcohol Concentration (BAC) Limits

Blood level concentration limits are the legal thresholds found to prevent accidents related to consuming alcohol because these levels are related to decrease driving skills and slow reactions. BAC defines the percentage amount of alcohol in a person's bloodstream. For example, 0.08% means that there are 0.08 g of alcohol per 100 mL of blood. 0.08% BAC was the legal limit before it was lowered to 0.05% by the National Transportation Safety Board in research done by Fell et al. [37], where they state strong evidence for recommending this limit. The risk of having a car crash increases significantly at 0.05 BAC and above. The chance of death in a single-vehicle accident by those with a blood alcohol concentration between 0.05 and 0.079 is said to be 7 to 21 times greater than that of drivers with a BAC of 0.00. Many countries of the world have been successful in implementing measures that target lowering the BAC limit from 0.08 to 0.05.

Discussion

Control Theory

Control theory in engineering focuses on achieving desired outcomes by regulating systems. In this project, the alcohol sensor acts as a control mechanism for the ignition system, where the DC motor will only operate if the driver's alcohol level is within the accepted limits. Control theory will help design the system to respond quickly and accurately to the readings.

- The MQ3 alcohol sensor will be the input in a control loop, it will detect the BAC and send the data to the microcontroller.
- The microcontroller will process the blood alcohol concentration (BAC) levels and control whether to activate or deactivate the DC motor.
- DC motor, which will play the ignition system role, serves as the actuator, where its operation will be based on the BAC readings.

And other features that will be discussed later.

Behavioral Deterrence Theory

Behavioral deterrence theory explains how the alcohol sensor integrated with the ignition system can discourage drivers from attempting to drive while they are intoxicated. When the driver knows that his/her car will not start or it will stop if alcohol is detected, they are more likely to avoid drinking alcohol before and while driving.

- Voice buzzers will provide immediate feedback to the driver when detecting high alcohol levels. This serves as a clear deterrent.
- LCD will communicate and deliver the system's status to the driver. For example, "Alcohol Detected" or "Ignition Will Be Disabled"

Immediate negative feedback reduces risky behaviors, which aligns with the theory's principles.

Systems Theory

Systems theory views the entire setup from sensors, control units, display, buzzer, motor, GPS/GSM, etc., as a unified system, where each part has its crucial contribution in achieving the major goal: safe driving.

- MQ3 and heart monitoring sensor, microcontroller, and DC motor are the core components to detect BAC, heartbeat rate, process data, and control the ignition.
- GPS/GSM adds another layer for potentially integrating with the system by alerting and tracking the vehicle when it's necessary.
- LCD Display and voice buzzer integrate seamlessly with the microcontroller's decision-making processes to inform the driver. In other words, they both serve as a user interface.

When viewing the project as an interconnected system, the team will ensure that each component functions optimally. This theory will ensure effective communication between all the components, in addition to reliability in operation and cohesive interaction to achieve the goal of the project.

Conclusion

This review shows multiple theories, technological solutions, and components used in existing progress. Most reviews used similar methods that include the use of the GSM, GPS module, Internet of Things, alarm siren using buzzers, and other components, revealing an advanced understanding of how to further limit drunk driving, coupled with speed excess. For a more accurate system and variety of applications rather than only sensing and detecting alcohol, the integration of other procedures will enhance the system as a car guardian system that will sense alcohol, check the heartbeat for accuracy and slow down the engine before stopping the vehicle to avoid any accident that can occur for the reason of a sudden engine locking. In addition, more applications will contribute to the system, making it more useful in vehicle safety.

3 Methodology

Introduction

The methodology chapter of this project will outline the systematic approach undertaken in designing and implementing an intelligent driver monitoring and alert system. This project aims to ensure driver and road safety by integrating software and hardware components. To achieve the desired results, the project consists of various stages, including hardware selection, circuit design, simulation, programming, and system testing, where each stage plays a crucial role in ensuring that the final prototype is effective.

This chapter describes the procedures followed in developing the system step by step, divided into three different parts: the programming section, the experimental part that highlights the components used, the instruments employed, and the simulation side includes the software tools utilized in detail. Key stages of the development process include the design of a workflow shown in Fig. 12 and block diagram Fig. 13, the circuit building and simulation, and writing the software code.

The project relies on an Arduino Nano microcontroller as the central processing unit, which interfaces with various sensors such as an alcohol sensor, heart rate sensor, and ultrasonic sensor, in addition to a buzzer, GPS, and GSM modules, and a screen display. All these components work together to monitor the driver's condition and control any unwanted or dangerous situation, for example, alcohol influence or abnormal heart rates. If it detects any risk, the system triggers alerts and notifies a designated emergency contact through SMS notifications. The workflow of the project is structured into distinct phases to ensure a logical and coherent development process. Moreover, circuit simulation is one of the main steps in this project, as it allows a virtual representation of the circuit to guarantee the accuracy and feasibility of the design before the physical implementation. Once the circuit is verified, the actual hardware assembly is performed. Programming the microcontroller is also a critical part of this project. Arduino IDE was used to write and upload the code onto the Arduino board. The code includes the sensor's interface, data processing, decision-making algorithms, and communication protocols. To evaluate the performance, accuracy, and response time, the code was tested at various stages.

This chapter also includes a block diagram, which serves as a high-level overview of the system's functionality. In addition, a detailed circuit schematic is presented in this chapter, which illustrates the electrical connections between all the components. The project establishes an efficient and effective development process by following this structured approach. The subsequent sections of this chapter will provide an in-depth explanation of each step. The workflow of the project is shown in the figure below for better understanding. It starts with the idea and the research done about the whole concept, including the gap. Then, it was divided into two parts, the simulation and experiment, where Proteus 8 was used to combine all the components into one circuit, and the real-time application of the project.

The block diagram provides a high-level structural representation of the project. It emphasizes the input–output relationships and illustrates how sensors are feeding

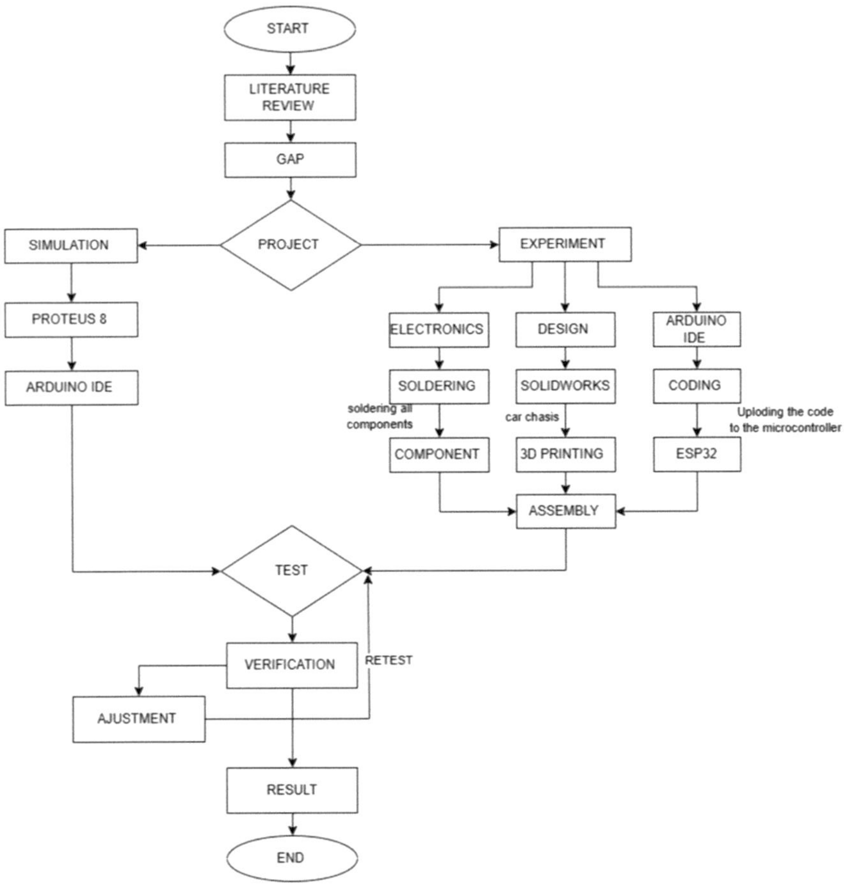

Fig. 12 Workflow

data into the microcontroller, which, in turn, makes decisions based on the code uploaded. In other words, this block diagram serves as a simplified functional overview to make it easier to understand the role of each component within the Guardian KS project.

Experimental Part

See Table 3.

Components Overview

The development of this project is based on several key hardware components that enable various functionalities. The following section provides a detailed overview of each component used:

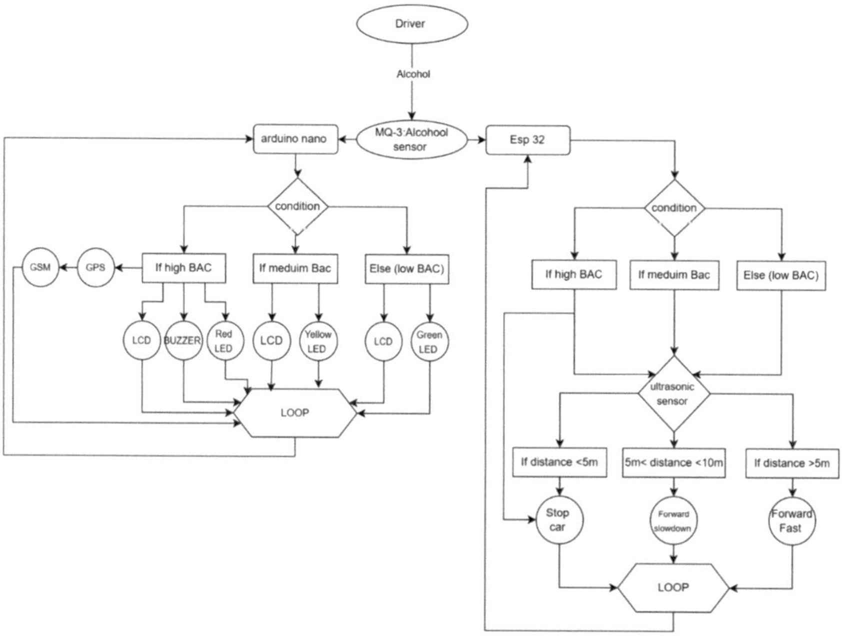

Fig. 13 Block diagram

MQ3 Alcohol Sensor

MQ3 sensor shown in Fig. 14 is an alcohol sensor that is used to detect the presence of alcohol in the breath of the driver. It provides an analog output that can be processed to determine the blood alcohol concentration BAC level.

Connections of the MQ-3 Sensor:

- VCC Pin (power supply): Connect this pin to a 5 V supply from the microcontroller to power the sensor and enable its operation.
- GND Pin (ground): Connect this pin to the ground (GND) of the microcontroller.
- DO Pin (digital output): Connect this pin directly to a microcontroller's digital input pin for processing.

When alcohol vapors in the driver's breath come into contact with the surface of the sensor, the resistance decreases, which leads to a change in the output voltage. This result is then read by the microcontroller (Arduino Nano in this project) to determine the alcohol concentration. The MQ3 sensor should be preheated before using it to stabilize its readings, and its sensitivity can be adjusted using an onboard potentiometer. Since the allowed BAC value differs from one country to another, and there are still many opinions about it, the average was taken while proceeding in this project; the results are divided into three parts. The first one is: BAC value < 0.05% is considered low, 0.05% <BAC value< 0.08% is considered medium, and the third

Table 3 Name of all the components used in the project with a brief explanation about each of them

Components	Explanation
16 × 2 LCD display I2C module	A display unit that shows text output with reduced wire complexity using I2C communication
Arduino buzzer board module	Produces sound alerts in response to specific triggers or conditions
Arduino nano clone CH340 V3 USB	A compact microcontroller board compatible with arduino IDE, featuring a CH340 USB-to-serial converter
Colored LED lights	Provides visual indicators
ESP32 WROOM 320 WIFI, bluetooth development module	A powerful microcontroller with built-in Wi-Fi and bluetooth capabilities
L298 DC and step motor driver module	Controls the speed and direction of DC and stepper motors
Large breadboard	A reusable prototyping platform for assembling and testing electronic circuits without soldering
Lithium batteries, battery socket	Supplies portable power to electronic circuits
MAX30102 heart rate oximeter sensor module	Monitors heart rate and oxygen saturation levels using optical sensors
MQ3 alcohol sensor	Detects alcohol concentration in the air, commonly used for breathalyzer applications
Neo-6 m GPS module	Provides real-time geographic location tracking using satellite communication
Rover 4*4 land hull set	A chassis with four-wheel drive
SIM800L GSM module	Enables mobile communication, including SMS and call functionalities
Ultrasonic sensor HC-SR04	Measures distance by emitting ultrasonic waves and detecting their reflection time

Fig. 14 MQ3 alcohol sensor

part is BAC value> 0.08%, which is considered high. (Note: For example, 0.08% means that there are 0.08 grams of alcohol per 100 ml of blood according to Fell et al. [37]).

Neo-6M GPS Module

The Neo-6 M GPS Module is a popular GPS receiver module. It is used in location-based applications. It provides real-time location data, including latitude, longitude, altitude, and time, via a serial interface. In this project, the team extracted both latitude and longitude from this Module.

Connections of the Neo-6 M GPS Module:

- VCC (power supply): This pin provides the power to the GPS module (5 V).
- GND (ground): This pin is connected to the ground (GND) of the circuit.
- TX (transmit): The TX pin defines the serial output from the GPS module. It is connected to the RX pin of the microcontroller to receive GPS data. It then sends data in the form of NMEA sentences, which contain the latitude and longitude of the driver.
- RX (receive): Connect the RX pin of the GPS module to the TX pin of your microcontroller.

The GPS module shown in Fig. 15 in this project will be powered the whole time, even if the alcohol sensor is not triggered, but it will send the location information only after detecting a high level of BAC value.

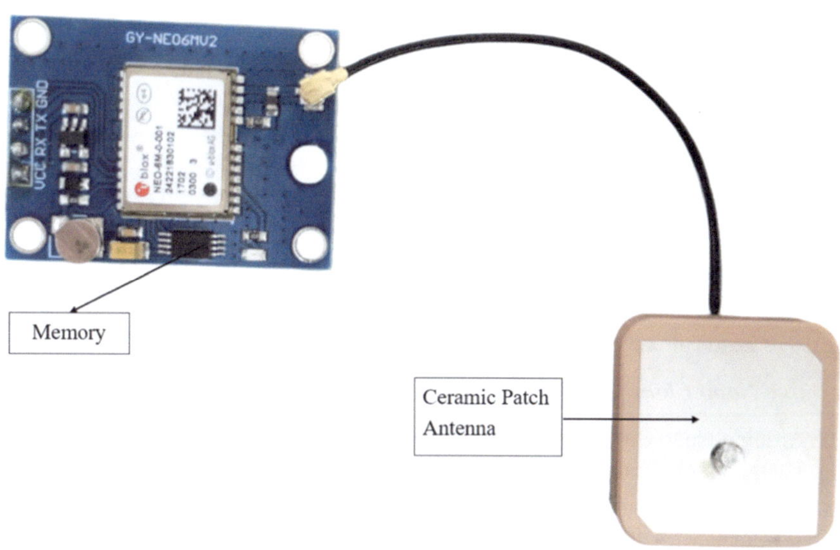

Fig. 15 Neo-6 M GPS module

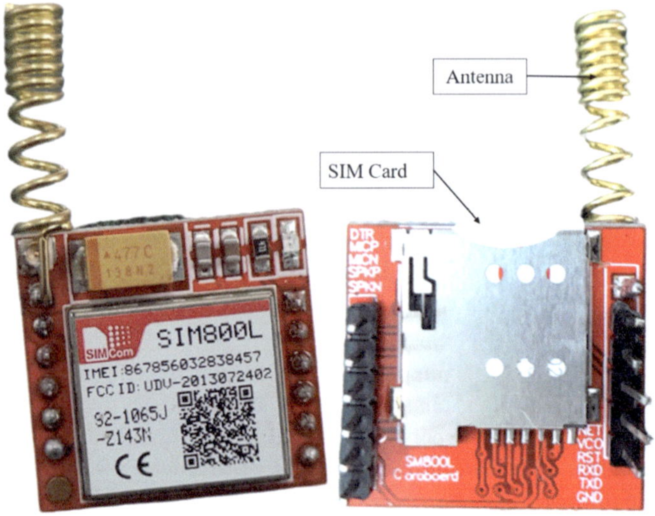

Fig. 16 SIM800L GSM module

SIM800L GSM Module

The SIM800L GSM Module is a low-cost GSM module that allows communication over mobile networks, as shown in Fig. 16. In this project, it is used to send and receive SMS messages.

Connections of SIM800L GSM Module:

- VCC (power supply): This pin is connected to the power supply (3.3 V).
- GND (ground): This pin is connected to the ground of the microcontroller.
- TX (transmitter): transmits data from the SIM800L module to the microcontroller. It is connected to the RX pin of the microcontroller.
- RX (receiver): is used to receive data from the microcontroller. It is connected to the TX pin of the microcontroller.

The GPS Module in this project is also powered the whole time. It will be triggered after the data from the GPS module is received. The next step is sending an emergency message that is already written in code to the desired person.

Arduino Nano Clone CH340 V3 USB

The Arduino Nano Clone CH340 V3 USB is a breadboard-friendly microcontroller, Fig. 17. It is an affordable alternative to the original Arduino Nano, and it is based on the ATmega329P, which makes it very suitable for this project. This microcontroller uses a CH340 USB-to-serial converter chip, which allows communication between the computer and the board via USB. The compact size of the Arduino, its low power consumption, and its multiple input/output pins make it an ideal option for the integration of the components used in the project. In other words, the Arduino

Fig. 17 Arduino nano clone CH340 V3 USB [38]

Nano Clone CH340 V3 USB serves as the central microcontroller; it coordinates all sensor readings, processes data, and controls various components' output.

In this project, the Arduino Nano Clone CH340 V3 USB is responsible for collecting data from the MQ3 alcohol sensor, MAX30100 heart rate oximeter, ultrasonic sensor HC-SR04, and GPS module, then it analyzes this data to detect if there is any unsafe driving condition.

HC-SR04 Ultrasonic Sensor

The HC-SR04 ultrasonic sensor shown in Fig. 18 is used in this project for distance measurement and obstacle detection. It emits an ultrasonic sound wave and then measures the time it takes for the wave to bounce back after hitting an obstacle. The ultrasonic is highly reliable, it has a range of up to 400 cm, and its accuracy is about 3 mm; all these make it an ideal option for real-time detection of any obstacle while driving.

Connection of the HC-SR04 ultrasonic sensor.

- VCC (Power supply): connect it to 5 V.
- GND (Ground connection): connect it to GND.
- Trig: Digital Pin, it triggers a signal for sending ultrasonic pulses.
- Echo**:** Digital Pin, it receives the reflected signal and measures the distance.

The Arduino sends a HIGH signal to the Trig pin for 10 microseconds, which triggers the sensor to emit an ultrasonic pulse (40 kHz). This pulse travels through the air, and if it encounters an obstacle, it bounces back to the sensor. Then the Echo pin receives the reflected wave and sends a signal to the Arduino to the time taken for the wave to return.

It uses the following formula: Distance (Time $\times$ Speed of Sound/2 or $\frac{\text{time}}{58}$) [39].

- Time: measured by the ultrasonic sensor in microseconds (μs).
- Speed of sound of the object in the range of the sensors.

The value is divided by two because of the reflection of the sensor on the object that doubles the travelled distance.

Fig. 18 HC-SR04 ultrasonic sensor

In this project, the ultrasonic sensor is placed on the top of the car. Whenever the distance is close to 1 m (100 cm), the car will slow down until it stops, while preventing any unwanted results like accidents.

16 × 2 LCD Display I2C Module

The 16 × 2 LCD Display with I2C module is used in this project to provide real-time visual feedback. It displays key information about alcohol level and heart rate. Unlike traditional parallel LCDs that require multiple pins, the I2C (Inter-Integrated Circuit) module reduces the number of pins needed for communication significantly, which allows easier wiring and more available pins for other components that will be used. The Arduino Nano will communicate with the LCD shown in Fig. 19 using only two wires (SDA and SCL). The system will continuously update the display with the MQ3 alcohol sensor and MAX30100 heart rate sensor readings, which will ensure that the driver receives real-time information about potential risks.

Buzzer Module

The buzzer module shown in Fig. 20 is an essential component of this project. It acts as an audible alert system to notify the driver of any potential dangers. It produces sound signals when it is triggered by the Arduino Nano, which happens when the BAC value is high. The buzzer Module ensures immediate attention is drawn to critical warnings.

Connection of Buzzer Module:

- VCC (Power supply): 5 V.
- GND (Ground connection): GND.

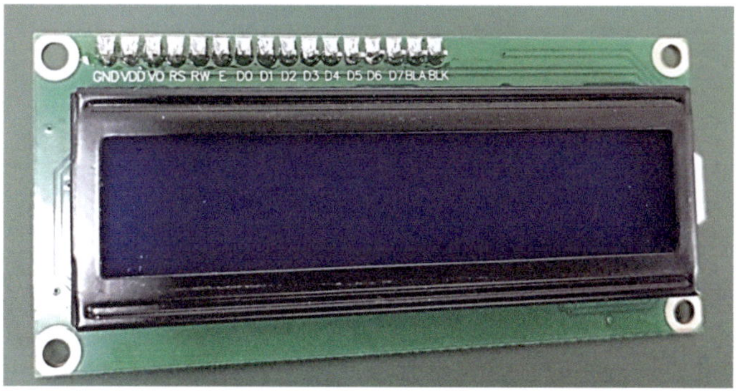

Fig. 19 16 × 2 LCD display I2C module

Fig. 20 Buzzer module

- I/O: Digital Pin.

MAX30100 Heart Rate Sensor

The MAX30100 sensor shown in Fig. 21 is an integrated circuit (I2C)-based low-power plug-and-play biometric sensor [40]. It computes pulse oximetry (SpO2) and heart rate (HR) signals by combining two light-emitting diodes (LEDs), a photodetector, and an optimized optic and low-noise analog signal [41].

The beat per minute (BPM) and oxygen level in the blood of the driver are determined by placing a finger on top of the sensor. The sensor measures, through a photodetector, the amount of light reflected by the infrared sensor (IR), Fig. 21 (1), and the red LED, Fig. 21 (2), on the finger. This principle determines the heart rate and the oxygen level in the blood. The pulse rate of an adult ranges between 60 and 100 BPM [42]. This value gives the state of the driver and helps to ensure safe driving, giving the project an advanced feature in addition to the BAC detection.

Sensor specification:

- Power supply: 3.3–5 v.
- Current: 600 μA (during measurement) and 0.7 μA (during sleep mode).
- Red LED wavelength: 660 nm.

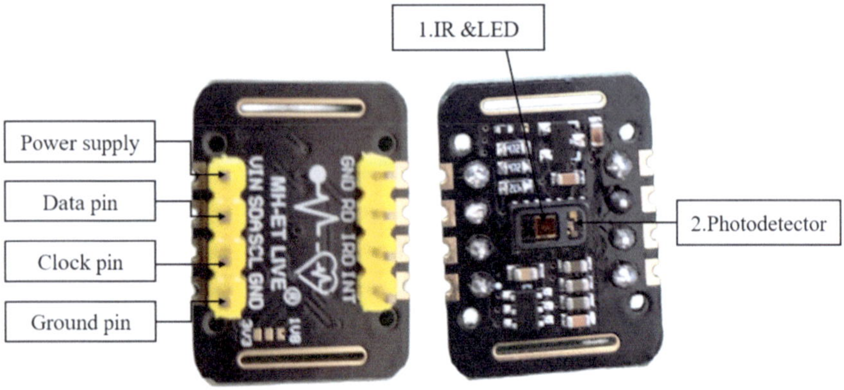

Fig. 21 MAX30100 heart rate sensor

- IR wavelength: 880 nm.

ESP 32 WROOM 32u Wi-Fi, Bluetooth Development

A microcontroller similar to Arduino and Raspberry Pi but with additional features like Wi-Fi, Bluetooth, or a camera. It offers a wide range of applications for numerous projects. It has 32 general-purpose input/output (GPIO) pins shown in Fig. 22, used to send and receive data and control external components or sensors connected to it. In this project, the ESP32 worked as the brain of the robot car, which is used to illustrate the working of the entire project in a real-time demonstration. It worked on 5 V powered by the motor driver and controls the motor driver, the ultrasonic sensor, and the output of MQ-3. Its GPIO pins: 5, 18, 19, 21, 22, 23 are, respectively, connected to the motor driver. The pins 25 and 26 connected to the ultrasonic sensor receive information about the obstacle within the range of the car and send this information to the motor driver, which can set the forward or stop mode of the car according to the condition in the code. The analog pin 34 is linked to the MQ-3 to read the state of the driver.

To ensure communication between the car and the Guardian KS system, the two microcontrollers, ESP32 and Arduino Nano, are grounded together. The Arduino IDE is the interface used to write, upload code, and communicate with the ESP32.

Microcontroller Specifications:

- Operating voltage: 3.3–3.6 V.
- Average operating current: 120–500 mA.
- Operating temperature: −40 to 105 °C.
- Frequency: 2.4 GHz.
- Interface: Pulse width modulation (PMW), GPIO, universal asynchronous receiver/transmitter (UART).
- Size: 19.2 × 18 × 3.2 mm.

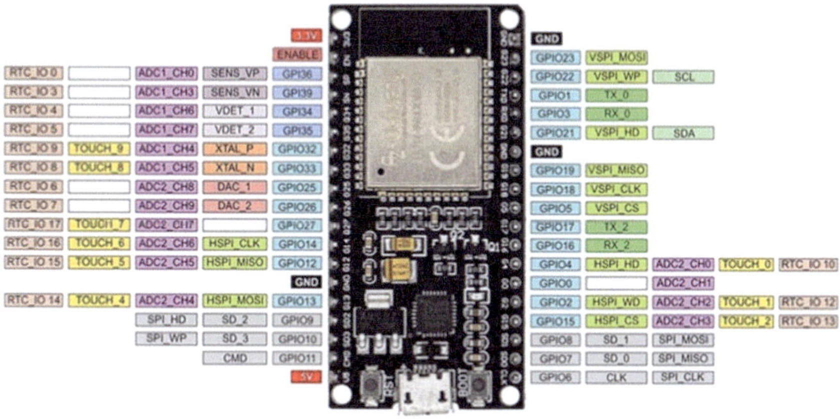

Fig. 22 ESP32 [43]

L298 DC Step Motor Module

Introduced previously, the L298 motor driver illustrated by Fig. 23 is the motor driver based on the L298N IC chip that monitors the speed and direction of the DC motors that run the car. It is normally operated on 12 V or more, but in this project, it's powered by four lithium batteries of 3.7 V. It has a voltage regulator that steps down the input voltage up to 5 V to supply the ESP32. It has six pins, of which two enable one and two, respectively: ENA and ENB control the speed of the DC motors and are connected to pins 5 and 18 of the microcontroller. The input pins IN1, IN2, IN3, and N4 are employed to control the direction of the motors, either forward and backward or right and left. Each motor is controlled by three pins: one enable's pin for the speed and two input pins for the directions. The motors are connected to the output pins as shown in Fig. 23.

This motor driver has two channels for two motors. Since it is needed to run four-wheel DC motors, connect the set of two in parallel, which helped in running four motors simultaneously.

Specifications of L298:

- Motor output voltage: 5–35 V.
- Motor output recommended voltage: 7–12 V.
- Continuous current per channel: 2 A.
- Max power dissipation: 25 W.

Rover 4 × 4 Land Hull Set

The rover 4 × 4 land hull set, Fig. 24, is a robot car chassis encompassing four DC gear motors, four wheels, and the ultrasonic sensor holder. Because of its shape that feet well for holding all the components of the system, this robot was chosen as the car of the project to demonstrate the working of the Guardian KS system. Its motors are controlled by the L298 motor driver, and the ultrasonic sensors detect the

Fig. 23 L298 motor driver

obstacles within its very close range and avoid collision by stopping the motor. The DC gear motor is the combination of the DC motor and the gearbox that holds the wheels; it operates on 12 V. It uses electromagnetic induction to produce the rotary motion that moves the wheels in the desired direction.

3.7 V Lithium Battery 18650 and Battery Socket

The 3.7 V lithium battery is a rechargeable battery, with lightweight, large capacity, and no memory effect that distinguishes them from other types of batteries [44]. In this project, it serves as a power supply. Four batteries of 3.7 V placed in series in the battery socket constitute the power supply to the DC motor that monitors the car. Lithium was chosen because of its rechargeable cycle ability and capacity.

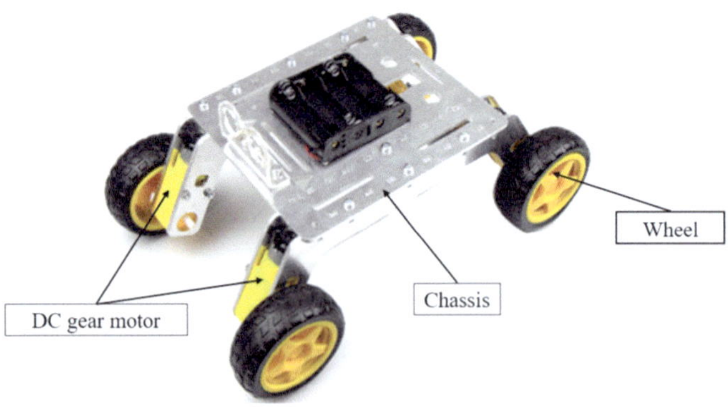

Fig. 24 Rover 4 × 4 land [43]

Specifications:

Colored Light Emitting Diodes (LEDs)

LEDs, known as light-emitting diodes, are used as an indicator of the state of BAC concentration. Three different colors revealing the BAC level are used: red, green, and yellow. The red LED Fig. 25 indicates the state of high BAC concentration greater than 0.08%, the yellow LED Fig. 26 shows the medium state between 0.06% and 0.08%. In this mode, the driver can drive, but he should pay attention to the road. The safe state is specified by the green LED in Fig. 27 with a BAC range lower than 0.06%.

Each LED is connected to one pin of the Arduino Nano microcontroller that defines its LOW or HIGH mode. They operate at 2.2 V and 20 mA. Because the microcontroller has several connected components in parallel, the current is sufficient to source the LEDs without resistance. However, the normal utilization with a 5 V power supply requires the resistance to step down the current up to the recommended one (16 to 18 mA).

Fig. 25 Red LED [45]

Fig. 26 Yellow LED [46]

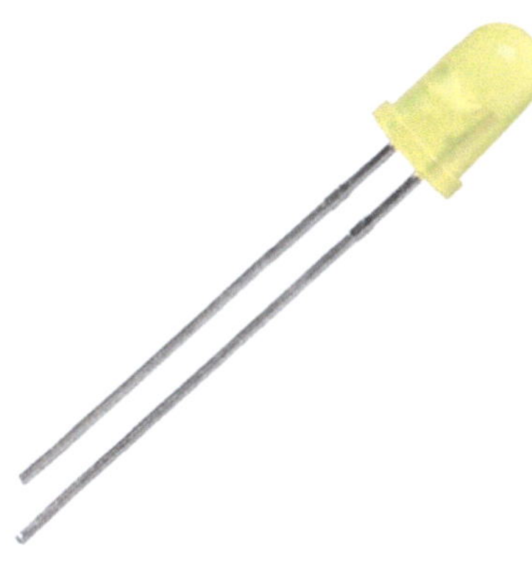

Fig. 27 Green LED [47]

Large Breadboard

A breadboard is a rectangular plastic board with tiny holes where electronics components can be inserted. It is used in electronics to prototype electronic circuits. The connections are not permanent, which makes it useful for projects for beginners in electronics. The holes are arranged vertically (columns) and horizontally (rows). The rows are electrically connected, forming the series configuration, and the columns define a parallel connection. The breadboard has letters, numbers, and a minus sign described in Fig. 28. These labels help in locating the holes' direction and building the circuit. Additionally, it has strips marked by lines: red for positive and blue for negative. These strips are called buses or rails and serve as a power supply to the breadboard. However, the right and left buses are not directly connected but can be linked using jumper wires or wires [48].

The breadboard offers the possibility to be combined, in need of more space and holes for the circuit through the pin shown on Fig. 28.

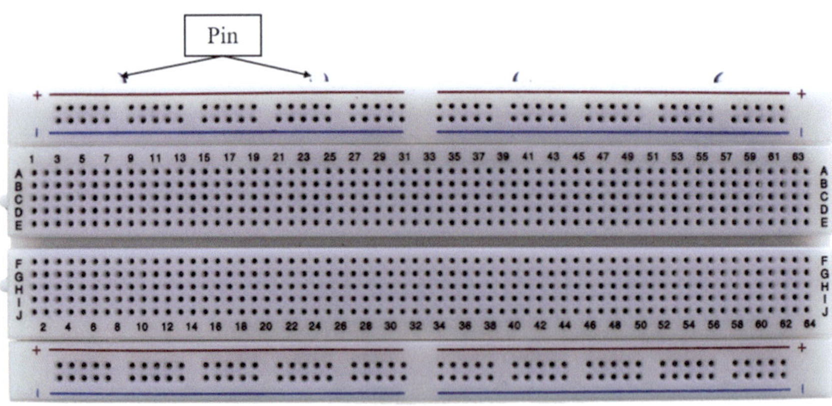

Fig. 28 Large breadboard [49]

Fig. 29 Soldering gun

Instruments and Tools Used

To build and test the project successfully, several essential instruments and tools were used. They are used to ensure proper assembly, connectivity, and troubleshooting of the circuit components.

Digital Multimeter (DMM)

In this project, the digital multimeter was used extensively to measure voltage and resistance across various components. It played an important role in ensuring correct voltage levels, detecting loose connections, and verifying power supply stability to the Arduino used and other components.

Soldering Gun

The soldering gun was used to create strong and permanent electrical connections between components and wires. For example, it was used to solder the wires of DC gear motors and some components with their pins (Fig. 29).

Jumper Wires and Normal Wires

Jumper wires were used for temporary and flexible connections between components on the breadboard. It allowed easy modifications and testing during prototyping. While normal wires were used for permanent wiring, like power connections and communication lines. But both types of wires were carefully selected based on length and insulation type, which is important to ensure reliable and safe electrical connections.

Simulation

Simulation is a representation of a system using software applications. It helps to emulate the working process of a system or process by providing a real-time scenario. In the experimentation procedure, the simulation of the project is important to capture the working of the project and perform calculations about the required component power, current, voltage, etc., as well as adjusting the default, or errors, and optimizing the system. Despite that, the simulation can also be used in cases where it is not possible to perform any experimentation that results in various causes, such as high cost, component limitations, etc. For this project, based on the needs, three simulation tools were employed: TinkerCad, Proteus 8 Professional, and Wokwi.

TinkerCad is a free and easy-to-use simulator application for 3D design, electronics circuits, and coding. The circuit feature of this application that supports the Arduino microcontroller directs the choice to it and helps in testing the virtual components of the project before their implementation in the real circuit. The limitation of TinkerCad regarding components, which limits its application to basic electronics and basic Arduino-based projects, lead to the selection of another software with a wide range of applications, Proteus 8 Professional. An example of the limitation of TinkerCad illustrated in Fig. 30 shows how the MQ-3 sensor, which constitutes one of the main components of the project, was replaced in TinkerCad simulation by the gas sensor MQ-2 that works quietly similar to MQ-3 but doesn't give the same result, accuracy, and performance as the alcohol sensor.

Proteus 8 Professional is powerful software used in circuit simulation and passive-based control (PBC) design layout. And microcontroller firmware development. Widely used by electronics engineers and designers, this software provides a user-friendly interface and a real-time analysis of the circuit [50]. Regarding the demand for additional components, the ability of Proteus to export component libraries has been helpful in the design of the circuit of the project. Proteus 8 in this project plays a crucial role in the design, test, and simulation of the circuit. It helps in calculating the desired power of the overall circuit, testing the circuit, and fixing all bugs found.

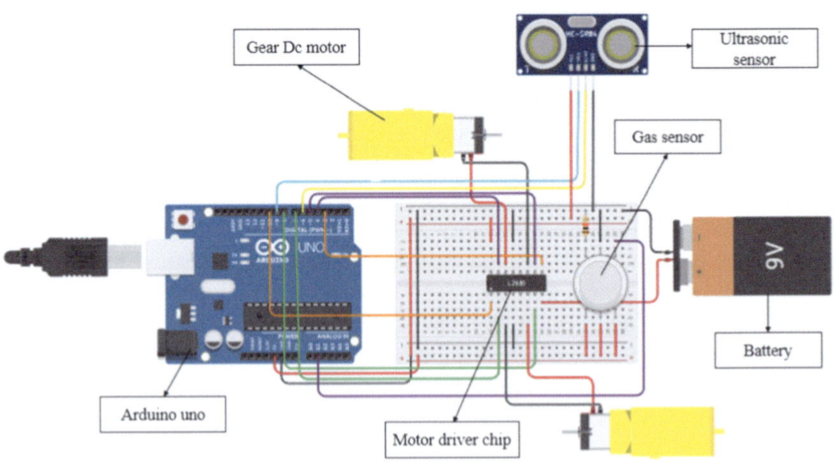

Fig. 30 TinkerCad component limitation

On the other hand, additional software was employed to perform some simulations on ESP32, the Wokwi simulator. Wokwi is an IoT and embedded system simulator that includes microcontrollers such as Arduino, ESP32, STM32, and Raspberry. The features that capture attention are the Arduino Nano and ESP32 simulations. However, it has limited components and is a restraint for the basic project. In this project, it helps in learning the basics of ESP32 that are not present in TinkerCad, but the completion of the circuit was done on Proteus 8 Professional, which provides advanced features.

Programming Section

The programming section is an important section that is crucial in both simulation and experimentation. The use of microcontrollers marks the importance of programming in this project. The most important tool used is Arduino IDE. Programming is the only way to command and communicate with the brain of the project Arduino Nano and ESP32, which control and command the overall project.

This system, based on Arduino, has been configured to monitor and analyze alcoholic levels, heart rates, and locations of drivers as well as to serve as an emergency service communications tool and interact with the ESP32 module. Inauguration of the system activates an LCD (16 $\times$ 2 with an I2C interface), which will display real-time feedback. A buzzer is initialized along with three LEDs (red, yellow, green), which will be activated according to the varying levels of alcohol detection. Input from the MQ3 alcohol sensor will determine the level of alcohol concentration, while serial communication has to be done with the GPS Module (Tiny GPS++), ESP32, and SIM module (GSM for SMS alerts) via Software Serial to get independent data transmission.

The system enables the MAX30105 heart rate sensor initialization for detecting heartbeat signals through infrared light. The program runs these three fundamental functionalities one after another in the main execution loop: heart rate monitoring, alcohol level detection, and GPS location tracking. The heart monitoring feature of the system reads infrared data points from the MAX30105 sensor to identify heartbeat signals, which it uses to calculate beats per minute (BPM). Heartbeat detection results in BPM calculation, followed by an average computation that accumulates recent readings from the last four measurements.

The alcohol sensor function utilizes MQ3 sensor-measured analog values that it processes into blood alcohol concentration (BAC) estimates. The system enables emergency mode when BAC exceeds 0.08, which includes red LED activation and buzzer activation, and SMS alert transmission to a pre-defined contact containing real-time GPS location data. The system then instructs the ESP32 module to shift the vehicle speed to lower levels until it stops to indicate dangerous driving conditions. A yellow LED will turn on when BAC levels range from 0.06 to 0.08, but the green LED stays on during safe BAC operation.

The GPS module tracks the driver's position by running continuous updates for their latitudinal and longitudinal data, which maintains reliable emergency location detection. Precritical situations require the SIM module to use AT commands to send

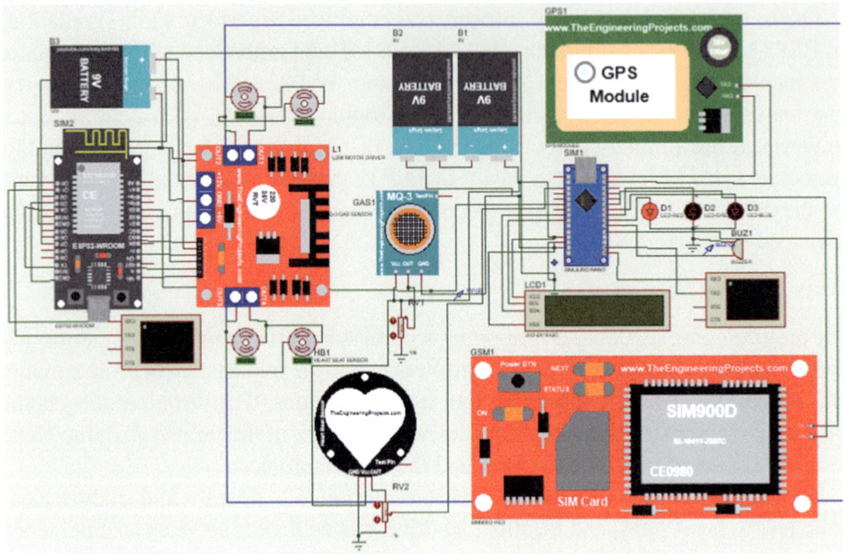

Fig. 31 Overall circuit and connected components

programmed SMS messages. During the whole program duration, serial communication enables live system monitoring through the serial monitor prints sensor data and current system conditions.

The overhead figure (Fig. 31) indicates the connection of all components to the microcontroller used. As described earlier in detail a simulation using Proteus 8 Professional was performed through this circuit, which shows the behavior and overview of the project. Despite this simulation, some confirmations and adjustments have been made throughout the experiment because of the component's quality and real-time performance, which slightly differs from the simulation.

In search of merging an important field of mechatronics in the project, the team make use of the SolidWorks skills taught in the CAD course to design a simple shape as the body of a car to cover the overall project for the good appearance and protection of components. Popular CAD software SolidWorks is used to design and simulate mechanical parts or other features useful in industries or the manufacturing of products. The Fig. 32 illustrates the drawing file of a shape on the SolidWorks software.

Summary

The chapter demonstrates how to create and deploy the Guardian KS project using hardware integration with software applications to boost road security. The methodology organizes its structured process into programming, experimental setup, and simulation sections. The system development follows specific phases of hardware selection, followed by circuit design, then simulation and coding, ultimately leading

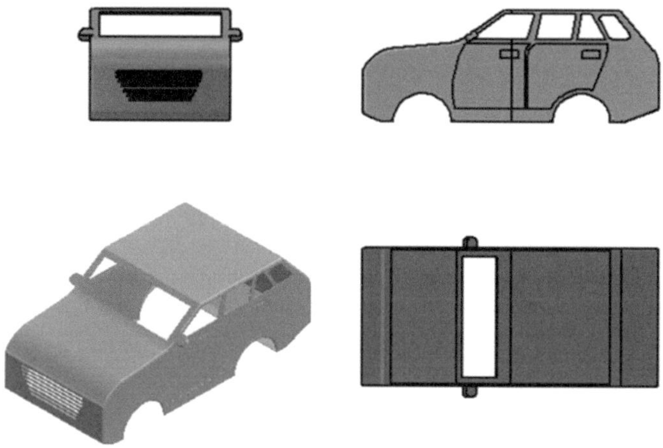

Fig. 32 Body shell

to testing, which secures both reliability and effectiveness for the final prototype design.

The main device of the project is an Arduino Nano microcontroller that works alongside alcohol sensor devices and heart rate detectors, as well as ultrasonic distance sensors. Different equipment, which includes a buzzer and GPS along with GSM modules and an LCD, collaborates to track driver states during critical conditions like alcohol intoxication or irregular heart rate activity. Alerts are activated by the system, which also generates SMS messages to notify users of detected risks. The procedural development guideline establishes a systematic workflow that starts with researching and concept development before using Proteus 8 for circuit simulations to verify feasibility before constructing physical devices. After validation, the hardware gets assembled at the same time as the programming process occurs with Arduino IDE. The system incorporates processing algorithms for sensors in addition to programming code for platform protocols. The system tests occur during multiple points of development with checks for proper functionality and response times to ensure accuracy and efficiency. A block diagram shows the system at a high level, but the detailed circuit schematic shows the electrical part concerning component interconnections. The formal approach to project development results in efficient and effective system development, which produces highly dependable driver monitoring and alert solution. Figure 31 shows the overall connected circuit from Proteus 8 Professional software, and the last part shows the design of the body shell for mechanical engineering integration and covering the circuit.

4 Results and Discussion

Introduction

The chapter will present findings about the Guardian KS project following implementation and testing procedures. The system utilizes alcohol sensors together with heart rate monitors, along with ultrasonic sensors and a GPS module to function as a whole. The main purpose of the project is to protect drivers through dangerous situations, identification, and prompt notification delivery. The research outcome arranges its findings into sections about system performance through evaluations of sensor data precision and accuracy, as well as response times and system integration capabilities, in addition to the alert system's operational effectiveness.

System Performance and Evaluation

Sensor Accuracy

The accuracy of the sensors used stands as the main evaluation point for this system. Testing each of the three main sensors took place individually using the alcohol sensor (MQ3), heart rate sensor (MAX30105), and ultrasonic sensor (HC-SR04) to determine their performance quality under real-world settings.

Analog signals from the alcohol sensor (MQ3) show the alcohol level present in air measurements. Evidence showed that the sensor produced results corresponding with a professional blood alcohol concentration evaluation and operated within expected parameters. The BAC sensor activated its alerts properly because its detection threshold was set at 0.08 percent, making it qualified for impaired driving detection.

The heart rate sensor MAX30105 operates with high efficiency to offer immediate readings of heart rate data. The measurements produced by the sensor matched manual pulse counting results within a satisfactory error range of $\pm$ 10 bpm.

Point-to-point distance measurements were obtained through the HC-SR04 ultrasonic sensor operated with high precision. The system accomplished successful vehicle stopping or speed change when the measured distance reached less than 50 cm with high precision.

System Integration and Response Time

Sensors required proper integration with the Arduino Nano microcontroller as a primary project component. The complete combined system had satisfactory performance levels. Delivery of signals through the GPS and GSM devices required several minutes due to both the GPS signal acquisition time and SMS communication delays. The GPS module needed "x" seconds to generate a reliable signal for navigation, so the SMS system triggered notifications after "x" seconds following system emergency detection. The system response period was measured under various operating conditions. The system activated the buzzer and switched on the red LED while sending text messages through SMS to a specific phone number when alcohol detection exceeded 0.08 BAC. The detection of unusual heart rates sometimes occurred based on how securely the finger stayed on the device sensor. The vehicle speed control through Ultrasonic Sensor Trigger required an average duration of "x" seconds before reacting to changes in distance to make necessary speed adjustments. Network issues and weak signal strength occasionally produced delayed GPS and SMS response times in the system performance.

System Computation and Conversion

The MQ3 pin A0 is connected to pin A1 of Arduino Nano and ESP32, reads an analog value varying from 0–1023. This value is a 10-bit analog-to-digital converter used to map the voltage of the devices. After performing different tests of BAC levels using ethanol that contains hand sanitizers, the team came up with different results and set a threshold of 0.08% as the limit of BAC for the driver to drive. This threshold value was defined using the formula below, in which alcohol represents the value read by the microcontroller from the MQ3 sensor pin; 1023 is the maximum selected value from the analog reading range, used as a divisor to get the rate that has been multiplied to obtain the BAC percentage. The multiplication of 0.123 is an arbitrary value assigned after performing multiple tests in order to get the desired BAC percentage. The graph below illustrates the variation of the BAC percentage related to the analog readings.

$$\text{BAC} = \left(\frac{\text{alcohol}}{1023}\right) \times 0.123$$

Variation of the BAC percentage related to the analog readings

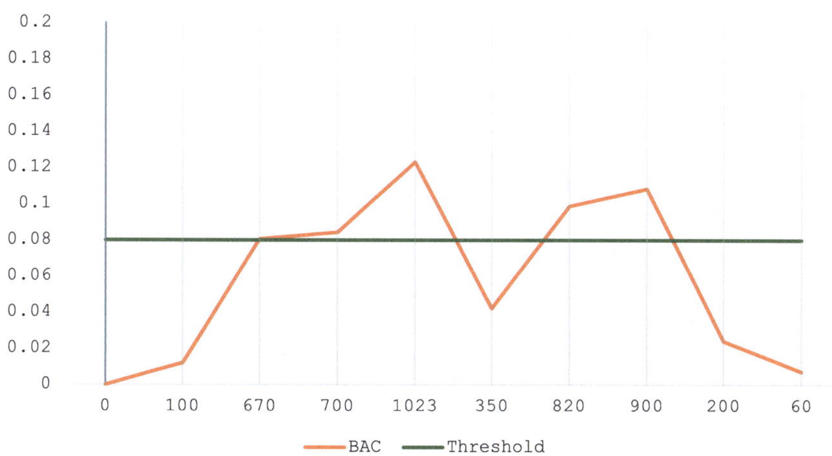

Result Interpretations and Discussions

Table 4 describes the condition state of the car resulting from different tests of the BAC percentage. The aim was to provide a suggested system with a suitable range for the driver and features to inform him about his current state, whether he can continue to drive or not. Most of the existing project measures the BAC level in mg/1000 ml; however, with component precision limitations, a formula shown above was settled to convert the values read from the alcohol sensor into percentage estimation. The system is allowed to start and run when the BAC level is less than 0.06%, illustrated by Fig. 33, with a green LED signaling the stable and preferable state: undrinked. An intermediate state shown in Fig. 33 was added between 0.06% and 0.08%. In this stage, the driver is still driving but has a percentage of alcohol in his blood and is requested to pay attention while driving via a message that appears on his screen and a yellow LED blinking. The last important state is the no motion state, where the driver's BAC surpasses the threshold of 0.08%. An alarm is activated by the buzzer, the red LED blinks to show the dangerous state, which automatically slows down the car and activates the location from the GPS, and sends an SMS to the driver's preferred relative through the GSM. These three conditions and numerous tests performed lead to an efficient and automatic working of the alcohol system recognition while focusing on maintaining the welfare of the driver on the road.

In addition to the BAC level from the alcohol sensor, an additional feature was implemented to confirm the state of the driver by testing the heartbeats. This function not only supports alcohol detection but also offers the system an enhancement in its aim of maintaining the safety of the driver. For the reason that the time was limited to deliver the project, the MAX30102 sensor was not exploited beyond basic BPM detection. However, it offers a wide range of features that can be used to enhance the system. The set of pictures in Fig. 34 illustrates the different variations of the bpm before gaiting stable readings (Table 5).

Table 4 Result of the experiment in comparison with the existing system

State of the car	Condition	Result	
		Existing method	Suggested method
Motion	BAC < 0.06%	Start	Start + green LED + screen message
	0.06% < BAC < 0.08%	–	Keep on + yellow LED with warning screen message
Rest	BAC > 0.08%	Stop alarm and SMS	Stop + red LED + warning stop screen message + alarm + SMS decreases the speed gradually

Fig. 33 BAC states: low, medium, and high

Fig. 34 Beat per minute reading variations

Table 5 Ultrasonic sensor data

State	Conditions	Results
Moving state	Distance > 100 cm or BAC < 0.08%	Move forward
	50 cm = < distance < 100	Slowdown
Rest	Distance < 50 cm or BAC > 0.08%	Stop the motors and reduce the speed gradually

Table 6 defines the working states of the robot car used for the approximation of real-time simulation of the alcohol ignition system in the real car. The aim of this additional feature in the project is to simulate the behavior of the overall project. The operating mode was divided into three states. In the first stage, when the distance measured by the formula described in chapter 3 is greater than 100 cm and the BAC level is less than the limit, the car can move without any interference. The slowdown state is a transitional situation where the HR-SR04 sensor measures the distance of the object ranging from 50 and 100 cm; the speed of the car should be reduced to avoid any collision with the object in the near range. The last mode is stopping motors where the distance is less than 50 cm and the alcohol percentage of the driver is greater than 0.08%. The same as the result of the test, all parts are connected with the BAC, and distances are required to move the car. This simulates how further this system will be included in the real car system to lock down the steering wheel. Although coming up with these conditions, some limitations are faced due to power consumption, and the sensors limit the development of the system.

One of the main features of this system was to alert by sending an SMS containing the location detected by the GPS through the GSM. The Figs. 35 and 36, shows a result of testing high BAC with confirmation of location and message sent to the selected number in the code, displayed by Arduino Serial monitor. As written, you can see BAC = “0.09%,” which surpasses the limit, and “SMS sent” the approval

Table 6 Loads analysis (power, voltage, and current)

Components	Individual voltage (v_1)	Current (A_1)	Individual power ($W_1 = V_1 \times A_1$)	Voltage in parallel (V_2)	Power in series ($W_2 = A_1 \times V_2$)
162 LCD display I2C module	5	0.022	0.11	5	11
Arduino buzzer board module	5	0.02	0.1	5	0.1
Arduino nano clone CH340 V3 USB	5	0.03	0.15	5	0.15
Colored LED lights (3)	6.6	0.01	0.066	5	0.5
ESP32 WROOM 320 WIFI, bluetooth development module	5	0.08	0.4	5	0.4
L298 DC and step motor driver module	12	0.53	6.36	5	2.65
MAX30102 heart rate oximeter sensor module	5	0.01	0.05	5	0.05
MQ3 alcohol sensor	5	0.16	0.8	5	0.8
Neo-6 m GPS module	3.3	0.05	0.165	5	0.25
SIM800L GSM module	4	2	8	4.7	9.4
Ultrasonic sensor HC-SR04	3.3	0.02	0.066	3.3	0.066
Total					14.026

after the message is sent, and the contents of the SMS: Emergency: the driver needs help, find him at: location.

Unpredicted Outcomes

Alcohol Sensor

The MQ3 sensor used presents a high accuracy of all alcohol contained in the air where it is placed, but it is inaccurate for precise applications. While working on

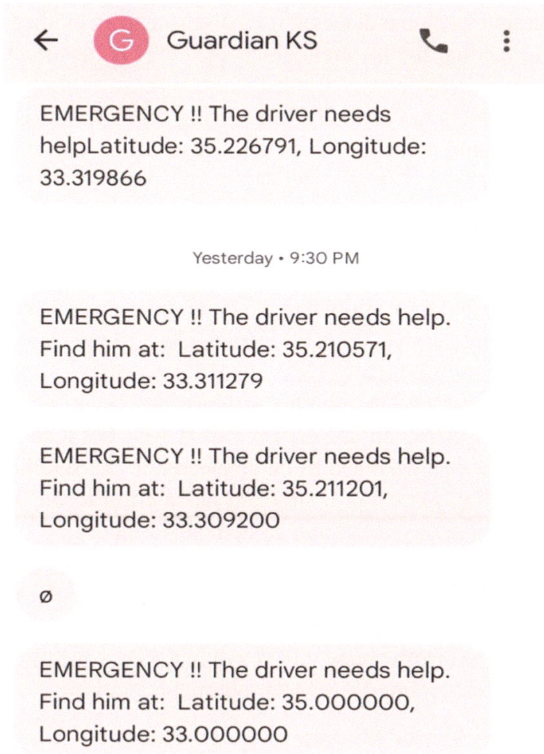

Fig. 35 GPS and GSM results

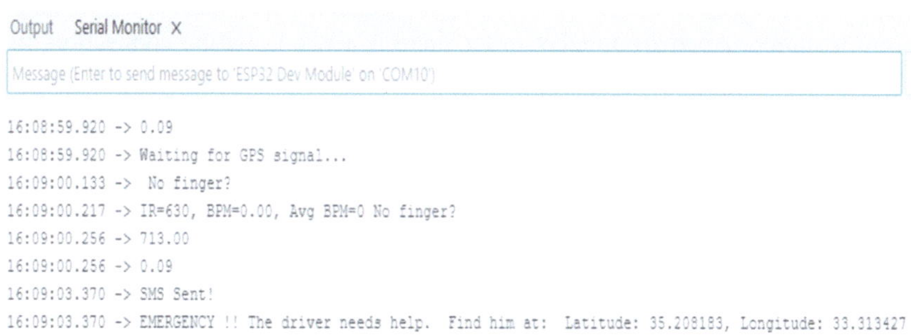

Fig. 36 Serial monitor GPS and GSM results

the project, the simple perfume detected by the sensor from clothes or spread close to it will result in reading the alcohol percentage. This is for the reason that MQ3 is not only used for alcohol in beer or wine, but can sense various types of alcohol, varying from low percentage to high. Considering this unexpected observation, a need for high-quality and specific sensors is suggested for more accuracy and target application.

GPS and GSM Modules

Neo-6 M GPS Module is an important element in the project, its aim already mentioned, which serves in location detection and presents an obstacle in enclosed areas. This impediment makes its usage outside, close to the clear sky. During the inside tests, no location was found, which made the working of the system incomplete, reaching the driver's relatives. This observation was an obstacle only in the testing stage that was performed in a closed area (LAB), but it shows that the GPS is used correctly, as it was designed to be connected to the satellite. Starting with the importance of being placed in a strategic place close to the sky, proper GPS positioning is essential to ensure reliable system performance.

SIM800L GSM Module presented irregular problems, the powering system due to insufficient current. The module was incapable of working when the current was not enough, which was detected by the number of blinks from the indicator LED. Apart from the current insufficiency, GSM needs a SIM card to operate, and this results in frequently recharging the SIM for optimal operation.

Heart Rate Oximeter Sensor Module

The MAX30102 sensor presents a delay during its working operation. Before collecting or reading the exact beat per minute, the sensor needs set-up time. This setup time affected the general working of the system, which caused a long delay before it started. Apart from delay, when there is a wrong wiring connection, the system is completely out of action, which affects the optimal performance, causing all components to be dependent on each other.

Loads Analysis

Load analysis is an important feature while dealing with different components. It helps in calculating each power consumed by individual devices as well as the total power of the system. This analysis is also useful when considering integrating other types of power supply, such as solar cells or external batteries. Table 6 summarizes individual voltage, current, and power along with power consumed by each component connected in parallel in the system and the overall power of the circuit from the formula power equal to voltage times current ($P = V \times I$). Since all components are connected in parallel, they have the same voltage and are taken as 5 V. This voltage comes from the motor driver that supplies the system from its 5 V output. The final analysis gives:

- Total power: 14.026 W.
- Supply voltage (from motor driver): 5 V.

- Currents: dependent on each device.

Summary

The Guardian KS project tested and validated its safety system, which combined monitoring of drivers' responses and actions to minimize accidents on the road. The system's functionality for detecting unsafe conditions depended on the MQ3 alcohol sensor and MAX30102 heart rate monitor, together with the HC-SR04 ultrasonic sensor and additional GPS and GSM modules to trigger proper alerts and actions. System tests verified that the sensors operated accurately in their specified ranges and that the defined alert thresholds allowed the system to identify secure from cautionary to hazardous states. Real-time data acquisition as well as decision-making capabilities, together with SMS alert functionality, made this approach practical for preventing accidents. The promising functionality of the system faced performance barriers that emerged from the components' instability and the power supply issues, alongside dependency on environmental factors. The MQ3 sensor showed excessive sensitivity toward non-alcoholic vapors, yet the GPS module had trouble acquiring signals inside buildings, and the GSM module needed steady power and monetary resources at all times. The heart rate sensor from MAX30102 needed time to initialize before operation and demanded precise positioning conditions for average performance. Proof has emerged that the system design allows the implementation of a smart safety module that supports driver health and shows promise to minimize traffic accidents. Further improvements in hardware components, together with enhanced software optimization, will make this prototype ready for use as a reliable system in intelligent transportation systems of the future. Proof of embedded systems and interdisciplinary cooperation exists through Guardian KS, which demonstrates how to create novel problem solutions for real-world issues.

5 Future Plans and Conclusion

Conclusion

The report established an overview of the Guardian KS project development as it developed a prototype embedded system, which identified impaired driving reasons among alcohol consumption or medical reasons before taking active measures to reduce incidents. The project development organized its information through structured chapters that demonstrated analytical progression from concept creation to execution and assessment stages. The motive for the project emerged from international concern about vehicle accidents that occur due to driver intoxication.

The Guardian KS project is a designed smart system integrating with vehicles to automatically respond to unsafe conditions to minimize risk. This paper documented project obstacles starting from funding shortages and sensor quality problems, alongside complete programming ignorance at the beginning. The MQ3 sensor, along with the MAX30102, produced irregular measurement results, which also depended on specific environmental factors. The team implemented and tested a functioning prototype by putting in substantial effort combined with self-taught knowledge acquisition.

The literature review documented academic findings regarding microcontrollers combined with alcohol detection systems and heart rate sensors, as well as GPS and GSM communication systems, and BAC legal limits. The project benefited from these findings, which led to critical hardware choices and multiple design choices during the development process. Detailed information about system architecture and programming methods, and the selection of components, along with wiring and programming structure, appears in the methodology section. A written document specified the precise project-related tasks for each module. The application needed a simulation environment for validating logic before conducting physical tests.

In the results and Discussion chapter, the performance evaluation of the system focused on three aspects: sensor precision, response delays, and hardware interconnection. The MQ3 sensor operated with a 0.08% BAC threshold calibration, whereas the MAX30102 measured heart rate with (x) bpm accuracy range. The system control utilized ultrasonic sensors to monitor motion according to detected obstacles near the device. The indoor performance of GPS and GSM modules was inconsistent, but they achieved optimal functionality in outdoor testing conditions.

In conclusion, the Guardian KS project represents a successful integration of theoretical knowledge and practical implementation. It's aimed at enhancing road safety through technological innovation. The system combines alcohol detection, heart rate monitoring, distance sensing, GPS location tracking, and GSM communication to provide a comprehensive safety mechanism for drivers. Through testing and iterative development, the project demonstrated functional accuracy in its core components. It also achieved its objective of alerting and responding to potentially hazardous conditions. Despite facing various technical and logistical challenges, the system was effectively designed, implemented, and evaluated. The project serves as a foundation for future enhancements and reflects the commitment to developing a real-world solution that addresses critical safety concerns in modern transportation.

- The Guardian KS project built a prototype system to detect drunk driving incidents alongside medical impairment conditions while initiating protective actions to decrease highway accidents. The project dedicated its efforts to enhance road safety by implementing alcohol sensors, heart rate monitors, alongside automated response technology into vehicle systems, because of worldwide traffic accident worries.
- The team managed to tackle shortages of funding as well as sensor problems and programming difficulties by investing great effort into self-learning.
- The literature review yielded essential understanding about microcontrollers, alcohol detection systems, heart rate sensors, as well as GPS and GSM communication systems, together with BAC limits, which led to essential hardware and design decisions.

- The methodology chapter presented the system architecture together with component selection and wiring, and assigned tasks for each module, where simulation testing came before physical verification.
- The results and discussion section analyzed three fundamental operational points: measurement accuracy, response speed, and hardware system connection. Routine quality checks for the MQ3 alcohol sensor adjusted its response at 0.08% BAC, while the MAX30102 sensor required precise accuracy parameters during tests. The system used ultrasonic sensors coupled with GPS/GSM modules that gave optimal results during outdoor operation.
- The Guardian KS project successfully implemented theoretical knowledge into practical use by achieving functional accuracy when responding to dangerous driving scenarios and creating a base for enhanced road safety improvements.

Future Work

The Guardian KS system sets new benchmarks for embedded system utilization in driver safety technology, although it exists as a prototype that requires additional development work. The present development allows future modifications and additions that will strengthen the system's accuracy and practical market usage potential.

The advancement of sensors through development represents one of the fundamental areas needing improvement. The MQ3 sensor detects alcohol effectively but shows insufficient selectivity to identify BAC concentrations with the required accuracy. False sensor alarms frequently occur due to their incorrect response to environmental alcohols like perfumes or sanitizers. The subsequent development phase of the system must integrate alcohol sensors designed precisely for human respiratory alcohol detection. For future commercial applications, the sensors will provide precise readings according to legal standards. The future development of sensor technology for MAX30102 should concentrate on adding blood oxygen saturation detection to its capabilities, as well as ECG pattern reading and fatigue measurement features that would provide a better assessment of driver health.

Developing an actual integration of the system inside a real vehicle stands as the subsequent logical development. The research utilized a Rover 4×4 hull set as a movement simulator that can achieve realistic testing results. The actual deployment of this system demands a connection to the car's electronic control units through CAN bus communication protocols. Control functions for the ignition system, along with brakes and steering, could be implemented by using data from the system through its integration with vehicle control units.

The integration of AI with machine learning stands as an essential development field. The system depends on preset threshold values together with linear decision-making mechanisms at this current stage. AI models operating within the Guardian KS system would acquire individual health data and driving characteristics from users over time, which enables it to enhance warning precision and lower nuisance alert percentages. Shifting warning alerts from the current guard system poses a problem to drivers who naturally experience high heart rates. The alert threshold would adjust according to the user's historical data through machine learning technology.

In addition, the development of a mobile app will enhance system usability as well as user-friendliness while making the system more interactive. A smartphone application functions as a command center that stores and records, sends automatic GPS location information to contacts, while letting users set monitoring parameters. The healthcare monitoring system integration into this application would allow it to provide continuous physical condition assessments for drivers, thus developing an interconnected system for road security as well as healthcare supervision.

The Guardian KS project extends beyond being a student because it serves as a starting point to create a life-saving system. Technical enhancement, along with physical vehicle experimentation and AI programming, and legal conformity, will transform this system into a marketable solution which links automated systems to safety measures.

References

1. Road traffic injuries. https://www.who.int/news-room/fact-sheets/detail/road-traffic-injuries. Accessed 22 Nov 2024
2. Haghpanahan H et al (2019) An evaluation of the effects of lowering blood alcohol concentration limits for drivers on the rates of road traffic accidents and alcohol consumption: a natural experiment. Lancet 393(10169):321–329. https://doi.org/10.1016/S0140-6736(18)32850-2
3. Drunk driving | Statistics and resources | NHTSA. https://www.nhtsa.gov/risky-driving/drunk-driving. Accessed 23 Nov 2024
4. What works: strategies to prevent alcohol-impaired driving | Impaired driving | CDC. https://www.cdc.gov/impaired-driving/strategies/index.html. Accessed 23 Nov 2024
5. Angın M, Albrka Ali SI (2021) Analysis of factors affecting road traffic accidents in North Cyprus. Eng Technol Appl Sci Res 11(6):7938–7943. https://doi.org/10.48084/ETASR.4547
6. Şafakli OV (2018) Statistical anatomy of traffic accidents in Northern Cyprus. Int J Acad Res Bus Soc Sci 5:497–508. https://doi.org/10.6007/IJARBSS/v8-i3/3945
7. Driving on North Cyprus | Ada rent a car North Cyprus. https://adarentacar.net/driving_on_north_cyprus. Accessed 23 Nov 2024
8. Goniewicz K, Goniewicz M, Pawłowski W, Fiedor P (2016) Road accident rates: strategies and programmes for improving road traffic safety. Eur J Trauma Emerg Surg 42(4):433–438. https://doi.org/10.1007/S00068-015-0544-6/TABLES/1
9. Hamamy F, Adib Danial A (2022) Alcohol detection using arduino with motor locking. J Eng Technol 10(1):103–109
10. Navya Sri M, Saritha N, Navya N, Jagadeeshan N Vehicle tracking and alcohol detector with engine locking system with GSM and GPS
11. Raja DP, Barkavi G, Aishwarya S, Keerthana K, Vasudevan V (2022) Alcohol detection and emergency alert system using IoT. In: Proceedings of 2022 6th international conference on intelligent computing and control systems ICICCS 2022, pp 395–400. https://doi.org/10.1109/ICICCS53718.2022.9788419
12. Jain H, Zope H, Jain H, Rakte I, Raut S (2023) Mobile phone detection and alcohol detection with engine locking for the prevention of car accidents. In: Proceedings of 2023 international conference on digital application, transformation and economy ICDATE 2023. https://doi.org/10.1109/ICDATE58146.2023.10248639.

13. Okengwu UA, Taiwo AA (2022) Design and implementation of in-vehicle alcohol detection and speed control system. Eur J Electr Eng Comput Sci 6(5):10–16. https://doi.org/10.24018/EJECE.2022.6.5.464
14. Interface an alcohol sensor MQ3 with arduino—ElectroVigyan
15. Owoeye S, Durodola F, Akinade A, Olaonipekun O, Alkali A (2022) 2022 5th information technology for education and development (ITED). https://doi.org/10.1109/ITED56637.2022.10051302
16. IEEE Xplore full-text PDF. https://ieeexplore.ieee.org/stamp/stamp.jsp?tp=&arnumber=10200473. Accessed 21 Oct 2024
17. Lopes A, Tiwari K, Singh Jainath S (2023) Alcohol detection and engine locking with SMS notification. Int J Res Anal Rev
18. Joshua I, Stanley U, Charles O, Peter A, Okokpujie K (2019) Development of alcohol triggered vehicle engine lock system. Int J Robot Autom 8(1):68–76. https://doi.org/10.11591/ijra.v8i1.pp68-76
19. Kommey B, Tamakloe E, Nunoo-Mensah H, Opoku D (2022) Smart vehicle ignition interlock: a car ignition interlock device for alcohol impaired driving 04(1):51–62. https://doi.org/10.24874/PES04.01.008
20. ESP8266 NodeMCU microcontroller | Download scientific diagram
21. Strobel D, Oswald D, Richter B, Schellenberg F, Paar C (2014) Microcontrollers as (In)Security devices for pervasive computing applications. Proc IEEE 102(8):1157–1173. https://doi.org/10.1109/JPROC.2014.2325397
22. Rahmad IF, Nababan EB, Tanti L, Triandi B, Ekadiansyah E, Fragastia VA (2019) Application of the alcohol sensor MQ-303A to detect alcohol levels on car driver. In: 2019 7th international conference on cyber and IT service management CITSM 2019. https://doi.org/10.1109/CITSM47753.2019.8965395
23. Naga Ramesh JV, Prakasa Rao Y, Kumari Francis S, John Samuel Babu G, Neerugatti V, Sharma S (2024) Next generation alcohol testing device for safer living. In: 2024 2nd international conference on embedded wireless systems and networks, pp 1–6. https://doi.org/10.1109/ICNEWS60873.2024.10730904
24. Pughazendi N, Sathishkumar R, Balaji S, Sathyavenkateshwaren S, Chander SS, Surendar V (2018) Heart attack and alcohol detection sensor monitoring in smart transportation system using internet of things. In: 2017 international conference on energy, communication, data analytics and soft computing, ICECDS 2017, pp 881–888. https://doi.org/10.1109/ICECDS.2017.8389564
25. Santos MAG, Munoz R, Olivares R, Filho PPR, Del Ser J, de Albuquerque VHC (2020) Online heart monitoring systems on the internet of health things environments: a survey, a reference model and an outlook. Inf Fusion 53:222–239. https://doi.org/10.1016/J.INFFUS.2019.06.004
26. Pendurthi HK, Kanneganti SS, Godavarthi J, Kavitha S, Gokarakonda HS (2021) Heart pulse monitoring and notification system using Arduino. In: Proceedings of international conference on artificial intelligence and smart system ICAIS 2021, pp 1271–1278. https://doi.org/10.1109/ICAIS50930.2021.9395825
27. Pulse rate (BPM) monitor using arduino & pulse sensor
28. IOT based stress detection and health monitoring system | Helix—the scientific explorer | Peer Rev Bimon Int J. https://www.helixscientific.pub/index.php/Home/article/view/127. Accessed 05 Nov 2024
29. Aziz T, Faisal TM, Ryu HG, Hossain MN (2021) Vehicle speed control and security system. In: 2021 international conference on electronics, information, and communication ICEIC 2021. https://doi.org/10.1109/ICEIC51217.2021.9369808
30. Khan SU, Alam N, Jan SU, Koo IS (2022) IoT-enabled vehicle speed monitoring system. Electron 11(4):614. https://doi.org/10.3390/ELECTRONICS11040614
31. Meenakshi V, Immanuel DG, Kavitha M, Kaveri VV, Selva Rex CRE (2021) Automatic speed controller of vehicles using arduino board. In: Proceedings of 2021 IEEE international conference on innovative computing, intelligent communication and smart electrical systems ICSES 2021. https://doi.org/10.1109/ICSES52305.2021.9633790

32. Pham HD, Drieberg M, Nguyen CC (2013) Development of vehicle tracking system using GPS and GSM modem. In: 2013 IEEE conference on open system ICOS 2013, pp 89–94. https://doi.org/10.1109/ICOS.2013.6735054
33. Mounika J, Charanjit N, Saitharun B, Vashista B (2024) Accident alert and vehicle tracking system using GPS and GSM. https://papers.ssrn.com/abstract=3869132. Accessed 06 Nov 2024
34. Kushwaha VS, Yadav D, Topinkatti A, Kumari A (2015) Car accident detection system using GPS and GSM. Int J Emerg Trend Eng Basic Sci 2(1):12–17. https://ijeebs.com/wp-content/uploads/2020/06/IJEEBS-2015-V2-I1-004.pdf
35. Basic block diagram of proposed vehicle tracking system | Download scientific diagram
36. Abinaya M, Devi RU (2014) Intelligent vehicle control using wireless embedded system in transportation system based on GSM and GPS technology. Int J Comput Sci Mob Comput 3(9):244–258. https://d1wqtxts1xzle7.cloudfront.net/34735893/V3I9201432-libre.pdf?1410768235=&response-content-disposition=inline%3B+filename%3DIntelligent_Vehicle_Control_Using_Wirele.pdf&Expires=1730870008&Signature=K1S8Ez5OfMC8DRuoJmN5izVnsl0vVhUAbIiIi2oSgZD8-i1WUBd
37. Fell JC, Voas RB (2014) The effectiveness of a 0.05 blood alcohol concentration (BAC) limit for driving in the United States. Addiction 109(6):869–874. https://doi.org/10.1111/ADD.12365
38. NANO V3.0 16MHz USB—ATmega328P—CH340—Clone—solder pins—compatible with Arduino. https://abc-rc.pl/en/products/nano-v3-0-16mhz-usb-atmega328pb-ch340-clone-solder-pins-compatible-with-arduino-9588.html?srsltid=AfmBOorNH_prW2ZBhZtdUj6sMoeBADpRSOK4onafahqIyqYmHdnGjuAL. Accessed 16 Jul 2025
39. Morgan EJ (2014) HCSR04 ultrasonic sensor
40. Interfacing MAX30100 pulse oximeter and heart rate sensor with arduino. https://lastminuteengineers.com/max30100-pulse-oximeter-heart-rate-sensor-arduino-tutorial/. Accessed 07 Mar 2025
41. General description. www.maximintegrated.com Accessed 07 Mar 2025
42. What is a normal pulse rate?—BHF. https://www.bhf.org.uk/informationsupport/heart-matters-magazine/medical/ask-the-experts/pulse-rate. Accessed 08 Mar 2025
43. ESP 32 Wroom-32 pin allocation—projects/programming—arduino forum. Accessed 08 Mar 2025. https://forum.arduino.cc/t/esp-32-wroom-32-pin-allocation/1228704/8
44. Advantages of the 18650 battery. https://www.acebattery.com/blogs/advantages-of-the-18650-battery. Accessed 08 Mar 2025
45. Lite-On2.6 V red LED 5 mm through hole, LTL-307P | RS. https://ph.rs-online.com/web/p/leds/1278393. Accessed 08 Mar 2025
46. Lite-On1.8 V yellow LED 5 mm LTL2R3KYD-EM | RS. https://za.rs-online.com/web/p/leds/1689438. Accessed 16 Mar 2025
47. 3 mm green LED | Evelta. https://evelta.com/3-mm-green-led/. Accessed 16 Mar 2025
48. "How to Use a Breadboard." Accessed: Mar. 08, 2025. [Online]. Available: https://www.sciencebuddies.org/science-fair-projects/references/how-to-use-a-breadboard
49. Electronics: introduction to breadboards—bald engineer. https://www.baldengineer.com/electronics-introduction-to-breadboards.html. Accessed 08 Mar 2025
50. The ultimate guide to using proteus 8 professional for electronics design—Ask.com. https://www.ask.com/news/ultimate-guide-using-proteus-professional-electronics-design. Accessed 25 Mar 2025

Genetic Algorithm-Based Stabilization of Rotary Inverted Pendulum by State Feedback

Fatma Nur Şen and Ulaş Beldek

Abstract A rotary inverted pendulum is a standard test bed in control theory due to its nonlinear and unstable characteristics. Driving the pendulum to an upright position, sustaining it motionless there, and carrying on signal tracking tasks at this position are the main control applications related to this test device. This paper presents a comparative study of four state feedback controller structures performing pole placement that theoretically stabilizes the pendulum at an upright position and in real-time application. The first state feedback controller is established based solely on the pendulum dynamics. In contrast, the other controllers are developed due to a multi-criteria optimization procedure by running a Genetic Algorithm search. The Quanser kit is used to obtain and implement the controllers and to test their performance in real-time.

Keywords Rotary inverted pendulum · State feedback control · Pole placement genetic algorithm · Multi-criteria optimization

1 Introduction

The inverted pendulum [1, 2] is a prevalent problem in control theory and is used to test different control strategies. There are different versions of the inverted pendulum, and the rotary inverted pendulum (RIP) is one of them. RIP has nonlinear and unstable characteristics, and hence it resembles other inverted pendulum types; however, the motions of both of the joints of RIP are revolute. RIP also has real-time applications such as vehicles, rockets, missile systems, and aircraft landing systems [3].

F. N. Şen (✉)
Mechanical Engineering, University of Turkish Aeronautical Association, Ankara, Türkiye
e-mail: fnsen@thk.edu.tr

U. Beldek
Mechatronic Engineering, Çankaya University, Ankara, Türkiye
e-mail: u.beldek@cankaya.edu.tr

A. Azizi (ed.), *Autonomous Robotics and Intelligent Systems*, Emerging Trends in Mechatronics, https://doi.org/10.1007/978-981-95-5867-4_5

Driving the pendulum to an upright position and holding the stick of the pendulum stabilized in that condition is a crucial benchmark problem in control theory. Generally, the RIP control structure comprises two modes, i.e., swing-up mode and hold (stabilization) mode. Initially, the pendulum is set at a standstill situation, with the pendulum stick being in an entirely downward position (180°). The swing-up mode brings RIP a nearly upright position from this starting point. In hold mode, RIP is stabilized at its upright position when the pendulum angle reaches 0°. Usually, the pendulum's reference signal tracking and disturbance rejection capability are tested at this condition. In real-time applications, switching from swing-up mode to the hold mode is realized when the pendulum angle reaches a critical threshold value as small as possible such as $\pm 15°$ or even smaller. This process is called as switching process.

First proposed by Furuta [4, 5], several control strategies have been applied for the control modes of the RIP. The common strategies are pole placement and Linear Quadratic Regulator [6, 7] for hold mode, robust control, and energy-based methods for swing-up mode [8]. Dynamical modeling [4] is, most of the time, the starting point for the choice of control strategies. Improvements in the RIP theoretical model with additional torque terms due to friction is the primary concern in [4]. These terms are then approximately identified with experimentation. Feedback linearization of the derived model is employed in [9] to compose the control signal that carries out a signal tracking of arm angle movement having sinusoidal nature. As the model structure has uncertainties, adaptive control strategies can also be a good candidate. The development of an adaptive controller to deal with the uncertainties due to variation in system parameters such as the mass and center of gravity is studied in a simulative manner in [10]. Also, in [10], simulations demonstrate the controller's capability to bring the RIP from any state to an unstable equilibrium point. Another adaptive controller [11] uses sliding mode control techniques coupled with a PID controller to easily handle parameter variations, similar to [10]. In [12], sliding mode control and state feedback (SF) technique that uses Linear Quadratic Regulator controllers are compared.

Meta-heuristic search methods are also utilized to develop controllers. Hamza et al. [13] represents two cascaded fuzzy PD controllers improved using Genetic Algorithm (GA) and particle swarm optimization techniques. The cost function evaluates time-domain performance indices to update the controller structures' gain values. Similarly, GA is also used in Hassanzadeh and Mobayen [14] and Hassanzadeh et al. [15] to develop a cascade control structure with feedback linearization for optimizing a cost function that minimizes angles and velocities of the arm and pendulum for step response. There is only a tiny difference between these studies: Hassanzadeh and Mobayen [14] uses angles and velocities, whereas Hassanzadeh et al. [15] only uses the angle values for the cost function. GA, particle swarm optimization, and ant colony optimization methods are compared with each other in [16], where PID parameters are optimized by these methods due to a multi-criteria cost function governing time-domain performance indices. Mobayen [17] also used GA to develop a PID controller. All these studies employed similar time-domain performance indices, which are an integral part of a multi-criteria optimization process. Frequency-domain analysis techniques like Nyquist criteria are also used to develop controller structures [18], even if they are not as common as the time-domain techniques.

Nonlinear controller forms are also applicable to the problem. Fuzzy logic and artificial neural networks are the most common of these forms. Krishen and Becerra [6] employs two fuzzy inference mechanisms, one for the pendulum and one for the arm variables, to compose the control input signal. Using two fuzzy inference mechanisms is critical to limit the number of fuzzy rules since increasing the number of fuzzy rules also increases memory consumption for computation which may cause lags in real-time applications. Oltean [19] employs a fuzzy PD controller for hold mode and a PD controller in the swing-up mode of operation. Neural Networks are also used both to control [20] and to identify [21] RIP. A scheme that utilizes the nonlinear adaptive neural network control method is established in [22] for the hold mode of operation. In [23], a review of all possible applicable techniques to RIP is presented, and pole placement with state feedback is considered as one of these techniques.

This study intends to design SF controllers for the hold mode of operation of the RIP. In the development process, a multi-criteria optimization procedure based on cost function concerning frequency-domain performance indices is employed. In this context, two SF controllers are developed by GA (another SF controller is also created as a variant of the developed more competent SF controller), and their performances are compared with a standard SF controller's performance both in real-time applications and in simulations. Additionally, our methodology enables us to put forward a bound for the SF gain values for stability considerations, and these bounds are used as an integral part of the optimization.

The multi-criteria optimization procedure enables to limit and adjust the SF gain values. This is not possible when a standard SF is preferred: Commonly, pole placement by SF is a static process: Once the closed-loop poles' locations are determined, the SF gain values are calculated automatically. However, multi-criteria optimization limits these gain values based on the determined performance criteria. Another significant focus of the multi-criteria optimization process is the relative position of the closed-loop poles of the system concerning the imaginary axis. It is intended to put these poles to the left open half of the s-plane (to the left of the imaginary axis) to sustain stability, which is also made possible with the help of a multi-criteria optimization procedure. Limiting the SF gain values and putting the closed-loop poles at the left open half plane are two conflicting situations with each other. There is a trade of to satisfy both of these criteria at the same time with sufficiency. Hence, this problem is a good candidate for multi-criteria optimization. Besides, the effect of the ratio of the imaginary part to the real part of the closed-loop poles and a relative stability criterion demonstrating the effect of putting the poles to the left of a negative real point in the s-pane are also studied. These two criteria are related to the damping ratio and the practical stability considerations of the nonlinear system, respectively. Once it is shown that the multi-criteria optimization procedure is successful for SF, one can easily set up a new multi-criteria optimization procedure to develop more sophisticated control algorithms with new criteria and with different controller structures. In this context, optimization gives the chance to manage the applied control method according to the preferred track.

Another contribution of this study is the usage of bounds obtained for SF gain vector components by Routh–Hurwitz criteria in the optimization process. The characteristic equation of the controlled system by SF pole placement at the unstable equilibrium point is derived first. Then Routh–Hurwitz criteria are applied to characteristic equations to achieve nonlinear inequalities giving regions of stability for the closed-loop system. These inequalities are employed in the optimization process to determine the suitable SF gain vector components and restrict the search domain of the GA to speed up the optimization process. Otherwise, one has to compute each eigenvalue of the closed-loop system one by one and determine whether the closed system is unstable or not, which is computationally a more time-consuming process. Instead, whether the nonlinear equations are satisfied or not for stability considerations is checked for decreasing the computation time.

Finally, switching from swing-up mode to hold mode is studied in real-time. The success of switching depends mainly on two parameters. The first parameter is the critical threshold pendulum angle value, which influences the success explicitly; the other parameter is a gain parameter that contributes to the speed of the pendulum in swing-up mode. This second parameter has an implicit effect on stability. In real-time applications adjusting these parameters have deep impact for sustaining stability. The relation between the components of SF controllers with the closed-loop pole locations and stabilization of the RIP at unstable equilibrium point is also monitored during the switching process.

2 Methodology

2.1 System Description

The RIP system "ROTPEN" is shown in Fig. 1 [24]. It consists of three major parts: the servo motor, the rotary arm, and the pendulum stick (arm). The servo motor supplies the necessary actuating motion signals to the arm. The rotary arm is attached to the load gear by a pin, and from the other end, it is connected to the metal shaft. This arm rotates on a horizontal plane. The pendulum is connected to the rotary arm, and it rotates 360° freely in a vertical plane.

According to Fig. 2, the Cartesian coordinates are used to determine the arm and pendulum locations. L_p, ℓ_p, and α are total lengths of the pendulum, the pendulum length from the arm's pivot to the center of mass of the pendulum, and the angle of the pendulum, respectively. θ is the rotation angle of the arm around the z-axis. L_r and ℓ_r are the arm's total length and the length from the pivot to the center of mass of the arm, respectively.

Fig. 1 SRV02 ROTPEN system [24]

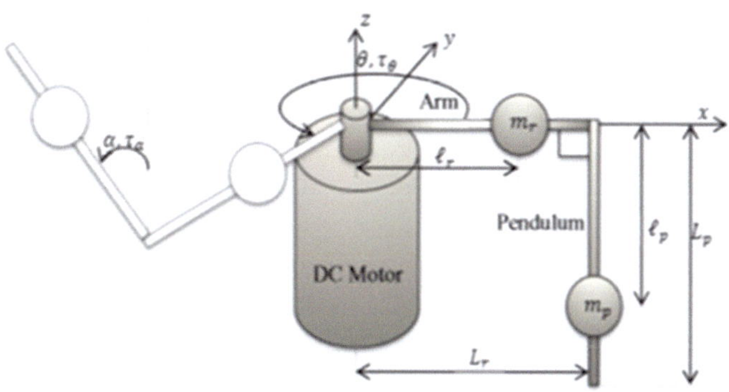

Fig. 2 Rotary inverted pendulum diagram [25]

2.2 System Dynamics

To obtain the dynamic equations [26] of the RIP center of masses of pendulum and arm are calculated, and these equations are obtained by the Euler–Lagrange formulation [27].

$$\begin{aligned}
&\ddot{\theta}\left(m_pL_r^2 + \sin^2\alpha\left(J_p + m_pl_p^2\right) + J_r + m_rl_r^2\right)\\
&-\ddot{\alpha}\left(m_pL_rl_p\cos\alpha\right) + \dot{\theta}\left(B_r + \left(J_p + m_pl_p^2\right)\dot{\alpha}\cos\alpha\sin\alpha\right)\\
&+\dot{\alpha}\left(\left(J_p + m_pl_p^2\right)\dot{\theta}\cos\alpha\sin\alpha + m_pL_rl_p\dot{\alpha}\sin\alpha\right) = \tau,
\end{aligned} \tag{1}$$

$$\begin{aligned}
\ddot{\theta}\left(-m_pL_r\ell_p\cos\alpha\right) + \ddot{\alpha}\left(J_p + m_p\ell_p^2\right) + \dot{\theta}\left(-\left(J_p + m_p\ell_p^2\right)\dot{\theta}\sin\alpha\cos\alpha\right)\\
- m_pg\ell_p\sin\alpha + B_p\dot{\alpha} = 0.
\end{aligned} \tag{2}$$

B_r and B_p are viscous damping coefficients. τ describes the torque applied on the arm, which is generated via servo motor defined by

$$\tau = \eta_g k_g \eta_m k_t\left(V_m - k_g k_m\dot{\theta}\right)/R_m \tag{3}$$

where η_g and η_m are gearbox efficiency and the motor efficiency, k_t, k_g, and k_m are the motor-torque constant, the gear ratio, and the back-emf constant, and R_m is the armature resistance and V_m is applied DC motor voltage.

The RIP is a Single Input Multiple Output (SIMO) system where the input is the applied DC motor voltage V_m, and the system states can be assigned as the arm angle ($x_1 = \theta$), inverted pendulum angle ($x_2 = \alpha$), and angular velocities of the arm and the inverted pendulum ($x_3 = \dot{\theta}$, $x_4 = \dot{\alpha}$). The outputs of the system are x_1 and x_2. The equilibrium point is defined as the upright position of the pendulum when the pendulum angle is 0° ($x_2 = 0$). The system parameters are shown in Table 1 [24, 38].

The system is linearized at the unstable equilibrium point with parameter settings as in Table 1, and corresponding state space matrices are obtained.

$$A = \begin{bmatrix} 0 & 0 & 1 & 0\\ 0 & 0 & 0 & 1\\ 0 & 53.29853 & -16.5724 & -0.6582\\ 0 & 108.2996 & -19.5897 & -1.3373 \end{bmatrix}$$

$$B = \begin{bmatrix} 0\\ 0\\ 29.8002\\ 35.2259 \end{bmatrix}$$

Table 1 System parameters

Symbol	Description	Value	Unit
m_p	Mass of pendulum	0.127	kg
L_p	The total length of the pendulum	0.337	m
l_p	Distance from pivot to center of mass	0.156	m
J_p	Pendulum moment of inertia about the center of mass	0.0012	kg.m^2
B_p	Pendulum viscous damping coefficient as seen at the pivot axis	0.0024	Nms./rad
m_r	Mass of rotary arm with two thumbscrews	0.257	kg
l_r	Rotary arm length from pivot to center of mass	0.0619	m
J_r	Rotary arm moment of inertia about its center of mass	9.98 10^{-4}	kg.m^2
m_b	Rotary arm viscous damping coefficient as seen at the pivot axis	0.0024	N.m.s/rad
B_r	Viscous damping coefficient of arm	0.0024	Nms/rad
L_r	The total length of the arm	0.256	m
R_m	Armature resistance	2.6	Ω
k_g	SVR02 system gear ratio	70	–
k_m	Back-emf constant	7.68×10^3	Vs/rad
k_t	Motor-torque constant	7.68×10^3	Nm/A
η_g	Gearbox efficiency	0.9	–
η_m	Motor efficiency	0.69	–

The eigenvalues of the systems are 0, –4.6674, –20.9277, 7.6853. As one of the poles of the system has a positive real part, the system is unstable.

2.3 Control Strategies and Development of Controllers

It is intended to stabilize the system using SF with pole placement around the unstable equilibrium point. For this purpose, K gain vector with components K_1, K_2, K_3, and K_4 is suggested as the controller structure.

$$K = \left[K_1 \; K_2 \; K_3 \; K_4 \right]$$

The components of the K vector, K_1 is the proportional gain of the arm angle (θ), K_2 is the proportional gain of the pendulum angle (α), and K_3 and K_4 are proportional gain values of the arm's and pendulum's angular velocities, respectively.

In literature, the K vector is habitually determined based on the desired closed-loop pole locations. However, this way of determining them has particular side effects.

Once the closed-loop pole locations are fixed, the *K* vector components are set automatically. Occasionally, some of the components of the *K* vector can have comparatively high values creating enormous control input signals, especially when the control of the RIP passes from swing-up mode to hold mode or at signal tracking application in real-time. In these cases, rapid RIP movements can be observed, even creating instability. For this reason, an optimization process can be set to put restrictions over the *K* vector components coupled with closed-loop system stability considerations to obtain SF controllers with unique characteristics. In this context, GA can be used for such a purpose. GA is a popular meta-heuristic search method that is generally applied to optimization problems. Darwin's theorem of natural selection inspires this method [28–32].

The procedures of GA are as follows:

- Generate the population randomly.
- Calculate the fitness values of each chromosome.
- Select the chromosomes for their fitness values.
- Mate the chromosomes to create the new generation.
- Investigate the solution.

This procedure continues until an optimal solution is supposed to be obtained. For generating a new generation, selection, crossover, and mutation operators are of significant importance. These genetic operators are a milestone to guarantee that the search continuously improves and, in general, better chromosomes are obtained and selected in new generations. Pseudo Code of the GA process used in the optimization process is shown in the Appendix section.

3 Implementation of Controllers

Four different SF controllers are established to stabilize the pendulum in the hold mode. The first controller is a standard SF controller. The second and third controllers are developed by GA search due to a multi-criteria optimization procedure, and the fourth controller is a variant of one of the developed controllers by GA. The GA cost function is taken as a weighted average of four different criteria: the general stability of the closed-loop system, the restrictions over the *K* vector components, the relative stability of the closed-loop system due to practical considerations, and the damping ratio limitation. In real-time applications, switching characteristics of three controllers are monitored at different parameter setting conditions. Two more competitive controllers' performances are compared for sinusoidal and square wave signal tracking due to arm angle changes. Besides integral square error (ISE) [33], and total applied square voltage (TASV) performance criteria are set to monitor the controllers' performance at a standstill position when there is no reference signal tracing operation in real-time application. Additionally, theoretically, all controllers' step response time-domain performance indices for arm angle are monitored in a simulation environment.

3.1 Standard SF Controller

The poles of the linearized systems are 0, −4.6674, −20.9277, and 7. 6853. As one of the poles has a positive real part, the system is unstable. If the system is completely controllable, SF control with pole placement can be applied to relocate the closed-loop poles in any desired location in s-plane. Thus, the controllability condition of the system is checked with the standard controllability test using the controllability matrix. The controllability matrix of the linearized system is composed, and this matrix's rank is found as four, which means that the linearized system is completely controllable as the system is also fourth-order. Thus, SF with pole placement is applicable. It is decided to put the dominant poles to the location having a damping ratio of $\zeta = 0.7$ and undamped natural frequency of $w_n = 4\text{rad/s}$; giving $s_{1,2} = -2.80 \mp j2.86$, and the remaining two closed-loop poles are selected as at −30 and −40 as in [34]. With these selections, the corresponding K vector is computed

$$K = \begin{bmatrix} -14.224 & 63.587 & -6.5397 & 7.1702 \end{bmatrix}$$

This controller is named as $SF_{standard}$. In [34], the gear and the motor's efficiency are taken into account, and for this reason, the K vector has totally different components.

3.2 Using Genetic Algorithm to Develop SF Controller

At first sight, it is impossible to estimate the $SF_{standard}$ controller's performance in real-time application as K vector components are blindly determined by setting the values of the closed-loop poles. If the pole locations are not set wisely, as a side effect, huge K vector components can be achieved that might end up with creation of enormous input signals causing rapid movements and poor overshoot and settling time performance. Besides, stability might also hardly be sustained in such circumstances as RIP passes from swing-up mode to hold mode. Thus, it is better to construct K vector components by considering the general stability of the system but with some limitations. Thus, a multi-criteria optimization process is carried out using GA. The GA's cost function is chosen to handle different criteria while adjusting the components of the K vector through optimization. These criteria are as follows: Criterion A (the general stability of the system): The closed-loop system's stability with the selected K vector components is monitored. If at least one of the closed-loop poles obtained due to the corresponding K vector has a positive real part (unstable pole), the related chromosome is punished with a very high penalty. This condition is accomplished in a more time-efficient way using Routh–Hurwitz criteria [35, 36]. Indeed, it is not intended to find the pole locations directly to decide on stability during the GA search. Instead, due to Routh–Hurwitz's criteria, four nonlinear inequalities are obtained [37], and with the selected K vector, it is checked whether these inequalities are satisfied or not. If all inequalities are satisfied, the corresponding K vector stabilizes the system making all closed-loop poles stable. As mentioned, this method is more time-efficient than finding the closed-loop pole locations and checking the stability in the GA search. Besides, some of the inequalities are simpler compared to the other ones. Hence, using these inequalities, it becomes possible to narrow the GA search space, and the initial population of the GA algorithm is selected randomly but according to some of these constraints, which consequently decreases the GA search computation time. Criterion B (restrictions over the K vector components): This criterion is linked with the restriction of the K vector components. In different GA searches, the summation of K vector components' absolute values is restricted to the threshold value of 60. If a K vector exceeds this threshold value, the corresponding chromosome is punished to an extent it exceeds the threshold. Criterion C (the relative stability of the closed-loop system due to practical considerations): Even if all poles of a linearized system are stable when a pole is very close to the imaginary axis, it has the potential to make the system unstable in practical real-time applications. For this reason, a new cost contribution is added as an extra term to guarantee that all the stable poles are relatively far away from the imaginary axis. Hence, if a chromosome puts some poles between the imaginary axis and the $s = -2$ line in the s-plane, the corresponding chromosome is further punished. This condition restricts the dominant poles. All poles should be replaced to the left of the $s = -2$ line by the chromosome to get rid of this penalty. Criterion D (the damping ratio limitation): The final condition is correlated with absolute values of real and imaginary parts of stable poles. If a stable pole has some imaginary parts, the imaginary part/absolute (real part) ratio should be smaller than a threshold value to reduce the effect of undamped oscillations. This threshold value is chosen as 0.25. Hence, the absolute value of real parts of a pole should be four times higher than the absolute value of the imaginary part of the same pole obtained by a K vector; otherwise, the corresponding chromosome is punished. It is assumed that this criterion will provide chromosomes having better overshoot performance. Two different optimization runs are carried out to understand the effects of these conditions. In the first run, Criterion A and Criterion B are valid for the cost function, whereas in the second run, all four criteria are available to contribute to the cost function.

The GA search parameters are determined as follows:

- Population size = 200.
- Number of generation = 500.
- Number of chromosome = 400.
- Ratio of crossover = 0.9.
- Ratio of reproduction = 0.06.
- Ratio of mutation = 0.04.

The GA code to calculate the chromosomes' fitness values is composed in Matlab® as the reciprocal of their cost values. When Criterion A and Criterion B are used to compose the costs, the GA search results the K vector as

$$K = \begin{bmatrix} -6.9181 & 45.5623 & -4.7962 & 5.7963 \end{bmatrix}$$

and corresponding locations of the closed-loop eigenvalue are obtained as −60.4847, −12.4166, −3.1303 + 1.6235j, and −3.1303 − 1.6235j. This SF controller is named an SFGA601. When all conditions are used to compose the cost, the K vector is obtained as

$$K = \begin{bmatrix} -13.909 & 77.1822 & -10.3866 & 11.274 \end{bmatrix}$$

and corresponding locations of the closed-loop eigenvalue are obtained as −82.8702, −10.3285 + 2.569j, −10.3285 − 2.569j, and −2. This SF controller is named an SFGA602. The absolute sum of components of SFGA602 exceeds 60 which is the threshold value according to Criterion B. However, the dominant eigenvalue is set to −2 which is consistent with Criterion C. This is how a multi-criteria optimization process trades of between different criteria. While one of them is partially fulfilled, some others are fulfilled totally.

$SF_{standard}$, SF_{GA601}, and SF_{GA602} are compared in different circumstances. Primarily, the real-time operating RIP, which is driven by Simulink®, is used as the environment to check controllers' real-time performance. In real-time applications, the RIP starts from the downward position. It swings up until the pendulum reaches nearly an upward position (α value reaches ±12°). After that, the pendulum changes the mode of control and gets into hold mode, where the computed SF vector stabilizes the pendulum at an upright position and there performance tests are carried out. Secondly, the previously established SF controllers and a variant of SF_{GA601} are also simulated and their capabilities are measured theoretically using time-domain performance indices by Simulink®.

3.3 Switching Process, Stability, and Reference Input Tracking Capability in Real-Time

The stability and the performance of the controllers while passing from swing-up mode to hold mode are important success indicators. In the real-time application, the pendulums start from the downward position, and in swing-up mode, which is governed by the default algorithm of Quanser, it is brought up to a nearly upward position. When the pendulum angle reaches the threshold value, the swing-up mode is switched to the hold mode, and the SF controllers developed in Sects. 3.1 and 3.2 drive the pendulum and try to keep the pendulum as stable as possible. In real-time applications, the sampling frequency is chosen to be 1000 Hertz. This value is sufficiently high as the Nyquist rate for the control system is considered.

RIP is a mechanical system, its natural frequencies are considerably low, and time constants corresponding to stable modes are relatively high. Hence, a sampling ratio of 1000 Hertz is enough to represent the dynamic system response properly. Unfortunately, SF_{GA602} is not able to stabilize the system in real-time application. When the control passes from swing-up mode to hold mode at the threshold angle $\alpha \pm 12°$, SF_{GA602} causes very rapid movements due to its components, and stability is not sustained. Especially, the second component (the gain for the pendulum angle) and fourth component (the gain for the pendulum angular velocity) of SF_{GA602} are responsible for this result in real-time. The second component of SF_{GA602} has a value of 77.1822, whereas it is only 45.5623 for SF_{GA601}. While passing from the swing-up mode to hold mode at a critical pendulum angle $\alpha \pm 12$, the second component of SF_{GA602} contributes formation of comparably larger applied input signal, causing very rapid movements. Similarly, the fourth component of SF_{GA602} has a value of 11.2741, whereas it is only 5.7963 for SF_{GA601}. Hence, if the pendulum angular velocity value is still very high when the critical pendulum angle is reached, it also creates a massive applied input signal.

Due to those two conditions, in switching process, the RIP makes sudden movements causing the RIP to pass from hold mode to swing-up mode reversely from the other side without stabilizing the pendulum at upright position. That is, if the RIP passes from swing-up mode to hold mode at the critical pendulum angle value $\alpha = 12°$, due to the rapid movement it passes from the hold mode to swing-up mode subsequently at the critical pendulum angle value $\alpha = -12°$.

Except for the K vector components; there are parameters that are significant at switching process, two of which are assumed to be prominent. One of these parameters is the critical threshold pendulum angle value. This parameter influences the switching process explicitly. The other parameter is a gain factor that adjusts the force given to the RIP in the swing-up mode. This parameter is implicitly responsible from the pendulum angular velocity value when the critical threshold pendulum value is reached. If gain factor is comparably high, more power is applied to the RIP causing higher pendulum angular velocities around the critical threshold pendulum angle value. Reversely, if gain factor is not adjusted appropriately and chosen as a comparably low value, the swing-up mode might never pass to the hold mode as the power applied to the system becomes short of to create necessary applied input signal that brings the RIP to the critical threshold pendulum angle value. For this reason, the effect of these parameters in switching process is critical. To observe the effect of these parameters in switching process, five critical threshold pendulum angle values ($\alpha = \pm 15$, $\alpha = \pm 12$ nominal value, $\alpha = \pm 10$, $\alpha \pm 8$, and $\alpha = \pm 6°$), two different swing-up gain factor values (nominal value 4 and 3) are set, and the success of the controllers to sustain stability of RIP in the upright positions is monitored. Table 2 summarizes these results.

Table 2 Effect of system parameters in switching process performance

Swing-up gain factor = 4 (nominal value)			
Critical threshold pendulum angle α (in degrees)	$SF_{standard}$	SF_{GA601}	SF_{GA602}
±6	U	U	U
±8	S	S	U
±10	S	S	U
±12 (nominal value)	S	S	U
±15	S	S	S
Swing-up gain factor = 3			
Critical threshold pendulum angle α (in degrees)	$SF_{standard}$	SF_{GA601}	SF_{GA602}
±6	U	U	U
±8	S	S	S
±10	S	S	S
±12 (nominal value)	S	S	S
±15	S	S	S

It is observed that whatever the swing-up gain factor is chosen, all the controllers are unsuccessful when the critical threshold pendulum angle value is chosen as $\alpha = \pm 6°$. At these cases, all the controllers are unable to sustain stability of the RIP at the upright position due to comparably fast movements observed between comparably small pendulum angle gaps. As the angle gap between limit values are smaller compared to other cases, RIP returns back to swing-up mode from hold mode more rapidly. $SF_{standard}$ and SF_{GA601} are able to sustain stability for all other cases whereas SF_{GA602} is successful when swing-up gain factor is decreased to three or the critical threshold pendulum angle is 15°. However, although the stability is sustained, strong vibrations are observed when RIP operates at hold mode. Such vibrations are not observed when other two controllers are in use. Thus, using SF_{GA602} for performance measures is not a good idea. Besides, for this reason, limiting the *K* vector by multi-criteria optimization is not only important to stabilize the RIP but also it is critical to remove the vibrations when the pendulum is stabilized and to get rid of sudden movements during the switching process. Due to these reasons, for a fair comparison, only arm angle, pendulum angle, and input voltage signals obtained by $SF_{standard}$ and SF_{GA601} real-time applications are considered, and these results are shown in Figs. 3, 4, 5, 6, 7 and 8 [37].

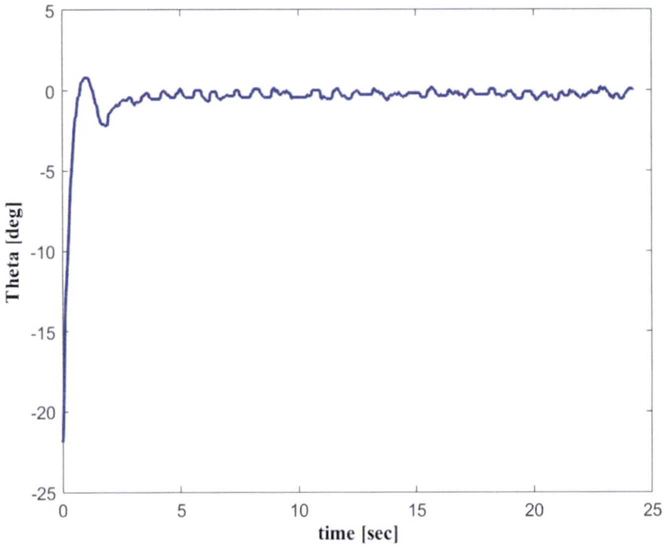

Fig. 3 Arm angle for $SF_{standard}$

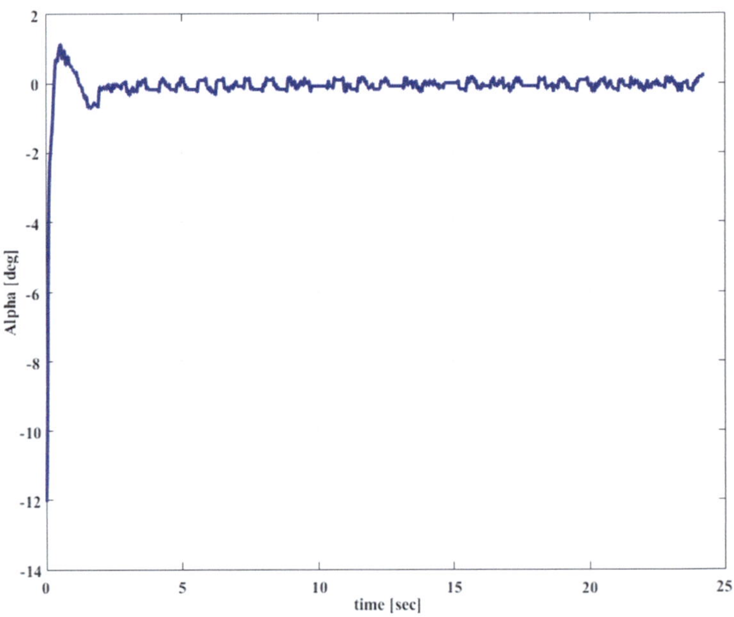

Fig. 4 Pendulum angle for $SF_{standard}$

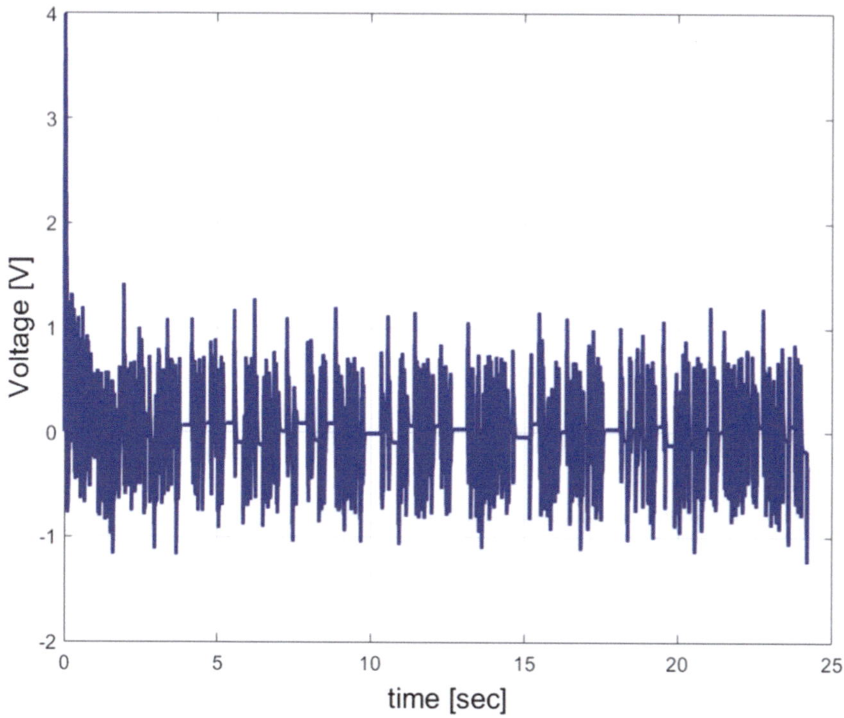

Fig. 5 Voltage for $SF_{standard}$

As seen from Figs. 3, 4, 5, 6, 7 and 8, both controllers can stabilize the pendulum, and during the switching process, instability isn't observed in the real-time application. The reference input tracking capability is also an important test to check the controllers' performance. Hence, real-time reference signals are applied to the system after the controllers stabilize the pendulum at an upright position. For this purpose, square wave and sinusoidal wave reference signals are applied to the arm angle (θ), and the controller tracing capacity is monitored. The reference input tracing performance $SF_{standard}$ is given in Figs. 9 and 10 for square wave and sinusoidal wave reference signals. The reference input tracing performance for SF_{GA601} is given in Figs. 11 and 12 for square wave and sinusoidal wave reference signals.

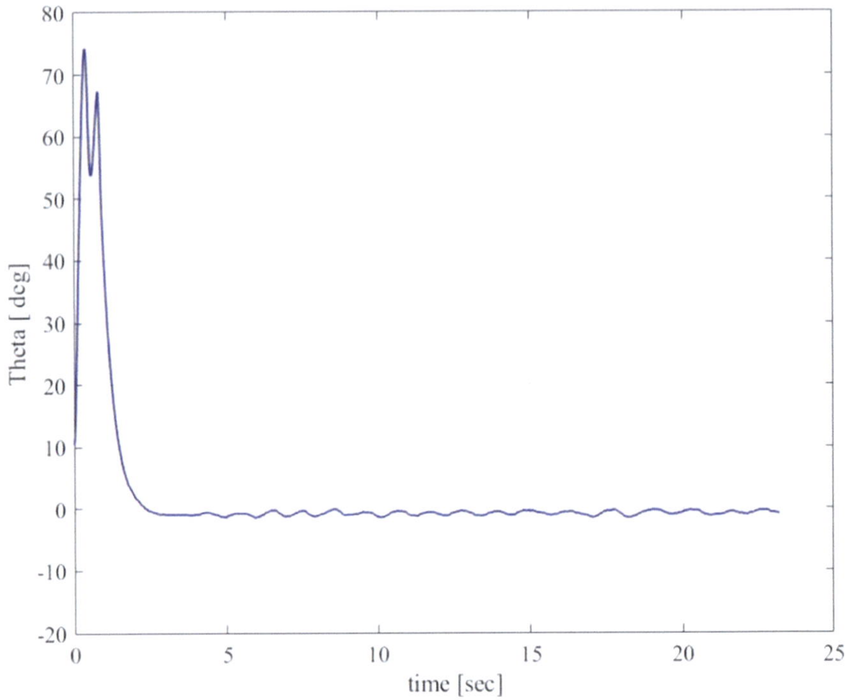

Fig. 6 Arm angle for SF_{GA601}

From Figs. 9, 10, 11 and 12, it can be concluded that SF_{GA601} outperforms $SF_{standard}$ as $SF_{standard}$ cannot track the reference signals very well.

3.4 Performance Index Measurement of Controllers in Real-Time

The performance of controllers is also tested when they are at a standstill upward position. One method to quantitatively measure controllers' performance is to use performance indices. Two performance indices are selected for this reason: Integral Square Error (ISE) for arm angle and total applied square voltage (TASV) to show the power of applied input voltage. ISE of arm angle is essential to show how much the arm angle deviates from its reference value. Similarly, TASV is a measure of

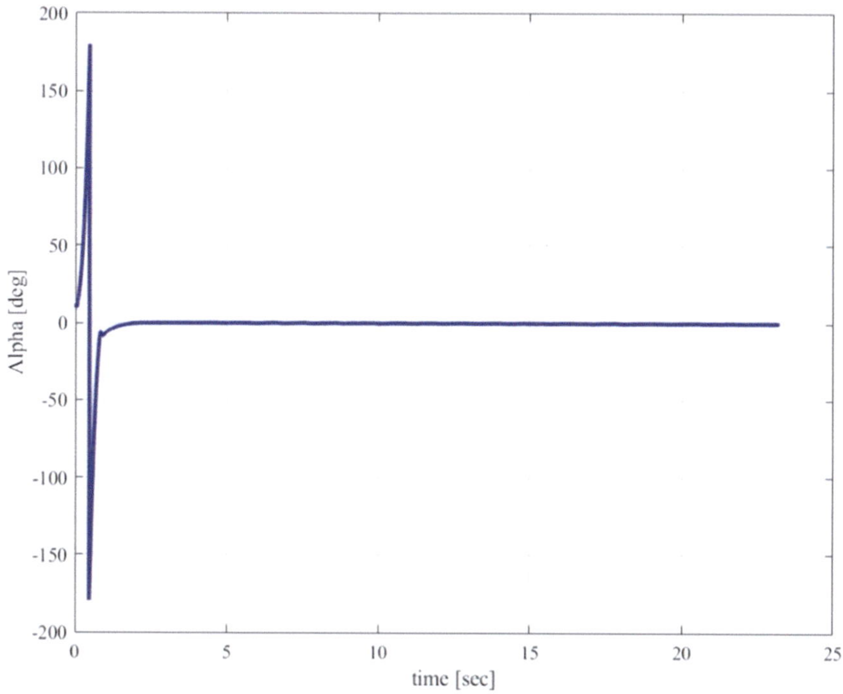

Fig. 7 Pendulum angle for SF_{GA601}

how much the controller changes the voltage waveform to stabilize and keep the pendulum at an upright position. The formulation of ISE in general is

$$ISE = \int_{t=0}^{\infty} e^2(t)dt \tag{4}$$

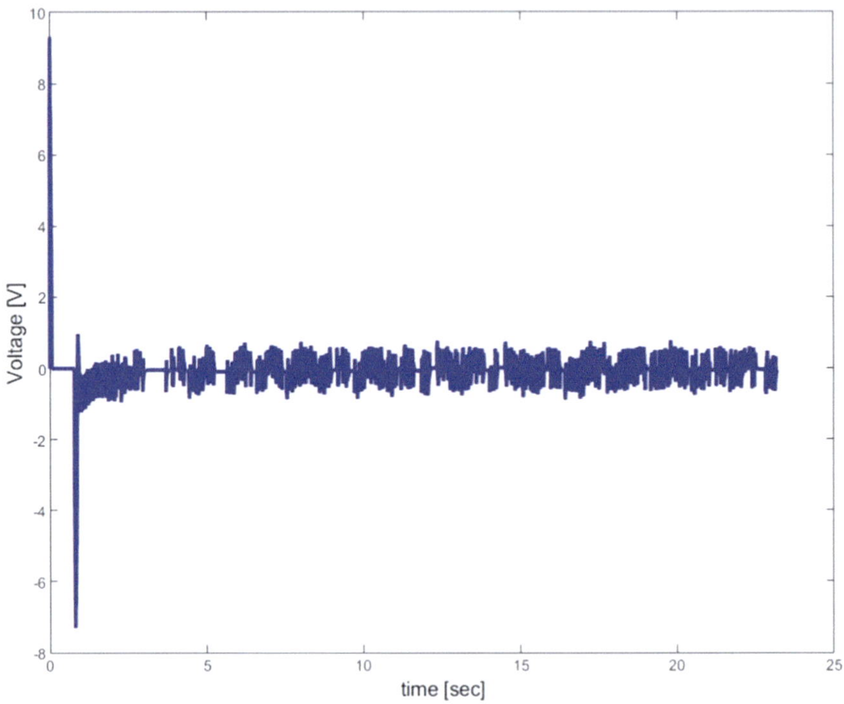

Fig. 8 Voltage for SF_{GA601}

In the real-time application, the sampling frequency is 1000 Hz (hence sampling period is 0.001 s). The last 10000 data are recorded and used to calculate performance indices, and TASV for input voltage and ISE of arm angle are formulated, respectively.

$$TASV = \sum_{i=1}^{10000} \text{Voltage}_i^2 \times 0.001 \tag{5}$$

$$ISE = \sum_{i=1}^{10000} (\theta_i - 0)^2 \times 0.001$$

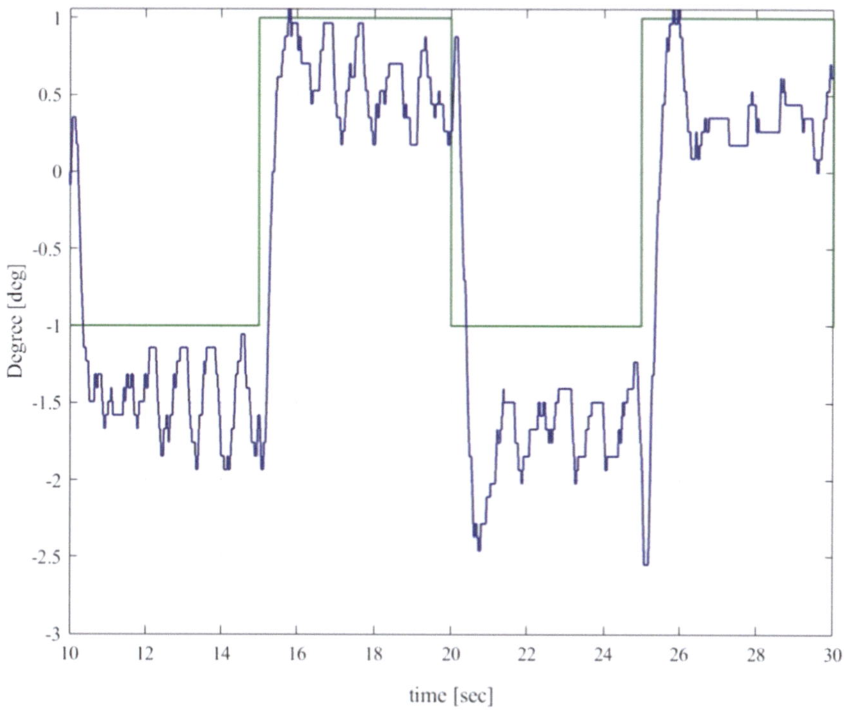

Fig. 9 Reference signal in square wave (green) and the arm angle of the system (blue) when f = 0.1 Hz by $SF_{standard}$

According to Table 3, SF_{GA601} uses less power as its TASV value is smaller compared to $SF_{standard}$. Therefore, SF_{GA601} balances the unstable system by relatively less effort compared $SF_{standard}$. Besides its ISE value for the arm angle is also relatively small. On the other hand, $SF_{standard}$ is still a good choice as it has a better ISE value with a slightly higher TASV.

4 Simulations

It is also intended to demonstrate the theoretical significance of developed controllers. Thus, the corresponding time-domain performance indices achieved by SF_{GA601}, SF_{GA602}, and $SF_{standard}$ are monitored and compared by simulation. A variant of SF_{GA601} ($SF_{GA601var}$), which is later derived artificially without any optimization, is also compared with the other controllers. The simulation is carried out in Simulink®. The state space representation of the linearized plant model is used assuming the

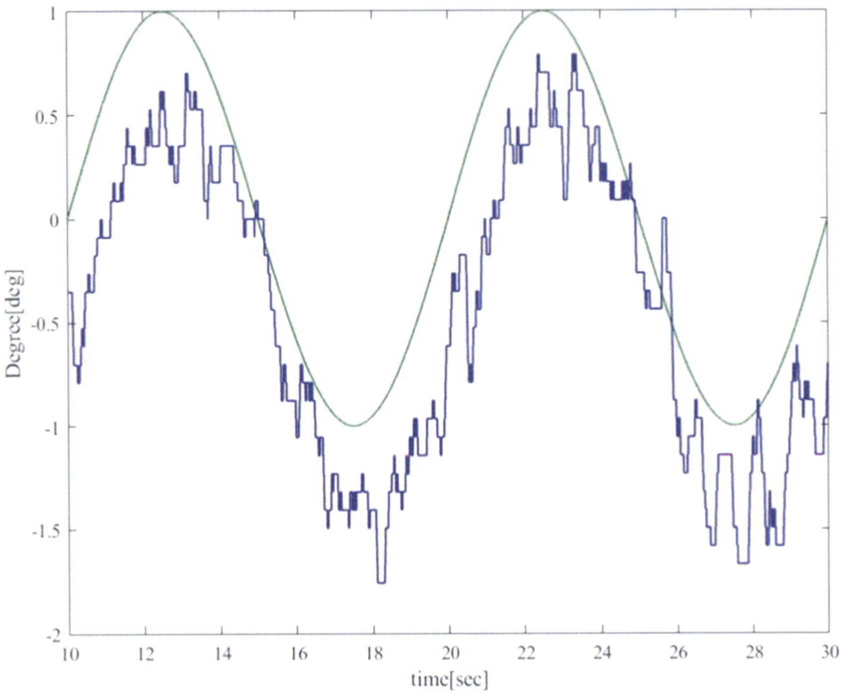

Fig. 10 Reference signal in sinusoidal wave (green) and the arm angle of the system

pendulum is at the upright position initially, and then with the corresponding SF controllers, the plant model is controlled to track a reference signal of a unit step change of 10° in arm angle. The results are tabulated in Table 4. In the simulations, to obtain a fair comparison, the pre-filter coefficients are also employed with SF controllers to guarantee that output signals reach steady-state conditions with a minimum steady-state error.

To compare the performance of the controllers, the arm angle, pendulum angle, and the control signal (input voltage) variations are demonstrated for $SF_{standard}$, SF_{GA601}, and SF_{GA602} in Fig. 13.

As seen from Table 4, $SF_{standard}$ is the fastest one based on rising time comparisons, but it creates a high overshoot, and hence it takes a long time for the response to settle down. SF_{GA601} has a minimal overshoot. The applied input signal has lower maximum and minimum levels (closer to zero Volt) due to lower K vector components. Besides, it has comparably fast rise time and best settling time values. It also exhibits better undershoot performance compared to other SF controllers. The third controller (SF_{GA602}) has the best overshoot performance; however, it has high K vector components causing significant maximum and minimum input voltage signals.

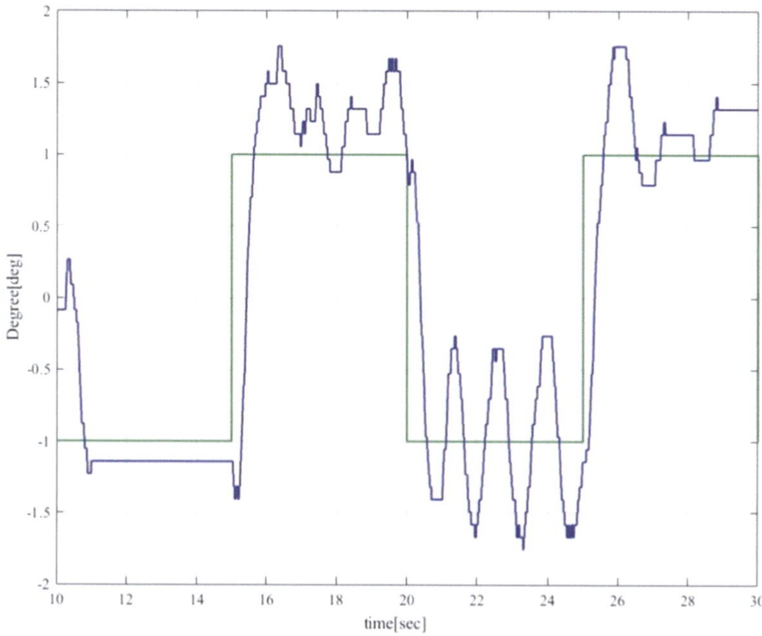

Fig. 11 Reference signal in square wave (green) and the arm angle of the system (blue) when f = 0.1 Hz by SF_{GA601}

Due to use of Criterion C at the optimization process, the dominant closed-loop pole obtained for SF_{GA602} is at location $s = -2$. This dominant pole is much closer to imaginary axis than any other closed-loop pole obtained. Hence, it has a slower settling and rise time. The dominant poles of the $SF_{standard}$ and SF_{GA601} controllers are complex conjugate pole pairs. The overshoot performance is generally related with the dominant poles' imaginary part to real part ratio. In this context, the SF_{GA601} has better performance in terms of overshoot compared to $SF_{standard}$.

GA search with multi-criteria optimization presents new opportunities. It produced SF controllers having comparably high and reliable time-domain performance indices with slightly lower *K* vector components which enable accomplishing the control process with the lower levels of applied input signal. A standard SF controller where the closed-loop pole locations are set directly lacks these properties: There is no possibility to control *K* vector components. Additionally, applying GA is just a starting point to introducing more competent controllers. For this reason, SF_{GA601}, which introduced comparably the best performance concerning other controllers, is modified without optimization, and its variant $SF_{GA601var}$ is introduced in Table 4. Non-dominant pole locations of $SF_{GA601var}$ and SF_{GA601} are the same; however, the dominant poles of $SF_{GA601var}$ have no imaginary parts, whereas

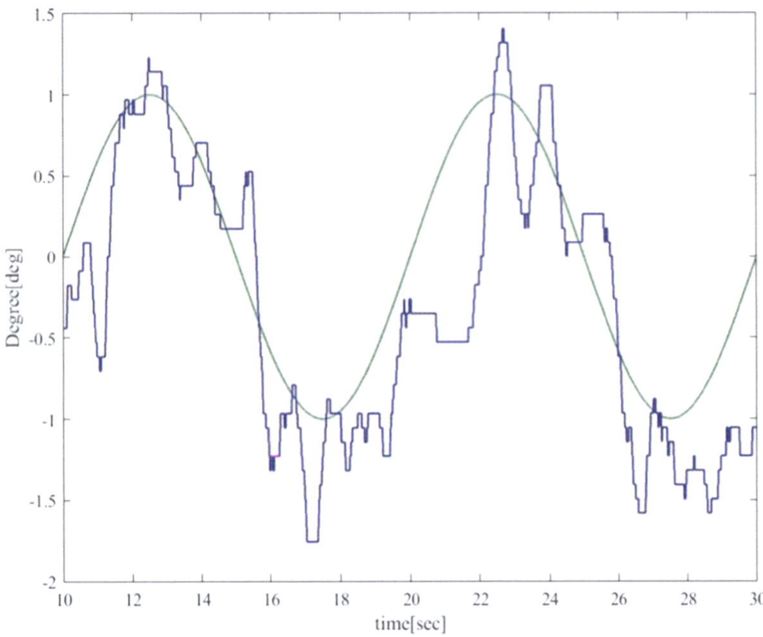

Fig. 12 Reference signal in the sinusoidal wave (green) and the arm angle of the system (blue) when f = 0.1 Hz by SF_{GA601}

Table 3 Results of the performance indices

Case	ISE results	TASV results
$SF_{standard}$	1.7146	0.8070
SF_{GA60}	2.1810	0.5166

the dominant poles of SF_{GA601} have imaginary parts. Thus, $SF_{GA601var}$ has better overshoot performance compared to SF_{GA601} and it possesses even lower applied input signal levels. However, as a side effect, $SF_{GA601var}$ has a slower response as it has higher rise time and settling time values. In Fig. 14, these two controllers' arm angle, pendulum angle, and input signal variations are given.

5 Discussion

This study's primary purpose is to design SF controllers by the GA that optimizes a multi-criteria cost function off-line and compare their performance with an SF controller that is obtained solely due to a pole placement procedure. The RIP system,

Table 4 Comparison of developed controllers

SF controllers	$SF_{standard}$	SF_{GA601}
Gain vector K	[−14.224, 63.587, −6.5397, 7.1702]	[−6.9181, 45.5623, −4.7962, 5.7963]
Poles	−30, −40, −2.80 ± j2.86	−60.4847, −12.4166,j −3.1303 ± 1.6235
% OS	6.5798	0.2939
Undershoot	20.7626	11.9699
Applied max and min input voltages (Volt)	0.4053, −2.4822	0.2161, −1.2073
Settling time	1.4955	1.3084
Rise time	0.4105	0.7129
SF controllers	SF_{GA602}	$SF_{GA601var}$
Gain vector K	[−13.909, 77.1822, −10.3866, 11.2741]	[−5.4517, 44.1711, −4.6358 5.6606]
Poles	−82.8702, −2, −10.3285 ± j2.569	−60.4847, −12.4166, −3.1303, −3.1303
% OS	0.0145	0
Undershoot	14.6709	9.4822
Applied max and min input voltages (Volt)	0.2823, −2.4279	0.1694, −0.9515
Settling time	2.0637	1.8899
Rise time	1.0741	1.0005

which demonstrates nonlinear, unsteady, and complex characteristics, is selected as a test bed. When the system is linearized, SF can directly be applied as the control structure to stabilize the system at an unstable equilibrium point(s). However, this is a blind effort as nobody knows where the closed-loop poles should be obtain the best performances at different applications. Hence, using a multi-criteria optimization gives us clues about where the optimal positions of the closed-loop poles should be. RIP problem requires two different modes of control operations in real-time applications: the swing-up and hold mode. The standard Quanser algorithm is chosen for the swing-up mode, whereas the developed SF controllers are in use for the hold mode. The SF controllers try to stabilize the RIP at the unstable equilibrium point(s) in the hold mode. The success of the controllers in the hold mode of operation depends on many system parameters. One of the system's essential control parameters is the critical pendulum angle value, where the system switches from the swing-up mode to the hold mode of operation. This angle is generally chosen at a value between ±6 and ±15°. If this angle is not chosen wisely, a sudden impact of the high K vector components at switching process might make the RIP unstable due to the pendulum's rapid movements. Another parameter that influences the success of switching process is the swing-up gain factor which is used to adjust the pendulum angular velocity in swing-up mode. If the angular velocity of the pendulum is high

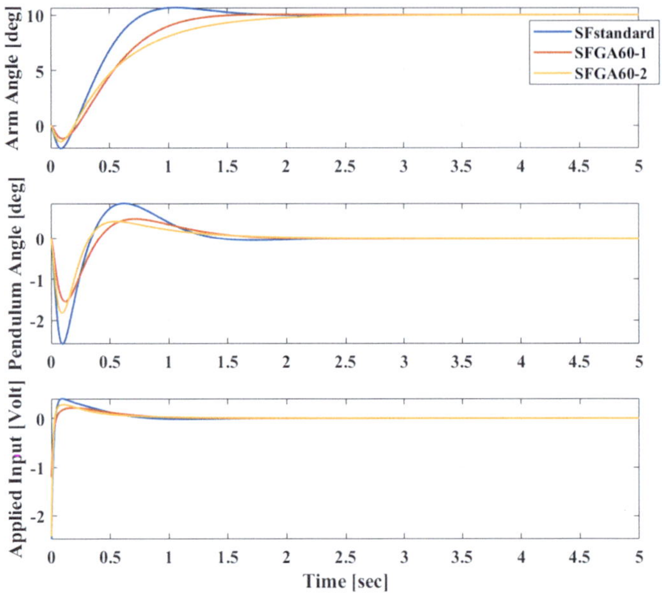

Fig. 13 Arm angle, pendulum angle, and applied input signals were obtained for reference tracking in simulations for $SF_{standard}$, SF_{GA601}, and SF_{GA602}

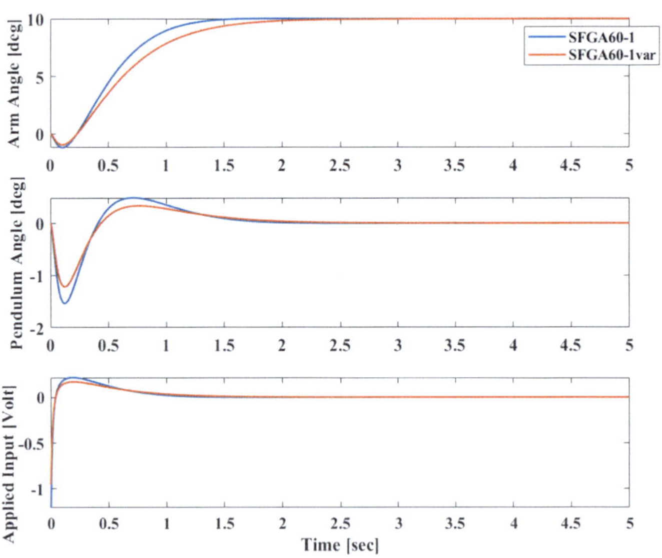

Fig. 14 Arm angle, pendulum angle, and applied input signals were obtained for reference tracking in simulations for SF_{GA601} and $SF_{GA601var}$

at the switching process, again comparably rapid movements of the RIP can be observed causing instability. On the contrary, very low swing-up gain factor values might create comparably insufficient control signals that have the potential to be short of providing the necessary force to make the RIP pass from swing-up mode to hold mode. Due to these reasons, many SF controllers developed off-line, which seem to be sufficient to stabilize the pendulum, can be unsuccessful in the real-time application at first sight. However, by changing the critical threshold pendulum angle value and swing-up gain factor at switching process, they can also stabilize the RIP. Two SF controllers obtained by multi-criteria optimization procedure are demonstrated in this study. They are SF_{GA601} and SF_{GA602}. Theoretically, both are successful. However, SF_{GA601} stabilizes the pendulum in the real-time application without any parameter change. SF_{GA601} also demonstrates better reference input tracking capability than the $SF_{standard}$, which is derived solely using a standard pole placement procedure by SF. When the two controllers' performance index values are compared, $SF_{standard}$ is observed to be slightly better in terms of ISE for arm angle, whereas SF_{GA601} is slightly better in terms of TASV. However, none of the controllers has significant dominance over the other. In simulations, SF_{GA601}, SF_{GA602}, and $SF_{standard}$ are also compared in terms of time-domain performance indices. This time SF_{GA601} overperforms the other controllers in most of the time-domain performance indices. A variant of SF_{GA601} is also derived to observe the effect of parameters changes in SF controllers. The new derived controller $SF_{GA601var}$ is similar to SF_{GA601} with one significant difference: $SF_{GA601var}$ puts nearly all the closed-loop poles in the same locations as SF_{GA601} but the dominant closed-loop poles attained by $SF_{GA601var}$ do not possess imaginary parts whereas the dominant closed-loop poles attained by SF_{GA601} have imaginary parts. Thus, $SF_{GA601var}$ has better overshoot and undershoot performance, but as a side effect, it has poor rise time and settling time characteristics.

The controllers developed using multi-criteria optimization have several advantages: Primarily, the SF gain vector components can be limited, resulting in better performance while switching from swing-up mode to hold mode of operation in real-time. Besides, rapid movements that can cause loss of stability in the real-time application are prevented. Secondly, time-domain performance indices can be improved.

The drawback of the proposed method is also being an optimization process. Every optimization process requires parameter adjustment to reach to promising and satisfactory solutions. Hence, the weights of multi-criteria cost function and GA parameters should be wisely selected. As SF is a strong and easily implemented method, it is chosen as the control structure in this study. Additionally, GA is chosen as the optimization technique because its cost function facilitates the usage of multi-criteria optimization straightforwardly. The selection of the controller structure and optimization algorithm may spark the question of whether they are the most competent ones for the RIP system. The answer is no. Every meta-heuristic optimization algorithm depends on some parameter settings, and depending on different optimization runs, one may obtain different and better solutions. The statistical analysis of different meta-heuristic search methods was not the concern of this study. Besides, nonlinear methods like fuzzy logic and neural networks might also be employed instead of

SF, and they might be more competent theoretically. However, SF can be easily implemented to real-time working RIP system ROTPEN as it has a small number of variables that will not cause computation problems in real-time. In contrast, nonlinear methods require more optimization time off-line and more computation time online in the real-time application, which may create lagging in the input signal causing problems. To sum up, our primary concern is the ease of using the optimization algorithm and simple structure of the controller both in simulations and in the real-time application.

As future work, disturbance conditions will be added in the real-time application and simulation, and the system's response will be examined under these conditions. Furthermore, the effects of swing-up mode parameters and the SF gain vector components in steady state and transient responses should be investigated deeply.

Declarations This manuscript is related with pre-print. The pre-print's doi: https://doi.org/https://doi.org/10.21203/rs.3.rs-2373423/v1.

Appendix

The Pseudo Code of the optimization:

1: Input Problem Data Set
2: Parameter Setting (Crossover Ratio, Mutation Ratio, Reproduction Ratio etc.)
3: Randomly Create Initial Population (use constraints over the K vector to diminish the search space)
4: Set i = 1 for First Generation
5: **for** i = 1 to Last Generation **do**
6: Calculate Fitness with Cost Functions
7: Sort Population according to Fitness Values of Chromosomes
8: If i = Last Generation get out from for Loop
9: Select Chromosome Using Roulette Wheel Method
10: Apply Crossover Operation
11: Apply Mutation Operation
12: Apply Reproduction Operation (elitist chromosome is also selected)
13: New Population is obtained
14: **end for**
15: Output is the best chromosome of the last generation.

References

1. Prakash B, Roy BK, Biswas RK (2017) Design, implementation and comparison of different controllers for a rotary Inverted Pendulum. In 1st IEEE International conference on power electronics, intelligent control and energy systems ICPEICES 2016, pp 1–6. https://doi.org/10.1109/ICPEICES.2016.7853165
2. Sarkar TT, Dewan L (2017) Pole-placement, PID and genetic algo rithm based stabilization of inverted pendulum. In: 8th international conference on computing, communication, and networking technologies ICCCNT 2017. https://doi.org/10.1109/ICCCNT.2017.8204047
3. Nath K, Dewan L (2018) Control of a rotary inverted pendulum via adaptive techniques. In: 2017 international conference on emerging trends in computer & communication technologies ICETCCT2017. https://doi.org/10.1109/ICETCCT.2017.8280315
4. Acosta JA (2009) Furuta's pendulum: a conservative nonlinear model for theory and practice. Math Probl Eng 2010. https://doi.org/10.1155/2010/742894.
5. Furuta K, Yamakita M, Kobayashi S (1992) Swing-up control of inverted pendulum using pseudo-state feedback. Proc Inst Mech Eng Part I J Syst Control Eng 206(1992):263–269. https://doi.org/10.1243/pime_proc_1992_206_341_02
6. Krishen J, Becerra VM (2006) Efficient fuzzy control of a rotary inverted pendulum based on LQR mapping, in: IEEE Int. Symp. Intell. Control- Proc., IEEE, 2006: pp. 2701–2706. https://doi.org/10.1109/CACSD-CCA-ISIC.2006.4777066
7. Ben Hazem Z, Fotuhi MJ, Bingül Z (2020) Development of a fuzzy LQR and fuzzy-LQG stability control for a double link rotary inverted pendulum. J Franklin Inst 357(2020):10529–10556. https://doi.org/10.1016/j.jfranklin.2020.08.030
8. Chou KY, Chen YP (2014) Energy based swing-up controller design using phase plane method for rotary inverted pendulum. In: 2014 13th international conference on control, automation, robotics and vision, ICARCV 2014, IEEE, pp 975–979. https://doi.org/10.1109/ICARCV.2014.7064438
9. Aguilar-Avelar C, Moreno-Valenzuela J (2014) A feedback linearization con troller for trajectory tracking of the Furuta pendulum. In: Proceedings of the American control conference, American automatic control council, pp 4543–4548. https://doi.org/10.1109/ACC.2014.6858724
10. Chen YF, Huang AC (2014) Adaptive control of rotary inverted pendulum system with time-varying uncertainties. Nonlinear Dyn 76:95–102. https://doi.org/10.1007/s11071-013-1112-4
11. Kuo TC, Huang YJ, Hong BW (2009) Adaptive PID with sliding mode control for the rotary inverted pendulum system. In: IEEE/ASME international conferences on advanced intelligence mechatronics, AIM, IEEE, pp 1804–1809. https://doi.org/10.1109/AIM.2009.5229784
12. Mathew NJ, Rao KK, Sivakumaran N (2013) Swing up and stabilization control of a rotary inverted pendulum. In: IFAC proceedings volume, IFAC, pp 654–659. https://doi.org/10.3182/20131218-3-IN-2045.00128
13. Hamza MF, Yap HJ, Choudhury IA (2015) Genetic algorithm and particle swarm optimization based cascade interval type 2 fuzzy PD controller for rotary inverted pendulum system. Math Probl Eng. https://doi.org/10.1155/2015/695965
14. Hassanzadeh I, Mobayen S (2008) GA based input-output feedback linearization controller for rotary inverted pendulum system. In: Proceeding 5th International Symposium on Mechatronics and its Applications, ISMA 2008, pp 25–30. https://doi.org/10.1109/ISMA.2008.4648838
15. Hassanzadeh I, Mobayen S, Harifi A (2008) Input-output feedback linearization cascade controller using genetic algorithm for rotary inverted pendulum system. Am J Appl Sci 5:1322–1328. https://doi.org/10.3844/ajassp.2008.1322.1328
16. Hassanzadeh I, Mobayen S (2011) Controller design for rotary inverted pendulum system using evolutionary algorithms. Math Probl Eng. https://doi.org/10.1155/2011/572424

17. Mobayen S (2017) Design of a novel genetic algorithm stabilizer for swing-up stabilization of rotational inverted pendulum system. In: ECTI-CON 2017- 2017 14th international conference on electrical engineering, electronics, computer, telecommunications, and information technology, pp 860–864. https://doi.org/10.1109/ECTICon.2017.8096374
18. Dwivedi P, Pandey S, Junghare A (2017) Performance analysis and experi mental validation of 2-DOF fractional-order controller for underactu ated rotary inverted pendulum. Arab J Sci Eng 42:5121–5145. https://doi.org/10.1007/s13369-017-2618-8
19. Oltean S-E (2014) Swing-up and stabilization of the rotational inverted pen dulum using PD and fuzzy-PD controllers. Procedia Technol 12:57–64. https://doi.org/10.1016/j.protcy.2013.12.456
20. Quyen ND, Van Thuyen N, Hoc NQ, Hien ND (2012) Rotary inverted pendulum and control of rotary inverted pendulum by artificial neural network. Int Conf Theor Phys 243–249. http://www.iop.vast.ac.vn/theor/conferences/nctp/proc/37/243_249.pdf
21. Chandran D, Krishna B, George VI, Thirunavukkarasu I (2015) System identification of rotary double inverted pendulum using artificial neural networks. In: 2015 international conference on industrial instrumentation and control, IEEE, pp 612–617. https://doi.org/10.1109/IIC.2015.7150815
22. Yang X, Zheng X (2018) Swing-up and stabilization control design for an underactuated rotary inverted pendulum system: theory and experiments. IEEE Trans Ind Electron 65:7229–7238. https://doi.org/10.1109/TIE.2018.2793214
23. Hamza MF, Yap HJ, Choudhury IA, Isa AI, Zimit AY, Kumbasar T (2019) Current development on using Rotary Inverted Pendulum as a benchmark for testing linear and nonlinear control algorithms. Mech Syst Signal Process 116:347–369. https://doi.org/10.1016/j.ymssp.2018.06.054
24. User Manual Inverted Pendulum Experiment (2012)
25. Al-jodah A (2013) Experimental verification and comparison of different stabilizing
26. Duart JL, Montero B, Ospina PA, Gonz´alez E (2017) Dynamic modeling and simulation of a rotational inverted pendulum. J Phys Conf Ser 792(2017):012081. https://doi.org/10.1088/1742-6596/792/1/012081
27. Furuta K, Yamakita M, Kobayashi S (1991) Swing up control of inverted pendulum. In: IECON Proceedings of industrial electronics conference, pp 2193–2198. https://doi.org/10.1109/IECON.1991.239008
28. Haupt RL, Haupt SE (2004) Practical genetic algorithms, 2nd edn. A John Wiley & Sons Inc., Publication, Hoboken, New Jersey
29. Kolcu I (1996) An investigation of genetic algorithms and genetic programming
30. Liang H, Zou J, Zuo K, Khan MJ (2020) An improved genetic algorithm optimization fuzzy controller applied to the wellhead back pressure control system. Mech Syst Signal Process 142:106708. https://doi.org/10.1016/J.YMSSP.2020.106708
31. Amar J, Nagase K (2021) Genetic-algorithm-based global design optimization of tree-type robotic systems involving exponential coordinates. Mech Syst Signal Process 156:107461. https://doi.org/10.1016/J.YMSSP.2020.107461
32. Li D, Deng L, Cai Z (2020) Intelligent vehicle network system and smart city management based on genetic algorithms and image perception. Mech Syst Signal Process 141:106623. https://doi.org/10.1016/J.YMSSP.2020.106623
33. Dorf RC, Bishop RH (2006) Modern control systems, 12th edn. Prentice Hall, Upper Saddle River, NJ
34. Apkarian J, Karam P, Levis M (2011) Instructor workbook, Markham, Ontario
35. Ogata K (2002) Modern control engineering, 5th edn. Prentice Hall, New Jersey
36. Nise NS (2011) Control systems engineering, 6th edn. John Wiley and Sons Inc., Hoboken, New Jersey
37. Şen FN (2019) Design and implementation of control algorithms for stabilization of rotary inverted pendulum. Çankaya University
38. User Manual SRV02 Rotary Servo Base Unit (2011)

Tracked Rescue Robot for Emergency Search and Surveillance Tasks

Berke Yücetaş, Serhan Gelens, Baran Turğay, and Onur Çelik Çelikezen

1 Introduction

This section discusses the study's difficulties, aims, importance, limits, and associated explanations. First, in an age where technology is more vital in solving complicated challenges, rescue robots are emerging as game changers in emergency response and catastrophe recovery. Rescue robots, for example, are game-changing advances in emergency response that aid in search and rescue efforts during catastrophes and crises. Furthermore, these contemporary robots are outfitted with a variety of technologies, including advanced sensors and cameras, allowing them to explore areas that are too risky or hazardous for people, search for survivors, and conduct critical jobs that can't be completed directly. Furthermore, by operating in locations such as collapsed structures, inclement weather, or poisonous environments, these robots improve the safety and efficiency of rescue efforts. Furthermore, their ability to offer real-time (i.e., immediately) information and execute difficult tasks has made them an essential tool in modern emergency management, revolutionizing how the group responds to and recovers from disasters.

Project Objectives

First and foremost, the fundamental goal of a rescue robot is, of course, to help in emergency circumstances by completing duties that improve the efficiency of search and rescue operations.

B. Yücetaş · S. Gelens (✉) · B. Turğay · O. Ç. Çelikezen
Department of Mechatronic Engineering, Near East University, Nicosia, Turkey
e-mail: gelensserhan@outlook.com

Department of Mechanical Engineering, Near East University, Nicosia, Turkey

Research Center for Science, Technology and Engineering (BILTEM), Near East University, Nicosia, Turkey

A. Azizi (ed.), *Autonomous Robotics and Intelligent Systems*, Emerging Trends in Mechatronics, https://doi.org/10.1007/978-981-95-5867-4_6

Specifically, the objectives include:

- **Locate and Identify Survivors**: The robot's main goal is to locate humans who are imprisoned or in need of aid. Sensors, cameras, and other detecting technologies are frequently employed in such scenarios to look for debris, devastation, and potentially dangerous places.
- **Provide real-time Data**: The built and developed robot offers human rescuers with real-time data such as video feeds, ambient conditions, and structure integrity information, allowing them to make better judgments and coordinate their efforts.
- **Perform Dangerous Tasks**: The robot also performs jobs that would be too risky for human rescuers, such as crossing shaky structures and entering poisonous regions or areas which include hazardous materials.
- **Improve Situational Awareness**: By inspecting regions that are difficult or hazardous for humans to access, the robot helps to gain a better knowledge of the situation, which can then be utilized to guide rescue operations and activities.

Finally, the purpose of a rescue robot is to improve the efficiency, safety, and efficacy of rescue operations, saving hundreds or even thousands of lives and lowering the risks involved with emergency response.

Project Aim

- **Safety**: Contributes to human rescuer safety by completing perilous activities in hazardous circumstances such as collapsed structures, poisonous places, or natural catastrophes.
- **Efficiency**: As previously noted, these robots can work in circumstances that people would find difficult, such as severe temperatures or low vision, which would speed up search and rescue operations.
- **Data Collection**: The robots gather basic environmental information, such as structural soundness, temperature, and the presence of dangerous items, to help direct rescue operations.
- **Access**: Rescue robots may penetrate regions where people cannot, such as narrow passages or shaky structures, boosting the likelihood of discovering survivors.
- **Communication**: It gives real-time information to rescue crews, including live video feeds, to help them plan and make decisions during rescue operations.

The primary goal of rescue robots is to consistently maintain the efficacy and safety of rescue missions, saving lives and lowering the risk to human rescuers.

Challenges and Considerations

- **Environment and Terrain**: Rescue robots must cross difficult terrain, such as debris, rubble, and unstable ground. They also need to navigate limited locations, such as fallen structures, without becoming trapped or causing damage, which necessitates careful design and engineering.
- **Communication**: Contact in disaster situations is challenging owing to debris and destruction, and robots require dependable contact in order to convey data. In summary, real-time data is critical for coordinating rescue operations.

- **Power Supply**: Robots used in rescue missions require long-lasting batteries or efficient charging techniques to function well over time. Therefore, smart power management is critical for maximizing operational time while reducing the need for frequent recharge.
- **Search and Detection Capabilities**: Robots plainly require improved sensors, such as cameras, to efficiently identify injuries or victims. Rapid processing of sensor data is essential for successful rescue operations. To address these challenges, a multidisciplinary approach is necessary. This approach combines expertise in engineering, robotics, computer science, and disaster management. By advancing technology and enhancing design and functionality, the group can overcome these obstacles and improve the effectiveness of rescue robots in critical situations.

Project Timeline

Approximate timeframe for the process, starting from design right through to documentation.

The schedule will involve phases and could span different lengths of time based on how intricate the project is and factors like teamwork and available resources.

1. **Project Initiation (2 weeks)**:
    - Define project's scope, goals, and objectives.
    - Create a comprehensive project plan and solidify team's structure.
2. **Research and Requirements Gathering (1–3 weeks)**:
    - Gathering requirements entails considering user needs, intended capacity, efficiency, and safety standards.
    - Research existing rescue robots and design considerations.
3. **Conceptual Design (2–3 weeks)**:
    - Generate initial sketches, conceptualize the design, and develop 2D and 3D models using SolidWorks.
4. **Detailed Design and Assembly (3–4 weeks)**:
    - Created detailed engineering drawings.
5. **Simulation and Testing**:
6. **Refinement and Iteration**:
7. **Documentation and Reporting**:

Work Done

1. **Deliberation on Design Aspects**

The project began with long talks concerning the design of the rescue robot. Design components and structural considerations were developed through collaborative brainstorming sessions among group members. These conversations provided the foundation for the rescue robot concept.

2. **Literature Review**

A series of literature evaluations was carried out, with an emphasis on publications and research in the fields of rescue robots, teaching tools, and engineering designs. This assessment method included examining existing solutions and identifying new design changes based on previous research and novel techniques.

3. **Group Strategy Formulation**

The group held extended group discussions to develop a comprehensive plan to achieve the project goals, which included: collaborating to create a roadmap that included work assignments, responsibilities, dates, and milestones, as well as strategy discussions aimed at optimizing collaborative efforts and project workflow.

4. **Task Delegation and Timeline Establishment**

The team assigns responsibilities based on their skills and interests, in accordance with the strategy framework. In this way, the group established precise timelines and objectives to guarantee a methodical approach to design development. In summary, this phase stressed effective time management and meeting deadlines.

5. **Ordering the Parts, the Project Need**

The team was also successful in locating key components for the rescue robot and acquiring the required maintenance from a highly qualified source. This will allow the group to begin assembling the rescue robot, enhance its functioning, and do comprehensive testing.

6. **Drawing the Mechanical Part**

The team convened an intense brainstorming session to identify the project's foundations and polish the concepts, with the main objective of developing a physical design that would successfully represent the initiative by articulating the shared vision. As a result, it concentrated on clarifying every element addressed before evolving into a unique design.

7. **Assembling**

The group worked together to design the different components and meticulously constructed the tank, ensuring that every piece fit correctly and functioned properly.

8. **Testing**

The group has evaluated the project's durability and functionality by putting it to a range of demanding environments, and this intensive testing was intended to verify that the group did not experience real-world problems in the field. In addition, the group investigated camera performance, picture quality, and responsiveness, and tested Bluetooth connections for reliability and range. The group also created extensive graphs of the outcomes of the testing, which provide a clear visual indication of its capabilities.

Table 1 Budget of materials list

Name of the item	Unit	Unit cost	Amount ($)
Arduino Nano 328 FT232	1	5.84	5.84
M-F, F-F Jumper Cable	80	0.02	2.34
3S 20A BMS	1	1	3.94
Esp-32 Cam + OV2640	1	14.57	14.57
Arduino UNO R3	1	12.65	12.65
HC-05 Bluetooth module	1	4.94	4.94
L298N motor shield	1	1.84	1.84
6 V 250RPM DC Motor	4	0.98	3.92
Large breadboard	1	1.45	1.45
Prolific PL2303 USB-TTL	1	1.26	1.26
3D parts	1	39.45	39.45
Panasonic NCR18650b battery cell	3	10.52	31.56
Total			123.76

Budget of Materials List

Table 1 shows the budget of the materials used in the project (13.08.2024 TL-Dollar Exchange Rate).

2 Literature Review

The design and positioning of numerous sensors, wheel systems, mechanical arms, wireless communication mechanisms, and remote-control mechanisms are all covered in depth. The coordination of its numerous components to fulfill various functions is also explored, and suggested robot's movement may be controlled wirelessly, either from the control unit within the range of its transmitter or from a GSM cellular phone at any distance where a comparable mobile network is accessible. After building the robot, it was tested both indoors and outdoors and confirmed to perform well. For civil use, it can be used to detect gas leaks in homes or industries, to monitor shopping malls, parking lots, offices, industries, banks, museums, and so on, to aid in rescue operations during natural disasters and their aftermath, and to provide food or other necessities to people living in unreachable areas where human access is difficult or dangerous. It may also be used for military surveillance, such as spying on enemy bases, exploring undiscovered enemy territory, night security in a variety of areas, anti-intrusion defence, or as a mobile communication base during wartime with GSM connection from anywhere. The suggested robot's structure and design are adaptable and may be tweaked to take its capabilities to the next level [1].

The paper aims to provide an overview of the planning, development, and fabrication of a multipurpose agriculture robot capable of digging the soil, leveling the mud, and spraying water and fertilizer separately. The robot's entire system is powered by a battery, and thus uses alternative energy. Approximately 40% of the world's population selects agriculture as their first employment, and autonomous agriculture has recently gained popularity. The car contains a relay switch and Bluetooth. This language input enables a user to engage with a robot that is familiar to the majority of people.

The agricultural robot runs on solar energy. These robots have the benefit of being hands-free and allowing for quick data entry. In the agricultural autonomous vehicle industry, a concept is being researched to see if several light compact autonomous machines can outperform traditional huge tractors and human forces. As a result, the gadget will be developed to assist farmers in overcoming the aforementioned challenge. Its goal is to increase production while lowering labor requirements; the robot is designed to perform the necessary activities for farm administration [2]. Figure 1 depicts a multipurpose agricultural robot.

Mobile robots are widely employed in research and teaching. Although there are various commercial mobile robots available for these jobs, they are typically expensive, do not always fit the requirements of specific applications, and are extremely difficult to modify due to proprietary software and hardware. In this work, the group explains the design ideas and details the development and applications of a mobile robot known as ExaBot. The primary objective was to obtain a single multipurpose low-cost robot (more than 10 times less expensive than commercially available platforms) that could be utilized not just for research, but also for teaching and public outreach.

Fig. 1 Multipurpose agriculture robot [3]

Fig. 2 Mobile robot for educational purposes [5]

The body of the ExaBot, its sensors, actuators, processing units, and control board are all explained in depth. The software and printed circuit board produced for this project are open source, allowing the robotics community to utilize and improve the present version. Finally, several configurations of the ExaBot are described, demonstrating multiple applications that meet the parameters for which this robotic platform was created [4]. Figure 2 depicts an example of a mobile robot for teaching purposes.

The project's goal is to create a clever, multifunctional, human aid robotic dog that requires no maintenance. The major function of this dog is to lead visually impaired and elderly individuals to a predetermined location while avoiding obstacles and traffic. It is also intended to function as a sophisticated multifunctional human assistance and service robot capable of recognizing the user's spoken words, conversing with them, and acting in response to the spoken voice command. An Android smartphone recognizes voice instructions and transfers the information to the main MCU via a Bluetooth serial connection running the Bluetooth SPP protocol stack. When given a vocal instruction, the robotic dog will follow a human.

Contact-sensitive e-skin detects human finger contact and assists in responding to complicated user demands, such as time, date, and meteorological variables like light and temperature. The identical question may also be posed in a voice. It even allows users to create wake-up alarms. A built-in audio playback mechanism can play MP3 files. One of the music tracks is set as the alarm tone. It also serves as a regular guard at night, barking like any other dog whenever it detects any unusual activity. During the daytime, it may charge itself by roaming around within a specific zone to obtain the highest solar light, intelligently avoiding shady places, thereby relieving the user totally from maintenance.

It has a head, eyes, and tail like a real dog, which it utilizes to make specific gestures during human–robot contact [6]. Figure 3 depicts a smart multipurpose human assistance robot.

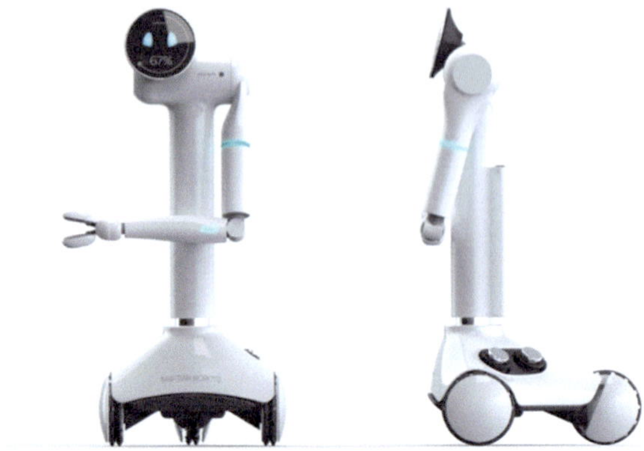

Fig. 3 Smart multipurpose human assistance robot [7]

The world's population grows by the day, as does the number of patients. People's life expectancy is rising as a result of technological advancements, yet the health industry has long struggled with a scarcity of sufficient doctors. The primary goal of this project was to create and construct a biomedical-based medical assistant robot known as "Docto-Bot" to address this issue. This study focuses on the design and development of a medical aid robot for the disabled and those in need. The primary purpose of such a robot is to reduce person-to-person interaction while also ensuring cleaning, sanitation, and assistance in hospitals and other comparable institutions such as quarantine.

This prototype robot includes a medication reminder and delivery system, an automatic hand sanitizer, and an IoT-based physiological monitoring system (body temperature, pulse rate, ECG, oxygen saturation level). A direct one-to-one server-based communication method and a user-end Android app maintenance system were created. It also had a controlling component that users could utilize both automatically and manually. Docto-Bot will play an important role in biomedical robot applications. Despite the fact that the gains presented in the study appear to be successful and advanced, there are still limitations [8]. Figure 4 depicts a medical assistant robot.

Navigating across unstructured settings is a basic ability of sentient beings, and so is of primary interest in the study. Navigation is a complicated job that requires the development of an internal representation of space, based on identifiable landmarks and powerful visual processing, capable of supporting continual self-localization ("I am here") as well as a representation of the objective ("I am going there"). Recent advances in artificial intelligence (AI) and associated technologies may make this possible. The number of robots deployed in the manufacturing business has grown fast, and this trend is expected to continue in the future, since autonomous robots have the ability to automate a wide range of labor-intensive jobs in the industrial setting, thereby increasing productivity.

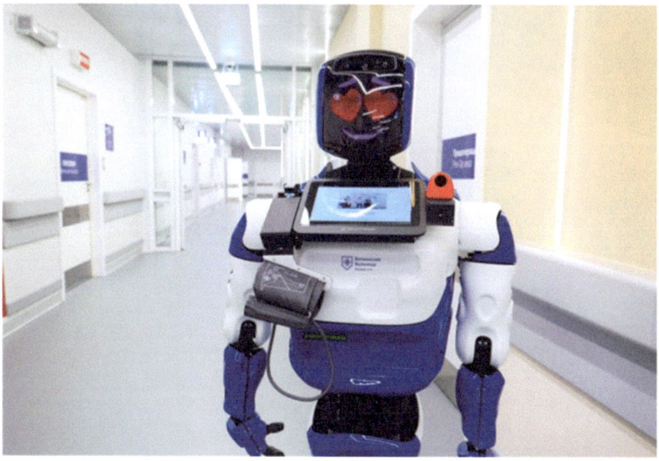

Fig. 4 Medical assistant robot [9]

There are several technological hurdles that must be overcome in order to build an autonomous multipurpose robotic platform. The major goal of this study is to solve the problem of autonomous robot navigation in the manufacturing environment. The robotic platform will be able to identify markings on the factory floor and independently move through the plant on the prescribed course, avoiding impediments along the way. In this study, the group employs a small number of sensors to lower the robotic platform's BOM cost while maximizing battery life.

In addition to all of these, they employed neural networks to detect landmarks and routes in the manufacturing environment, simultaneous localization and mapping (SLAM) to locate the robot, and a navigation algorithm to direct the robotic platform to its destination [10]. Figure 5 depicts autonomous factory robots.

Rescue teams rush against the clock to save lives in catastrophic calamities such as natural and mining disasters. The first 18–24 h are critical for rescue operations. For years, search and rescue operations have been carried out without the need for additional heavy equipment to help those trapped under the rubble. It is extremely difficult for rescue crews to enter through tight passageways, hunt for and image nearby locations, or measure coordinates. In such instances, the operating distance is critical to save people's lives. Many experiments have been conducted to move on in rocky or severe environments that humans find difficult to reach.

Kamegawa et al. created a robot called Kohga that resembles a snake and can crawl, climb, and pass through tiny openings. In the initial iteration of the rescue robot, the group employed a tank kit as its mobile foundation. It is tiny, moves rapidly, and has outstanding mobility in flat terrain. However, in a true catastrophe scenario, for example, under entirely fallen houses, this robot is less successful [12]. Figure 6 depicts the development of the snake-like rescue robot "*Kohga*".

Rescue robots play an important part in disaster recovery activities. A rescue robot locates the accident site, searches for survivors to aid in first aid treatment, and tells

Fig. 5 Autonomous factory robots [11]

Fig. 6 Development of the snake-like rescue robot "Kohga" [12]

the rescue team about environmental conditions and the potential for survival in the disaster-prone region. Several rescue robots are under development, and some have previously been deployed in emergency circumstances, but they have not functioned as expected. The industry's task is to produce rescue robots that are more advanced, intelligent, and have durable and dependable hardware to function in crucial zones and tackle all potential impediments (Baca et al., 2010). Swarm robots, or multi-robot teams that communicate and synchronize with one another, have the potential to expand the communication and sensor network. However, this concept is still in its early phases of research and may not appear in the mining business very soon [13]. Figure 7 depicts a coal mine search and rescue robot.

The Arduino microcontroller is open source and may be programmed and upgraded at any time. The original Arduino was debuted in 2005. The Arduino

Fig. 7 The robot designed for search and rescue operations of coal mines [14]

microcontroller was designed to let professionals and students construct interactive gadgets. Use sensors to interact with your environment. Arduino microcontrollers provide inputs and outputs that may be used to.

Arduino may generate output based on the information it receives. Arduino microcontrollers may also transmit and receive data over the Internet using HTTP requests. The Esp board is a simple microcontroller capable of connecting to the Internet. Esp Microcontrollers can connect to or function as Wi-Fi servers. The Arduino platform is divided into two components: hardware and software. The Arduino development board is the hardware that runs Arduino.

The Arduino Integrated Development Environment (IDE) is the software used to write code. Development Environment. These microcontrollers may be built with Atmel's 8-bit AVR microcontrollers or the 32-bit Atmel ARM. The Arduino IDE allows you to develop in C or C++ [15]. Figure 8 shows Arduino, NodeMcu ESP8266 ESP-12E CH340G WIFI Network Development, and an Arduino development board.

The Internet of Things (IoT) is a new technological system which combines a variety of information technologies and is a key component of the expanding modern service sector. The introduction of the Internet of Things is transforming traditional networks. Based on the Internet of Things, an Android smartphone is wirelessly connected to service-oriented robots, which are programmed via TCP sockets, allowing smartphones to control service robots. The system can operate the robot to sing, dance, and so on in response to commands, and the robot's qualities are visual and portable, with good motion control [17]. Figure 9 depicts a robot control concept based on a smartphone.

Since late 2019, the COVID-19 sickness has expanded worldwide. It creates serious issues for practically every country in the globe. According to World Health Organization (WHO) figures, there have been 57,639,631 confirmed COVID-19 cases and 1,373,294 fatalities worldwide until November 2020. Various tactics

Fig. 8 Arduino, NodeMcu ESP8266 ESP-12E CH340G WIFI Network Development, Arduino development board [16]

Fig. 9 Robot control design based on a smartphone [17]

and technologies, including robotics, are proposed to help in pandemic treatment and control. Previously, several studies briefly described the robotic uses during a pandemic.

Khan et al. divided robot applications in healthcare into several categories, including receptionist, nursing, ambulance, telemedicine, serving, cleaning, spraying/disinfectation, surgical, radiologist, rehabilitation, food, and outside delivery robots.

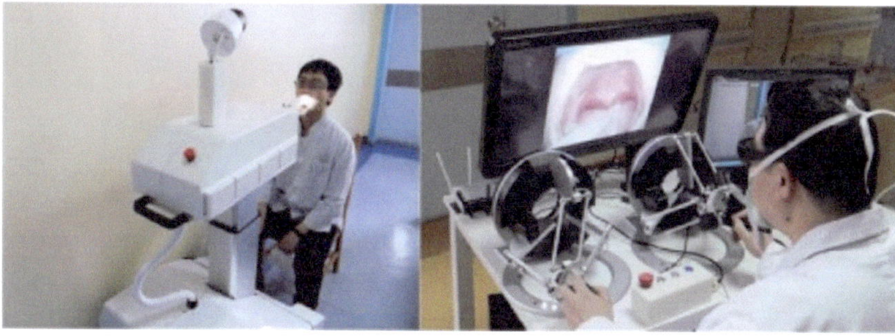

Fig. 10 A semi-automatic swab robot [18]

Zeng et al. evaluated robot applications based on their intended settings, such as hospitals, transportation, communities, recreation, airports, hotels, attractions, restaurants, and picturesque regions. Those assessments presented a summary of robotic achievements thus far. However, there is still a lack of comprehensive examination from a robotic technology standpoint. As a result, it is critical to address the important robotic technologies combating the epidemic and to suggest future research directions. Thus, in this research, a literature review is undertaken with the purpose of carefully analyzing the research achievements thus far [18]. Figure 10 depicts a semi-automatic swab robot.

Swarm robotics is concerned with the design, manufacture, and deployment of large groups of robots that collaborate to solve a problem or execute a task. It draws inspiration from natural self-organizing systems such as social insects, fish schools, and bird flocks, which exhibit emergent collective behavior based on basic local interaction rules. Swarm robotics often derives engineering ideas from the study of natural systems in order to create multi-robot systems with comparable capabilities.

This approach tries to create systems that are more resilient, fault-tolerant, and adaptable than individual robots and can better respond to changes in the environment [19]. Figure 11 depicts some of the robots often utilized in swarm robotics research.

The rise of artificial intelligence is mostly associated with software-driven robotic systems, such as mobile robots, unmanned aerial vehicles, and, to a greater extent, semi-autonomous autos.

Nonetheless, the enormous mismatch between the algorithmic and physical realms prevents present systems from attaining the desired objective of developing intelligent and user-friendly robots that can efficiently interact with and manage the human-centered world. The emerging discipline of machine intelligence (MI), which integrates robotics and artificial intelligence, seeks to create dependable and embodiment-aware AI systems. These systems are self-aware and comprehend their surroundings, allowing them to adapt to the interacting body in which they operate.

The integration of artificial intelligence (AI) and robotics into control, perception, and machine-learning systems is required for the realization of completely

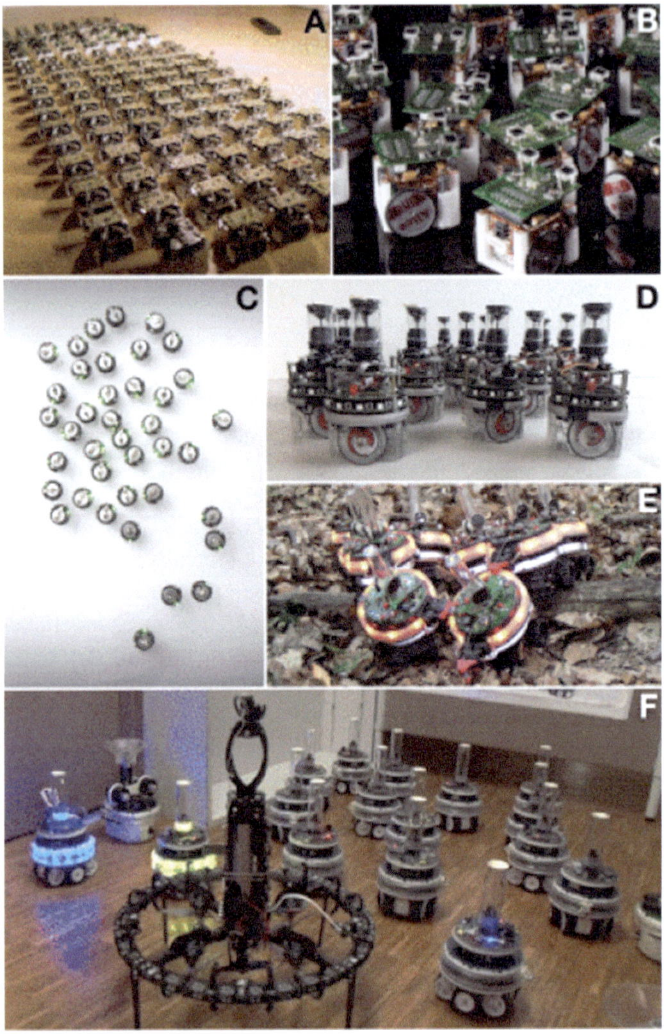

Fig. 11 Several robots commonly used in swarm robotics research include: **a** Jasmine (image from WikiMedia Commons); **b** Alice (image courtesy of Simon Garnier); **c** Kilobots (image courtesy of Massimo Berruti); **d** E-pucks; **e** Swarm-bots; and **f** Swarmanoid [19]

autonomous intelligent systems in the daily lives. This paper presents an outline of the historical evolution of machine intelligence, dating back to the eleventh century. It then goes on to explore the current state of robotics and artificial intelligence (AI), including [20].

The hotel sector began to deploy service robots, which are seen as the future labor. However, no attempt was made to investigate the dimensionality of service

Fig. 12 Hyundai motor group hotel delivery robot [22]

quality for service robots. This research seeks to investigate the impact of human–robot interaction from the perspectives of hoteliers and guests. Two studies have been undertaken in this regard. Study 1 conducts focus-group interviews with hotel managers from several departments to generate themes about guest-robot interactions and robot-delivered services. Based on the findings of Study 1, Study 2 conducts an experiment to investigate and evaluate hotel customers' opinions of the quality of services given by human personnel and service robots, as well as their collaborative services. Human staff services are viewed as superior to service robot services in terms of interaction quality and physical service environment. However, no substantial variation in outcome quality was seen [21]. Figure 12 depicts the Hyundai Motor Group Hotel Delivery Robot.

For more than 50 years, robotics has served as the cornerstone for sophisticated production. Robots and associated equipment are increasingly being employed in the military and law enforcement for a wide range of purposes as they get smaller, more reliable, and smarter. Mobile robotics is becoming increasingly important in military operations, from patrolling to dealing with potential bomb threats. Android smartphones are presently the most popular technology. Numerous Internet apps use the cell phones' built-in hardware, such as Bluetooth, Wi-Fi, and ZigBee technologies, to control other devices. The controller may be connected to the Bluetooth module using the UART protocol.

The robot's movements may be controlled via orders sent from Android [23]. Figure 13 depicts the block diagram of a military service robot.

The primary purpose of this research is to identify an application that might be used in the agile development of mobile apps. In this example, Dr. Mobile was selected as the primary approach for developing mobile apps, and the MIT App Inventor 2 tool was employed.

A real-world scenario is used to show the interaction between the Arduino UNO hardware development panel and a wireless environment via Bluetooth. With the Arduino hardware development board and its HC-05 Bluetooth module, it has been

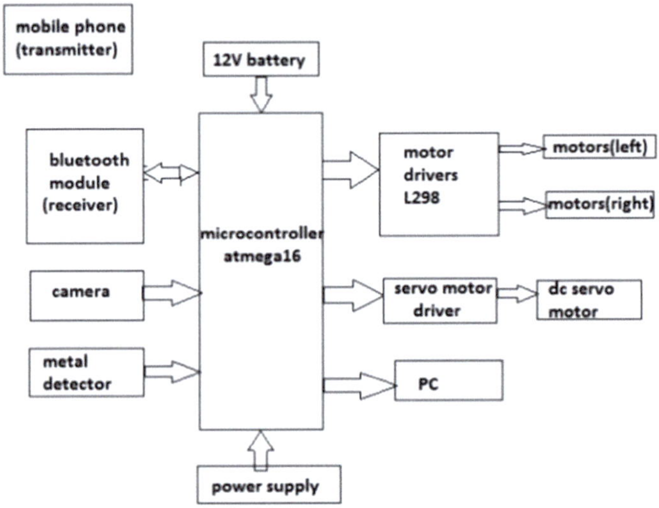

Fig. 13 Block diagram of military service robot [23]

discovered that the interaction between the application created and the Web App Inventor portal works flawlessly because the parameters are efficiently sent and received, and the response time is very quick [24]. Figure 14 depicts an example interface of MIT App design tab.

Every year, more advanced robots are produced for military uses. Currently, the US military has around 12,000 ground robots and 7,000 unmanned aerial vehicles. Forty-four other countries, including the United States, are studying unmanned military

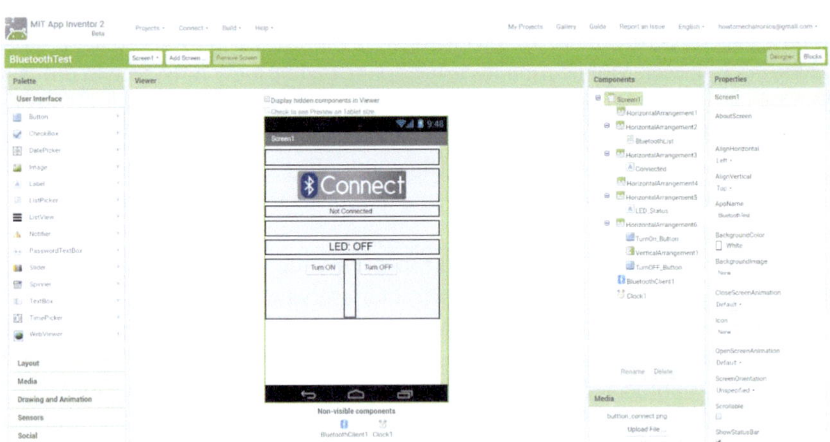

Fig. 14 App design tab [24]

Fig. 15 Prototype tank-based military drone using camera [25]

systems. Robotic systems can be veiled, purposely loud, or stealthy, and deceive their adversary in behavioral, technical, and physical ways.

Enabling total autonomy for military robots has important ethical implications. Instead of troops firing at each other on the battlefield, operators and robots will fight in the future. In the future, the group will propose a more accurate and exact method of shooting targets using wireless technology [25]. Figure 15 depicts a prototype tank-based military drone.

This study discusses the creation of an affordable robot that can assist the military by broadcasting video in close-quarters battle. A small, unarmed ground vehicle (SUGV) is a robot that uses a toy car as a platform and is equipped with an Arduino Uno controller, batteries, a Bluetooth module, and a smartphone. The DC motors of the wheels are linked to the Arduino controller, and using the Bluetooth HC-05 module, orders from the smartphone's accelerometer are transferred to the microcontroller, allowing the velocity and direction of the wheel to be controlled.

Robots can move forward, backward, left, and right, and their movement is controlled by the Android to Arduino software loaded on the smartphone.

The smartphone mounted on the platform utilizes the Skype video conferencing application to transmit live footage of its monitoring to the Skype program installed on the laptop over the Internet. Soldiers can take appropriate action based on the footage. Thus, SUGV helps to reduce the risk to a soldier's life [26]. Figure 16 depicts the SUGV drone after construction.

As the group all know, robots play an increasingly important role in both technology and the daily lives. A remote-operated robot is a wheeled device that can

Fig. 16 Developed SUGV drone [26]

be controlled from a distance, allowing users to travel and interact with their surroundings using wireless Internet access.

Typically, the robot uses a tablet or smartphone to give video and audio capabilities, as well as a visual depiction of faraway areas. The car can be controlled using an Arduino Bluetooth app, allowing users to move around and interact with their virtual world. The servo motor controls the camera's movement, while the Arduino board processes the data.

Arduino generates a specified connection, which the smartphone uses to receive a video. To visually capture the surroundings, an Arduino is utilized with a camera set at a distance. This gives consumers a more immersive experience while remotely managing the robot.

This project also included online real-time head movement control, which allows natural head movements to be mirrored in a distant environment [27]. Figure 17 depicts the live stream interface as visible from the URL.

Robot applications for human security are a hot issue in the scientific community. Many applications are being developed to improve human existence, ranging from the military to the medical fields. A mine-detecting robot is a ground-breaking military innovation and a life-saving device that can help all of humanity.

An Android application guides the robot as it scans the contaminated region, while a beat frequency oscillator serves as a mine-detecting sensor mounted on a servo motor in front of the vehicle. This work presents the design and implementation of a mine-detecting sensor in a robotic prototype. The suggested embedded system is built with Arduino technology and controlled by an Android application [28].

A tracked robot's embedded control system which is based on a single board computer. This system was created for educational purposes and quick prototyping of tiny mobile robot software. The challenges of power supply stability and network connection for visual feedback via the integrated camera, design adaptability, and chassis robustness are addressed.

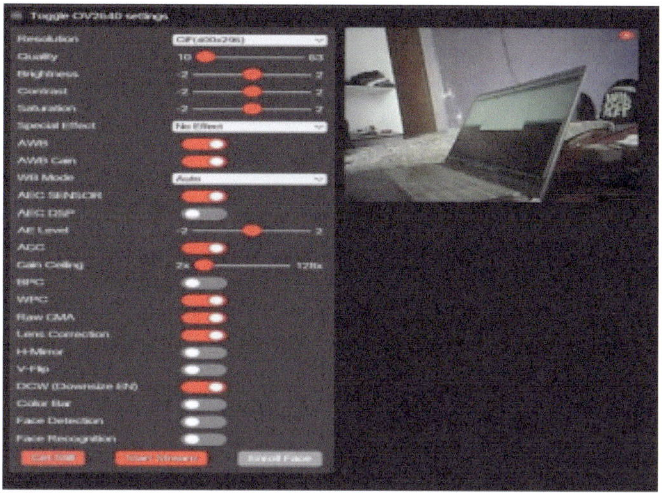

Fig. 17 Streaming content observed through the URL [27]

Data transmission methods, as well as the structure and functioning of the built program, are detailed. The proposed techniques would make it easier to utilize the robot throughout the learning process and may be applied to various types of mobile robots [29]. Figure 18 shows a tracked single caterpillar motor group.

This study discusses human surveillance utilizing IoT technology and robotics, namely an Arduino UNO microcontroller controlled by a smartphone and a PC. The goal is to create a spy robotic automobile capable of providing continuous observation in dangerous environments. The robot can capture real-time streaming both during

Fig. 18 Single-caterpillar motor group [29]

the day and at night using its wireless camera. The robot's motions are controlled manually by the user.

This robot reduces human interaction in a dangerous environment where constant observation and security are required. The entire system consists of several sensors, such as a PIR sensor, an ultrasonic sensor, and a gas sensor are interfaced with an Arduino board.

Spy robots monitor live streaming data and send it to the linked Android handset. The Android application may control the robot's navigation from a long distance via WIFI connectivity. The further progress of this concept will result in the use of spy robots in military and mining regions. Border areas require thorough and uninterrupted observation to ensure the country's safety and security from adversaries, and spy robots can be useful [30]. Figure 19 depicts cable connections and the robot's general interface.

The Arduino Mega 2560 microcontroller serves as the central processing unit of the robot. It manages the robot's movements, processes data from sensors, and facilitates communication with an Android device via Bluetooth. Wireless communication is achieved using the HC-05 Bluetooth module, enabling the user—through an Android phone—to remotely control the robot. A camera is integrated to capture images and provide live video streams of the surrounding environment. An ultrasonic sensor is employed to detect and avoid obstacles in the robot's path. A servo motor operates the robot's claw, allowing it to grasp objects for removal or transportation back to the base. The L298N motor driver controls the motion of the robot by regulating the DC motors. Power for the entire system is supplied by a battery pack. The complete circuit layout of the system is illustrated in the following diagram (Fig. 20).

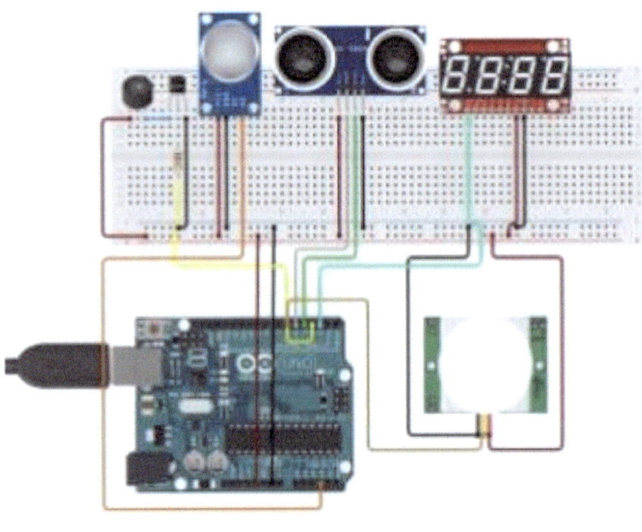

Fig. 19 Overall interface of the robot [30]

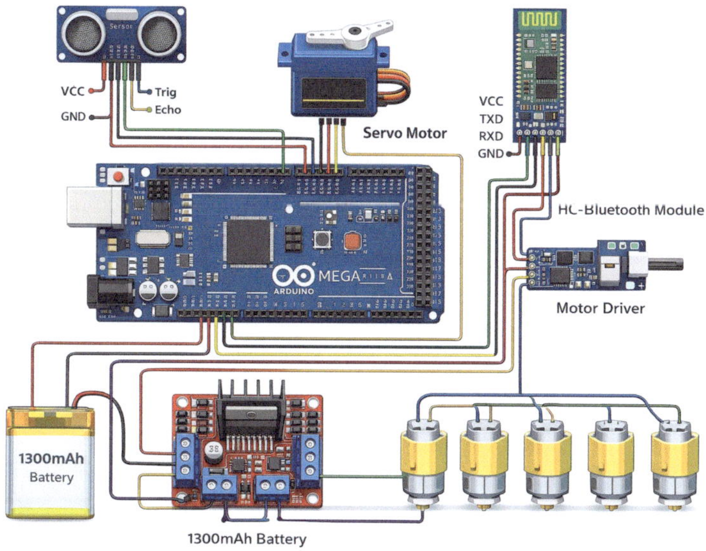

Fig. 20 Circuit diagram of the robot

The robot is operated using the Arduino Bluetooth Controller app on an Android device. This application enables the user to send specific command strings to control the robot's movements, such as moving forward or backward, turning left or right, and operating the claw to grasp objects. When an obstacle is detected, the robot automatically adjusts its direction to avoid a collision. Although the robot can react to obstacles, its overall operation is primarily manual, with the user directing its actions via the app. The user can navigate the robot to a desired location, pick up an object, and return it to the starting point. Figure 21 depicts Workflow Steps.

Summary

This literature review presents a critical analysis of the existing research concerning land surveillance drones, offering a comprehensive overview of significant studies, theoretical frameworks, and recent advancements in the field. The review examines a range of academic perspectives, emphasizing both historical and contemporary innovations that have shaped this area of inquiry. Synthesizing research from diverse sources, it highlights key findings, methodologies, and discussions relevant to the subject.

The primary goal of this chapter is to illustrate the gaps and discrepancies in the available literature in a methodical manner, identifying opportunities for further research. In this context, it comprehensively evaluates the effectiveness and limitations of various methodological approaches used in previous studies, and the review discusses how socioeconomic and technological changes have shaped the research

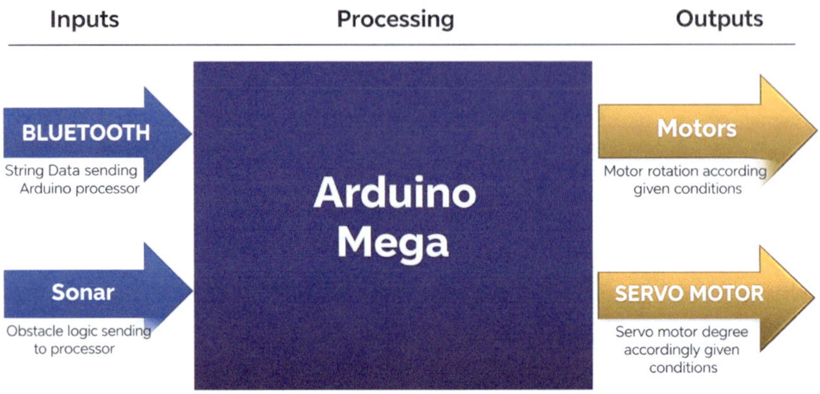

Fig. 21 Workflow steps of the robot

landscape, as well as how recent developments contribute to future research directions. However, this chapter critically examines diverse points of view, enabling a fair discussion of key problems and allowing for both current discussions and opposing viewpoints.

The findings stress the necessity of linking theoretical concepts to practical applications and show that this research has the potential to make a significant addition to the academic literature. In conclusion, this literature analysis establishes a sound theoretical foundation for the study, allows for the formation of a full knowledge of the issue, and clearly indicates the necessity for more research.

3 Methodology

Introduction

In recent years, robotics has emerged as a transformational technology that has transformed a variety of industries, including manufacturing and healthcare. The integration of robotic systems into everyday life applications is becoming more common as automation, artificial intelligence, and mechatronics advance rapidly. In this context, the initiative intends to help progress the area by building and implementing a practical robotic system.

During the project's first phase, extensive literature and market research were undertaken to develop a strong knowledge basis. Throughout this process, existing robotic devices were rigorously studied to establish their strengths and flaws, as well as the required assumptions to assure that the design would deliver considerable improvements, and the design process was then refined through iterative problem solving also collaborative cooperation with SolidWorks software, resulting in a detailed three-dimensional model.

Aside from the mechanical design, electronic components that would improve the overall operation of the system were carefully chosen, and the integration procedure was designed. Despite unforeseen problems throughout the assembly process, these challenges were effectively solved owing to extra research and detailed troubleshooting techniques. The project's final part involves programming the Arduino microcontroller using methodical learning and experimental investigations.

This thesis details the research, design, assembly, and coding stages of the project and offers an overall review of the process. Thus, it seeks to give the information, skills, and procedures required for the construction of an innovative robotic system, as well as to make a significant addition to academic and practical knowledge in the field of robotics.

Thesis Workflow

Figure 22 shows the workflow chart of the project.

Methodology

During the first part of the project, the group followed a thorough and painstaking approach to construct ground drones for surveillance and rescue. This approach began with a thorough literature analysis done in accordance with the important needs identified in prior reports obtained from Internet sources. Different elements of drone technology and its different uses in surveillance and rescue operations were extensively researched; as a result, the project team has a comprehensive understanding of the relevant sector. The primary goal of the project was to create a sophisticated drone that includes modern design concepts and new features while also resolving the issues noted in previous studies. In order to achieve this aim, existing technologies' capabilities, user requirements, and environmental concerns were thoroughly studied and balanced. Thus, the system was designed to function well in real-world scenarios, and the data gathered during the study process were meticulously documented and collated into an extensive report. Furthermore, regular discussion sessions were performed to promote team members' ideas and to benefit from suggestions from prior studies conducted on other research platforms. Thanks to this collaborative approach, the design process was constantly enhanced, and the robot's efficacy in crucial tasks was meant to be boosted.

In light of the information gathered, the robot design process was carried out with powerful SolidWorks software tools. This program enabled not only the depiction of new ideas in a three-dimensional environment, but also the early detection of possible difficulties throughout the development period. As the design process continued, prototype designs from prior studies were viewed as a valuable source of knowledge. These historical prototypes inspired the creation of the current design and helped to eliminate problems found in prior iterations.

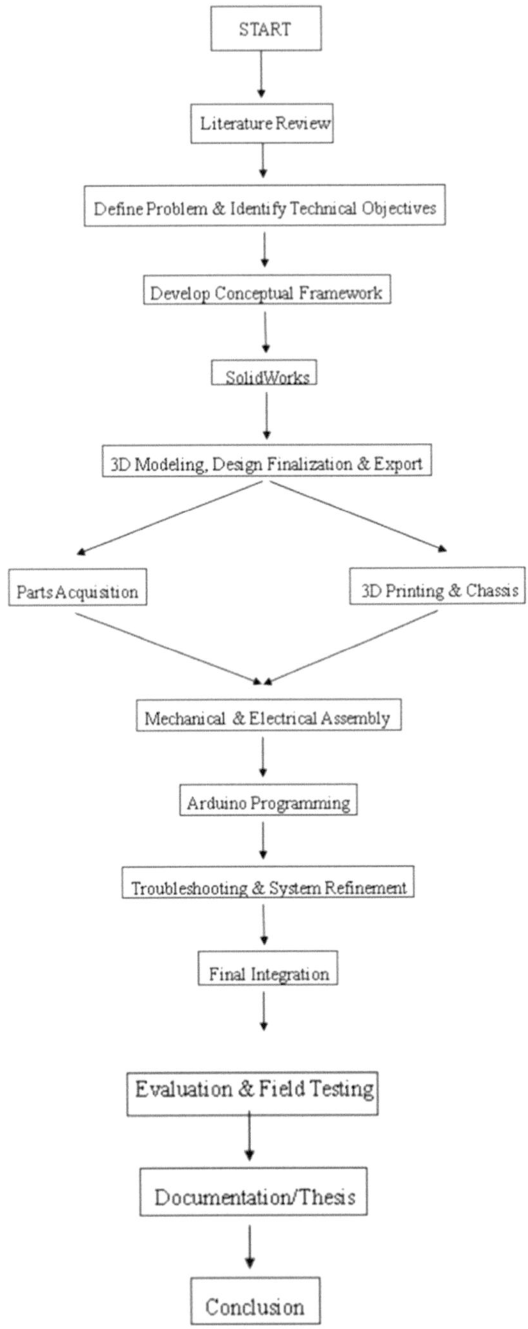

Fig. 22 Workflow chart

Fig. 23 3D Drawing of the project (SolidWorks)

As a result, the iterative design method improved not just the overall project quality but also the system's endurance and functionality (Fig. 23).

When the group began the product development journey, the project meticulously considered both the material selection and the associated manufacturing costs. The main goal was to design the robot to seamlessly integrate all the key components while minimizing errors at every stage.

To improve the design, the group went back to SolidWorks where the group made a series of critical adjustments, designing precise fittings for the moon gear and ensuring that the DC motors fit perfectly, resulting in smooth, frictionless operation. In addition to these steps, after extensive evaluation of various materials, the group concluded that Polylactic Acid (PLA) was the best choice for the robot. Known for its strength and reliability, PLA not only met the structural needs but also aligned with the commitment to sustainability, as illustrated in Fig. 24.

At this point, the group began converting the SolidWorks design files to STL files, which was critical for ensuring that the 3D printer could correctly comprehend and replicate the models. When the conversion was finished, the group immediately sent the STL files to the trusted manufacturer, who meticulously printed each model to the specifications. The entire process from conversion to final production took about twelve business days, ensuring that every detail was captured with precision and care, as shown in Fig. 25.

After a 12-day delay, the much-expected merchandise was eventually delivered. As the package was carefully opened, the team members examined all of the components with enthusiasm and some trepidation; it was discovered that some extra bolts would be required to finish the assembly promptly.

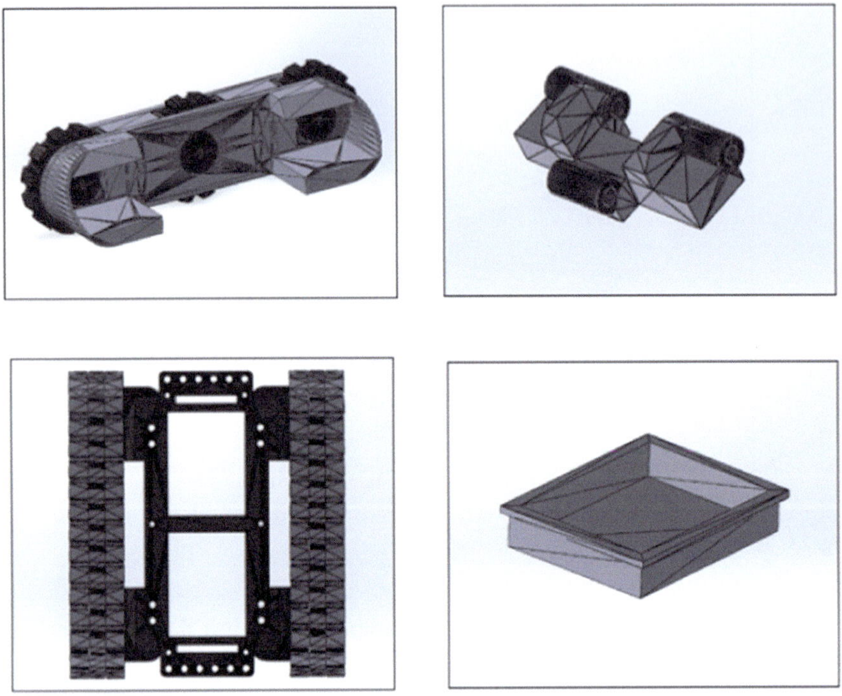

Fig. 24 Track frame and DC motor holders, tank track part, tank body, tray for electronic parts

To ensure a successful assembly, the size of the essential bolts was precisely measured. Based on the notes, the group headed to a nearby hardware store and were pleasantly delighted to find the goods the project needed.

For the construction of the inner and outer frame, the following parts were provided:

- 3 gears to ensure the efficiency of the movement mechanism,
- 4 M3 bolts, 25 mm long, used with nuts to fix the base assembly,
- 8 M3 bolts, each 20 mm long, with appropriate nuts to assemble the upper frame,
- 6 M3 spacers, each 30 mm long, to provide the necessary clearances of the structure,
- 24 M3 nuts to ensure that all components are securely and firmly fixed.

With all the necessary components provided, the team members were highly motivated and excited to start the assembly process. The parts used are shown in Fig. 26.

After gathering all of the essential components, the group carefully arranged the 3D-printed parts on the table, the excitement building as the surveillance robot began to take shape before the eyes.

The texture of the plastic felt cool against the fingertips as the group inspected each part, examining the precise details of the tracks and the sturdy frame. With a

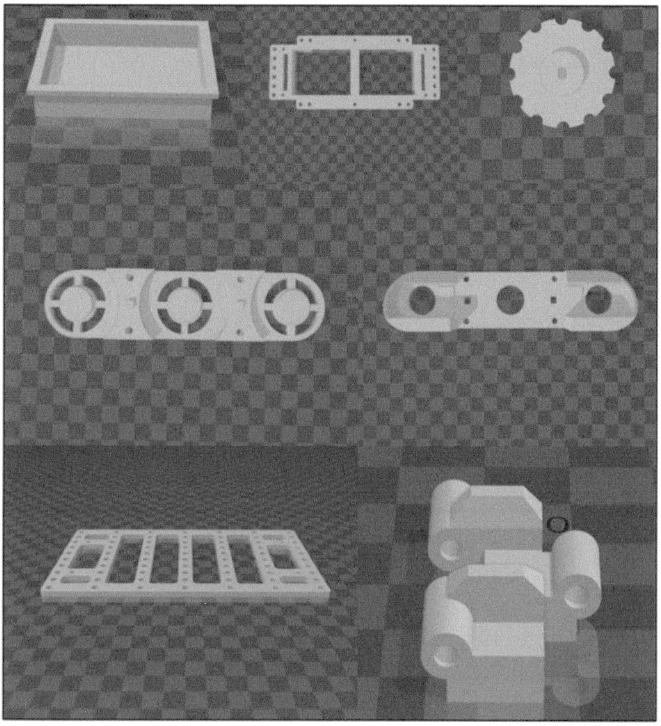

Fig. 25 STL file view of the parts

shared sense of excitement, the group discussed the plan of attack, deciding to kick off the assembly with the robust track frame that would serve as the foundation for the innovative creation, shown in Fig. 27.

Putting together a sturdy track frame proved to be a rewarding task. Although the group faced a few minor challenges during the assembly process, none were significant enough to derail the progress. The group had eight distinct components that needed to be meticulously aligned and secured with the bolts the group had acquired. Overall, the assembly unfolded as the group had envisioned to make steady progress toward the goal, as shown in Fig. 28.

The group successfully completed the construction of a robust track frame, meticulously fitting it to the tank's main base. The assembly process for the lower section was seamless, with all components aligning perfectly and functioning as intended. Each piece clicked into place with precision, allowing the group to move forward without encountering any complications, shown in Fig. 29.

The group opted for standard wood nails to fasten the tracks, as members felt they provided the most straightforward and efficient solution, allowing for assembly without the need for power tools. After carefully aligning the components, the group took the extra precaution of trimming the sharp ends of the nails, ensuring that there

Fig. 26 3D Printed parts

Fig. 27 Tools and parts used for the assembly process

Fig. 28 Track frame parts

were no exposed points to pose a risk to anyone using the finished product. This attention to detail reflects the commitment to safety and usability, shown in Fig. 30.

After successfully assembling the surveillance robot tracks, the group transferred them to the sturdy track frame for rigorous testing. As the group conducted the evaluations, it attentively examined every detail, hoping to identify any potential issues. To the satisfaction, everything functioned flawlessly; the gear meshed seamlessly with the track, allowing for smooth and precise turns. With this optimistic outcome, the group was ready to move on to the next assignment while also eager to continue the growth. Figure 31 depicts the course's results.

To build the upper frame of the surveillance robot, the group carefully placed sturdy bolts that were specifically designed to hold the weight and support the overall structure. In the center of this assembly, a central column was created that would house the Arduino and other key components, allowing for seamless integration.

Fig. 29 Assembled parts of track frame and the main base

Furthermore, to provide stability and avoid movement, the group tightened each bolt with two robust nuts, securely fixing the frame and eliminating the possibility of it sliding up or down. In all, the group added six solid bolts and twelve dependable nuts, precisely building the robot's upper section for maximum performance and longevity. Figure 32 is presented on the following page.

After patiently building all of the intricate mechanical components, the group focused the attention on precisely and carefully assembling the surveillance robot's tracks. Each component fit flawlessly, and the robot's mechanical skeleton stood proudly before us, a visible tribute to the efforts. Not only was it elegantly designed, but its solid construction also gave it the appearance of being ready for action as the group stared at the beautiful fruition of the efforts, see Fig. 33 on the following page.

The long-awaited electrical components have finally been received, and extensive tests have been carried out on each one. As a result of the checks, it was happily

Fig. 30 Assembly process of tracks

determined that there was no damage caused by the transportation process and that all components were operating at the expected performance level.

The electrical plan, which was meticulously prepared in line with the findings from the previous research and analysis, was then reviewed in detail. This process allowed significant improvements to be made to the previous design and allowed the developed system to best suit the specific requirements of the surveillance robot.

After this stage was completed, the complex installation process of the project's electrical system was started, and the integration of this system was considered an important step that will increase the overall functionality of the innovative design. The relevant electrical plan and components are presented in Fig. 34.

The programming procedure for the robotic system began with the integration of the ESP32-CAM module via Arduino UNO, and this powerful camera system effortlessly links to smartphones via a specially constructed Wi-Fi network, allowing users to view the camera image in real time. To guarantee continuous and dependable

Fig. 31 Testing process of tracks

Fig. 32 Assembly process of the top frame

Fig. 33 Assembled model

Fig. 34 Electrical components

operation, the camera module was fueled by a portable battery bank, which allowed the robot to be deployed in a variety of environments.

A secure Wi-Fi network was established that all users could readily access, and photographs captured from the robot's front were made available live. This novel setup not only allows for remote operation of the robot, but it also considerably boosts surveillance capabilities in tough areas where human access is restricted or risky.

The device in issue enables the operator to efficiently monitor the surroundings, making it a useful tool for exploration and security applications. Figure 35 provides a graphic illustration of the essential arrangement.

The group utilized the Arduino IDE for the development, which is packed with built-in capabilities that considerably expedited the process. Its user-friendly interface and extensive resource collection allowed it to focus on the project without being bogged down in technical intricacies, and when the group encountered a problem, it went to YouTube for brief lessons that assisted in identifying and implementing the precise feature the project needed. This mix of tools and resources made the coding process both instructional and pleasant, allowing to work quickly and accurately. Figure 36 depicts the software, which may be found on the next page.

After finishing the CAM code, the group proceeded with the installation of the 9 V DC motors into the rails, putting them into the locations the group had defined and then carefully soldered the motors in parallel, ensuring that they functioned together harmoniously. This approach was crucial to guarantee that both motors operated in unison and maintained the same direction, thus preventing any unwanted collisions or the possibility of one motor becoming stalled. The attention to detail in the setup aimed to create a seamless and efficient operation for the entire system. You can see the soldering process shown in Fig. 37.

Fig. 35 Coding process of ESP32-CAM module using Arduino UNO

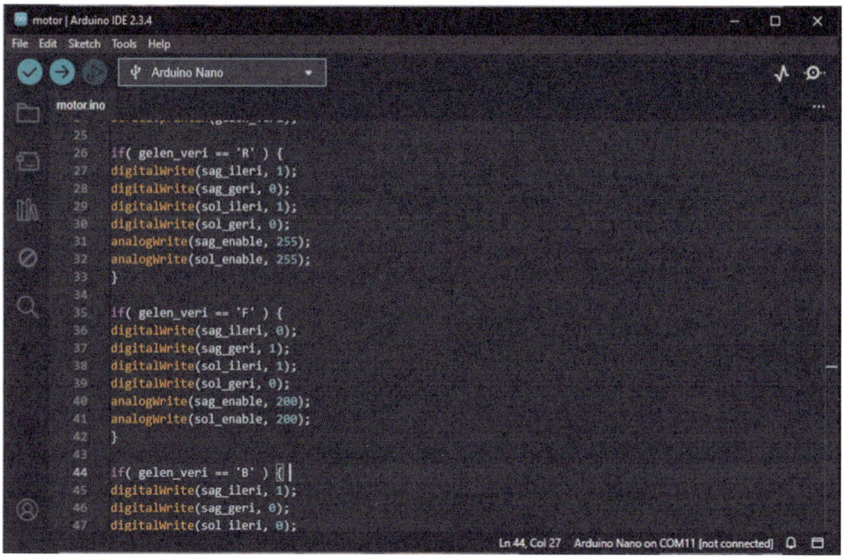

Fig. 36 Coding process using Arduino IDE software

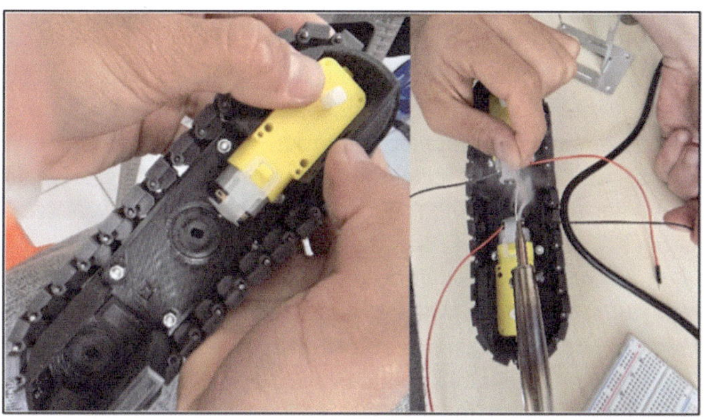

Fig. 37 Soldering process

After integrating the DC motors, the next step will be to mount them onto the motor shield L298N, which will serve as the primary interface to connect to the Arduino microcontroller. Using the Arduino IDE, the group will meticulously write and upload a program that ensures precise control and operation of the motors, allowing them to respond as intended.

To enable remote-control functionality, the group will establish a Bluetooth connection by incorporating the HC-05 module, which will allow a smartphone to communicate with the system wirelessly. The entire Arduino setup will be powered by a reliable 9 V battery, ensuring a sufficient energy supply for extended usage.

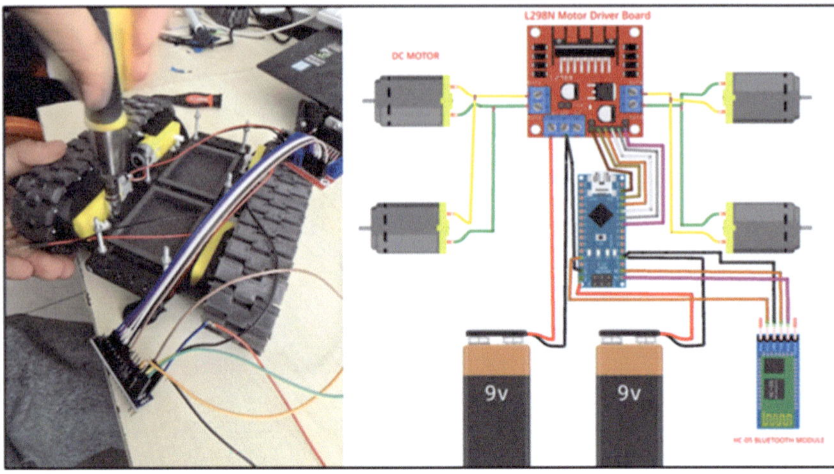

Fig. 38 Mounting process of motors

Once all components are properly connected and configured, the group will be poised to deploy this setup as part of the surveillance rescue robot, enhancing its operational capabilities in various scenarios. The whole process is shown in Fig. 38.

After installing the electrical equipment and meticulously completing the coding process, the project team focused on assembling the surveillance and rescue robot. While the components were properly built, extra care was taken to secure the cords with strong cable ties and create a nice and orderly interior structure.

To avoid any touch or entanglement hazards, the cables were strategically positioned on the upper portion of the robot frame. This technique was seen as a vital step in avoiding any long-term difficulties during surgery.

Furthermore, to make future maintenance and troubleshooting easier, the wires were color-coded: red for power cables, green for Arduino connections, and black for assembly components, so that this color coding not only offered visual order but it also made the system's general structure easier to grasp and navigate.

After the wiring was done, the ESP32-CAM module was carefully installed and securely connected to the power supply. As a consequence of the extensive testing, it was discovered with delight that all system components functioned well and that no faults occurred. The effective conclusion of the project was seen as both a significant milestone for the team members and a concrete result of the collaborative effort. Figure 39 shows the finished output.

Fig. 39 Final product

4 Result and Discussion

Discussion

The Surveillance and Rescue Ground UAV development process, as outlined in this project, clearly illustrates the viability and relevance of incorporating robotic technology into emergency response scenarios. The project's principal aim, to lessen reliance on human intervention in dangerous and limited access areas, has been successfully handled through precise mechanical design, electrical integration, and software applications.

In this sense, one of the project's most notable accomplishments is the effective development of real-time video transmission using the ESP32-CAM module. This feature not only improved situational awareness in the field, but it also aligned with the remote and wireless monitoring trends prevalent in modern surveillance systems. As underlined in the literature review, including Bluetooth and Wi-Fi connections in the project boosted usability and permitted the creation of user-friendly control interfaces.

Despite the early concerns with power consumption and motor efficiency, the choice to use 18,650 lithium-ion cells instead of typical 9 V batteries was a substantial improvement. This technique verifies the importance of energy efficiency, particularly in coal mine exploration robots and autonomous rescue systems, as stressed in the literature. This project solution exemplifies a tangible engineering technique that demonstrates theoretical planning via iterative design and experimental testing.

Compared to prior research and prototypes like the Kohga robot and tracked military drones, the created technology provides a more balanced and accessible alternative. For example, while the Kohga robot is highly maneuverable and flexible in tiny passages, this project focuses on qualities such as cost-effectiveness, dependability, and quick deployment, making the system more ideal for large-scale operations or educational applications. Furthermore, the Arduino-based control system's ease of use allows even individuals with little technical experience to utilize the robot efficiently in a short period of time. This successfully supports the accessibility principle, one of the project's primary aims.

However, the system's simplicity of use is not without drawbacks. The robot lacks modern SLAM techniques and artificial neural network-based autonomous navigation and mapping capabilities. However, given the project's scope and budget limits, it is reasonable to conclude that the established objectives were met.

Furthermore, the project's choice of reproducible components, as well as the creation of assembly procedures that are simpler and more repeatable, has resulted in considerable sustainability improvements. This strategy is also congruent with current robotics research, which emphasizes both high-performance solutions and sustainable and accessible technology.

As a result, the developed Surveillance and Rescue Ground UAV demonstrates the potential to produce effective solutions to urgent global problems by blending practical engineering applications with real-world needs and constraints. Although there are areas open to further development in automation, environmental durability, and data security, this study provides a meaningful and original contribution to the field of emergency response robotics.

Challenges

The employment of ground-based robots for ground surveillance provides substantial benefits in terms of monitoring and data collecting; nevertheless, some major hurdles must be overcome to assure the effectiveness and sustainability of such initiatives, and the most significant of these problems is the terrain on which the robot will operate. Ground robots, unlike airborne drones, are completely dependent on the surface they travel. Rough, rocky, or muddy terrain might limit the robot's movement, causing it to become immobile in some regions or reducing its effective coverage capability across wide areas. As a result, while designing the robot, it is critical to include a framework that can run constantly and reliably under a variety of ground conditions.

One of the most significant issues is the robot's mobility and durability. In addition to being able to navigate safely across various terrains, the robot must be resistant to adverse climatic conditions such as rain, dust, snow, and high temperatures. Protecting sensitive components like motors, sensors, and cameras from external factors is crucial for the device's long-term and dependable performance. Power management is another critical issue that must be addressed properly. Ground robots, like their flying cousins, have limited lifespans owing to battery capacity. Furthermore, ground robots consume more power since they carry greater payloads (such as sophisticated sensors and cameras). This needs the development of efficient energy management systems, particularly in distant and difficult-to-access regions where charging or battery replacement is not feasible.

Although 9 V batteries appeared to be sufficient in first theoretical evaluations, actual tests proved that they were insufficient to power four DC motors. To address this issue, a battery pack was created that was precisely manufactured in line with component data sheets to avoid damaging electrical components and wires. This battery pack is made up of three series-connected 18,650 lithium-ion batteries. During the design process, numerous scenarios that may improve or restrict the ease of replacement of the battery pack were considered, and two distinct battery pack solutions were established as a result: one with Battery Management System (BMS) integration and one with a battery holder. This enables people to select the best solution for their needs.

The security and privacy concerns associated with the data collected by land surveillance robots are of paramount importance. Surveillance in sensitive areas, such as private properties, agricultural lands, or industrial sites, raises questions about privacy. It is imperative to implement robust data encryption and access control mechanisms to safeguard collected data from unauthorized access or misuse. Furthermore, the robot's design must respect legal boundaries to prevent inadvertent infringements on privacy rights or violations of regulations. To enhance data security, the camera module (ESP32) is engineered to establish its own server, accessible solely from designated wireless connections.

To keep project expenses low, poor-quality Arduino models were used, which caused certain technical issues. Because these low-quality Arduino boards used different processors than the original versions, they were incompatible with most PC data reader sockets, making it difficult to upload, update, or alter software. To address this issue, studying and learning about processors and CPU design were encouraged within the scope of the project. The Arduino model used was compatible with the CH340 serial driver; therefore, it was feasible to successfully upload to the PC used for programming. Figure 40 shows New Battery Pack with 3 Series Batteries.

Lastly, operator training and ongoing technical support are essential to ensure that the robots operate efficiently and are maintained properly over time. Operators must be trained to handle the robots, manage data collection, and respond to any technical issues that may arise during operations. Similarly, a technical support team must be in place to troubleshoot and repair any mechanical or software issues that occur, ensuring minimal downtime. To achieve this, the control system was kept as simple as possible. Access to both the camera and the controls can be provided from any device with an Android or Windows operating system. Since this is not a complex control system, maintenance, repair, and accessibility are high. A 5-min training session will be sufficient for the user to understand and easily use the controls. Of course, a longer and more comprehensive training is required to understand and troubleshoot control issues.

By keeping the control system this simple, the costs that would be spent on a more complex control system (such as the DJI RC PRO Remote Controller costing 1477.56 $) have also been avoided.

Results

The implementation of the Surveillance and Rescue Land Drone has yielded promising and pragmatic outcomes that align with the initial objectives of the

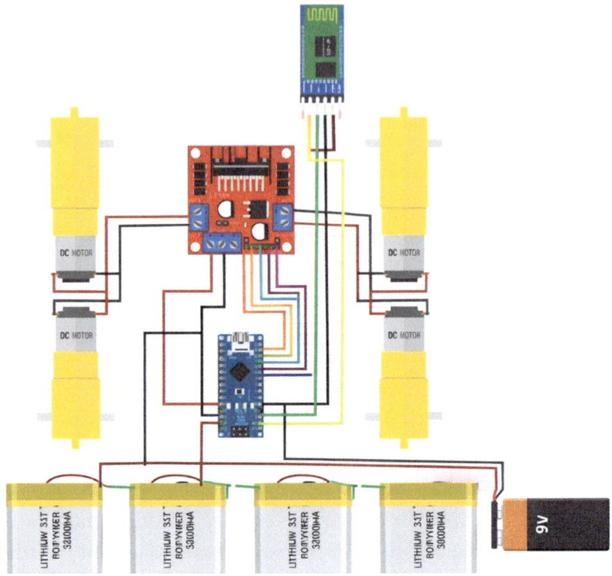

Fig. 40 New battery pack with 3 series batteries

project. The finalized prototype has effectively demonstrated its core functionalities, including the capability to navigate irregular terrain, provide real-time video streaming, and facilitate manual control via Bluetooth. These findings affirm the feasibility of developing an affordable, efficient, and reliable ground-based rescue robot utilizing accessible components and open-source platforms.

Functional Performance

- **Mobility**: The tank-track design ensured stable movement across different surfaces, including flat indoor areas and moderately uneven outdoor terrains. Testing confirmed that the robot could move forward and backward, as well as make left and right turns, while maintaining consistent motor coordination.
- **Wireless Camera System**: The ESP32-CAM module performed effectively, allowing users to connect to a dedicated Wi-Fi network and view a live video feed from the robot. This feature was successfully tested on both Android smartphones and computers. The camera provided sufficient resolution and real-time feedback, which improved situational awareness during simulated search and rescue operations. The range factor is illustrated in Fig. 41.
- **Control System**: The Bluetooth control system, which utilizes the HC-05 module, successfully transmitted commands to the Arduino-based setup. All programmed movements—forward, backward, left, right, and stop—were executed responsively with minimal latency. The control app was tested on several Android devices and consistently performed well. The range of the module has been tested, and the results can be seen in the graph below. The maximum reliable range is ~ 10 m

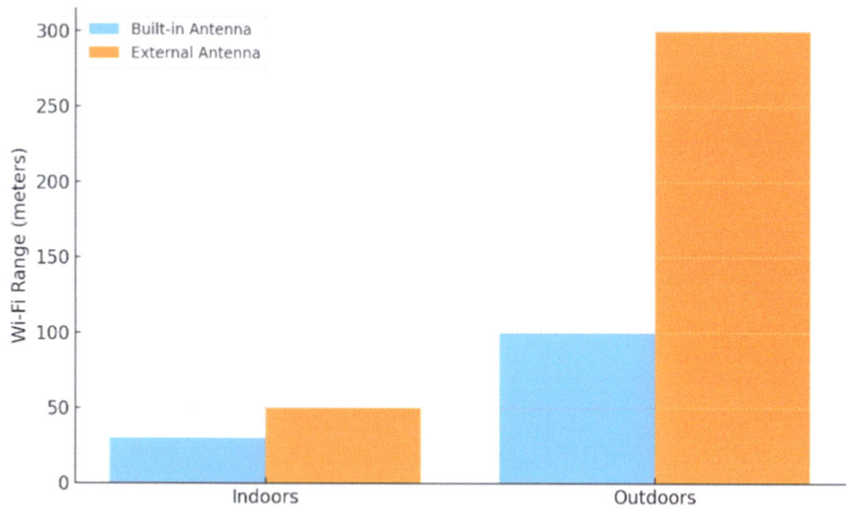

Fig. 41 ESP32-CAM Wi-Fi range comparison chart

(open space, line of sight). Indoor range (walls/interference): 5–8 m, depending on obstacles.

Signal drops significantly after 10 meters and becomes unstable. Interference (Wi-Fi, walls, etc.) can drastically reduce range. You can see the range graph in Figure 42 on the next page.

Fig. 42 HC-05 bluetooth module range graph

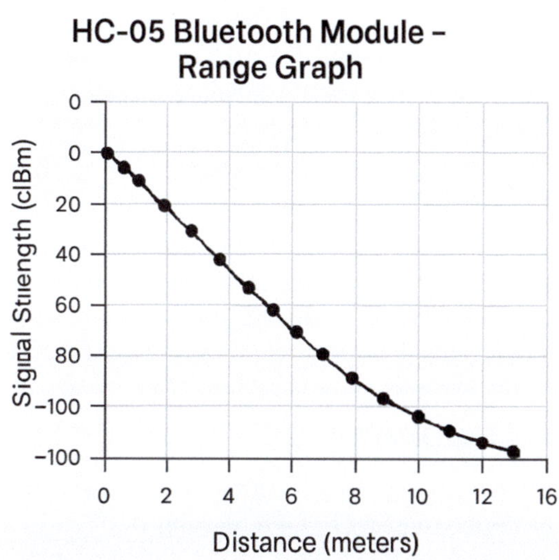

Fig. 43 3S battery charging circuit card

Hardware Integration

- **Electrical System**: All electronic components—including motors, drivers, microcontrollers, and sensors—were successfully integrated. The initial setup using a 9 V battery had limitations in terms of duration and performance. However, switching to a 3-cell battery pack significantly improved power efficiency and operational time. With the addition of the charger chip, the group can charge the battery directly on the tank, making it even more efficient. You can see the plan shown in Fig. 43.
- **Mechanical Structure**: The 3D-printed components showed precise fitting due to careful modeling and iterative improvements made in SolidWorks. Using PLA material delivered enough strength to withstand movement-related stresses, and the overall assembly demonstrated strong structural integrity throughout the testing phase.

Cost Efficiency

The entire system was developed within a budgetary constraint of approximately $123.76, demonstrating the feasibility of constructing a fully operational surveillance and rescue robot utilizing low-cost, commercially available components. This achievement underscores one of the principal objectives of the project: to provide a cost-effective alternative to high-priced commercial robotic solutions.

Limitations Observed

- The robot's operation is currently limited to manual control, with no autonomous obstacle avoidance or GPS tracking.
- The range of Bluetooth communication restricts usability in large-scale disaster zones.

- Battery life, though improved, still requires further optimization for extended field use.

Software Used

During the design and development phase of the rescue robot, the team relied exclusively on SolidWorks as its primary computer-aided design (CAD) software. All mechanical components on this strong platform, from the durable tank-type chassis to the novel camera housing and the robust structural frame, were rigorously developed and modeled in an integrated way. This targeted approach not only contributed to excellent dimensional accuracy but it also allowed for efficient modeling of mechanical interactions, providing a clear view of how each component would interact in real-world settings. SolidWorks software was used to conduct extensive analysis, optimize designs for performance, and develop high-quality representations of the robot's primary capabilities. This fluid and comprehensive design methodology aided the creative productivity and created an efficient preparation environment for the future production and prototype phases.

Finally, the effective and systematic use of SolidWorks software contributed to a clear and consistent embodiment of the project vision and provided a solid foundation for the successful development of the rescue robot. The corresponding model and design visualization are presented in Fig. 44 on the next page.

Working Steps of the System

- **Controller**: Sends control signals via Bluetooth.
- **HC-05 Bluetooth Module**: Receives the signal from the controller and transmits it to Arduino.
- **Arduino Nano**: Processes and analyzes incoming signals.

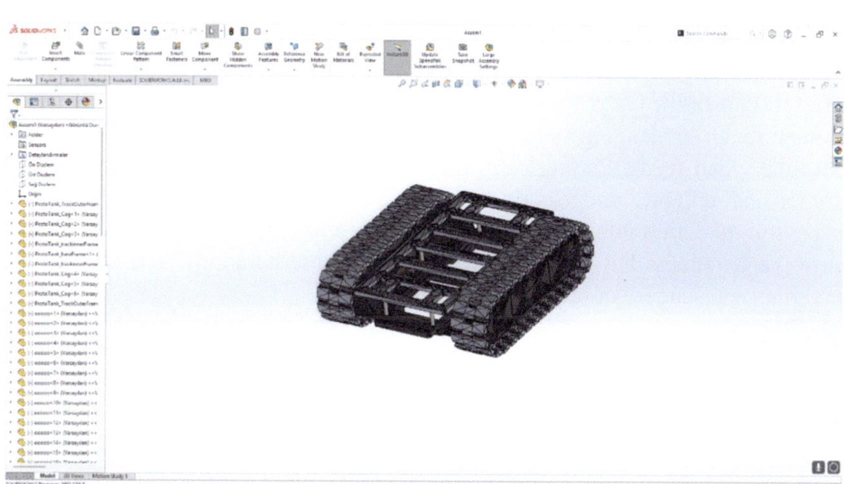

Fig. 44 3D drawing of the project in solidworks software

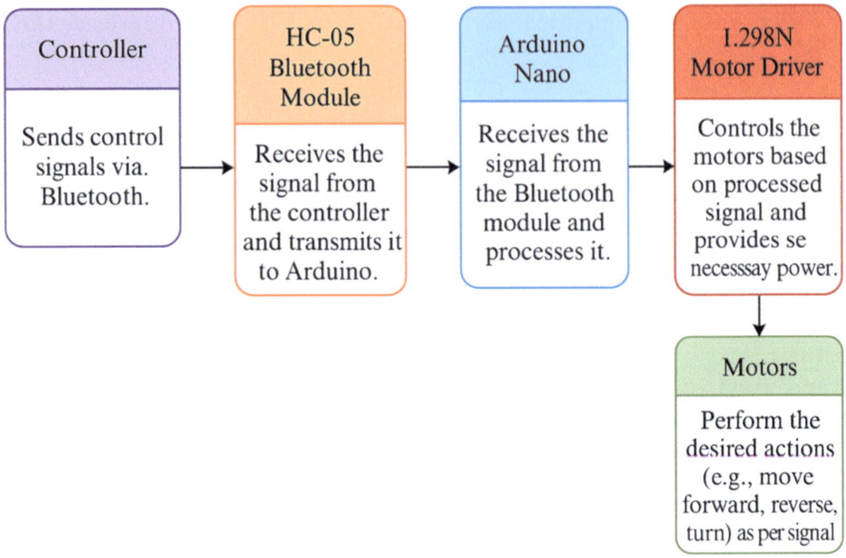

Fig. 45 Working steps of the system

- **L298N Motor Driver**: Controls the motors based on the processed signals and supplies the required power.

  In short, the signal flows like this:
  Controller → Bluetooth Module → Arduino → Motor Driver → Motors.
  Figure 45 shows Working Steps of the System.

Summary

In conclusion, the prototype not only met its design specifications but also excelled in demonstrating exceptional performance during rigorous practical testing. It showcased its capabilities as a versatile surveillance and rescue platform, adeptly navigating through challenging and unpredictable terrain while transmitting high-quality real-time video feeds. Furthermore, its responsiveness to remote commands was prompt and reliable, underscoring its effectiveness in critical situations. These successful outcomes not only validate the engineering principles that were applied during development but also pave the way for future iterations. You can see the surveillance rescue robot product shown in Fig. 46.

Fig. 46 Surveillance rescue robot

5 Conclusion

The project centers on the creation of a versatile land drone, meticulously engineered to navigate complex obstacles on rugged terrain using robust tank tracks while simultaneously collecting crucial data through its high-resolution camera. The primary beneficiaries of this innovative initiative will be search and rescue teams, although the drone's capabilities open doors to applications in military, logistics, entertainment, and experimental fields as well.

The major goal of this project is to create a sturdy and dependable device that can reach locations that human operators cannot, allowing for real-time data collection. In accordance with this lofty ambition, tank-type track systems have been added to the drone architecture to give excellent mobility even in tough and challenging terrains.

The development procedure began with a standard and optimized chassis design for electrical components, followed by precision-engineered plastic components, wheels, and tank tracks to ensure flawless motor integration. As a result, these designs were created using the sophisticated CAD program SolidWorks and implemented utilizing a reputable 3D printing service for assembly operations. The ESP32-CAM module, programmed with the Arduino IDE, was chosen as the display module. Two DC motors are fed and controlled by the motor driver card (motor shield), and the Arduino Nano microcontroller, which is picked owing to its compact size and flexible programmability, serves as the system's control center. The HC-05 Bluetooth module, which is compatible with Arduino, was utilized for remote control and monitoring.

During the first testing, 9V non-rechargeable batteries were employed as power sources. However, performance issues were identified as a result of the batteries' quick depletion and low voltage levels. In this regard, a more efficient power solution was devised, with the decision to utilize three series-connected 18,650 lithium-ion batteries for the motors and two more series-connected 18,650 lithium-ion batteries for the microcontroller. The ESP32-CAM module was powered by a portable power bank, which transmitted more current to the motors and improved overall system performance.

The adoption of lithium-ion batteries enhanced performance while also providing considerable cost-effectiveness and long-term benefits. In this context, the studies demonstrated that this power source increased battery life, improved performance, and facilitated charging in projects with DC or servo motors.

Because of its small size and adaptability, the Arduino Nano microcontroller is an excellent choice for this application. Furthermore, the test results demonstrated that the project objectives were mostly satisfied, and that the planned system was compatible with both design and operational goals, and the ESP32-CAM module takes and transmits high-definition real-time photos, and tank-type tracks allow obstacles to be readily overcome in a variety of terrain situations. The HC-05 Bluetooth module improves overall system dependability by enabling the safe and efficient transfer of control signals.

The drone is powered by four DC motors, with two positioned strategically on each side, connected to a motor shield circuit that simplifies voltage regulation and motor management. The group established a sophisticated system for motor management utilizing a compact Arduino Nano microcontroller, which serves as the brain of the operation. Additionally, a Bluetooth module was incorporated to facilitate seamless communication with the microcontroller, enabling remote control and monitoring.

During the initial testing phases, the group relied on 9V non-rechargeable batteries as the power source. However, the p encountered significant challenges, including rapid battery depletion and inadequate voltage levels. In response to these issues, the group pivoted to a more efficient solution: three series-connected lithium-ion batteries dedicated to powering the motors and an additional pair of series-connected lithium-ion batteries supporting the microcontroller. The ESP32 camera module was powered through a USB port linked to a portable power bank, resulting in enhanced current flow to the motors and allowing for greater operational speeds. This transition to rechargeable lithium-ion batteries not only improved performance but also led to considerable cost savings.

Future Works

The project serves as a platform for vast future enhancements. This allows the user to simply program desired functionality using the easy Arduino interface. Future advances might include incorporating more robotic devices into the drone's chassis, installing an upgraded radar system for obstacle identification, enhancing camera angles with a secondary camera mechanism, and employing more powerful DC motors to boost the drone's torque capabilities. These enhancements will expand the drone's adaptability and utility in a wide range of applications. Furthermore, the modular architecture enables future updates such as GPS integration, obstacle avoidance algorithms, and solar charging systems. Finally, while these functionalities were not part of the original project objective, they are useful enhancements for future applications.

References

1. Skillful project ideas on multipurpose robot for military. Accessed: Aug. 10, 2024. [Online]. Available: https://www.skyfilabs.com/project-ideas/multipurpose-robot-for-military
2. Aashish A, Kumar A, Deepak RS, Kusuma S, Sreekanth DV (2020) Review on multipurpose agriculture robot 8. https://doi.org/10.22214/ijraset.2020.5209
3. hq720.jpg (686×386). Accessed: Aug. 10, 2024. [Online]. Available: https://i.ytimg.com/vi/v85j1pIjMiQ/hq720.jpg?sqp=-oaymwEhCK4FEIIDSFryq4qpAxMIARUAAAAAGAEl AADIQj0AgKJD&rs=AOn4CLChdFvK9XIRXc4xXz0lCHEtocJZ6g
4. Pedre S, Nitsche M, Pessacg F, Caccavelli J, De Cristóforis P (2014) Design of a multi-purpose low-cost mobile robot for research and education. Lecture Notes in Computer Science (including subseries Lecture Notes in Artificial Intelligence and Lecture Notes in Bioinformatics), vol. 8717 LNAI, pp 185–196. https://doi.org/10.1007/978-3-319-10401-0_17
5. ReBeL EduMove: mobile robot for education purposes. Accessed: Aug. 10, 2024. [Online]. Available: https://rbtx.com/en-US/components/mobile-robots/rebel-edumove-mobile-robot-for-education-purposes
6. Sindhuja MAV, Anandan MR (2024) Smart multipurpose human assistance robot. Smart multi-purpose human assistance robot Asian. J Comput Sci Technol (AJCST 2(2):25–30. Accessed: Aug. 10, 2024. [Online]. Available: www.goniv.com
7. Domestic Robots are a new frontier for Industrial Designers: Whipsaw CEO, Dan Harden—Yanko Design. Accessed: Aug. 10, 2024. [Online]. Available: https://www.yankodesign.com/2021/08/09/domestic-robots-are-a-new-frontier-for-industrial-designers-whipsaw-ceo-dan-harden/
8. Hossain MA, Ebrahim Hossain M, Rahaman MA (2021) Multipurpose medical assistant robot (Docto-Bot) based on internet of things. Int J Electr Comput Eng (IJECE) 11(6):5558–5567. https://doi.org/10.11591/ijece.v11i6.pp5558-5567
9. Robot Medical Assistant will be serving hospital patients | PROMOBOT. Accessed: Aug. 10, 2024. [Online]. Available: https://promo-bot.ai/news/robot-medical-assistant/

10. Harapanahalli S, Mahony NO, Hernandez GV, Campbell S, Riordan D, Walsh J (2019) Autonomous Navigation of mobile robots in factory environment. Procedia Manuf 38:1524–1531. https://doi.org/10.1016/J.PROMFG.2020.01.134
11. Autonomous mobile robots could change metal fabrication. Accessed Aug. 10,2024.[Online].Available:https://www.thefabricator.com/thefabricator/article/shopmanagement/autonomous-mobile-robots-could-change-metal-fabrication
12. https://robotnews.wordpress.com/2006/03/21/snake-like-robot-kohga
13. Reddy AH, Kalyan B, Murthy ChSN (2015) Mine rescue robot system—a review. Procedia Earth Planetary Sci 11:457–462. https://doi.org/10.1016/J.PROEPS.2015.06.045
14. https://pdfs.semanticscholar.org/ea35/c97df0e95e1c35a9dbe9487415087faa7c16.pdf
15. Ismailov AS, Jo'rayev ZB (2022) Study of arduino microcontroller board. Sci Educ (Dordr) 3(3):172–179
16. D. V.-I. I. J. of F. and Nutritional and undefined 2022, "IOT Surveillance Robot Using ESP-32 Wi-Fi CAM & Arduino," researchgate.net
17. Lu X, Liu W, Wang H, Sun Q (2013) Robot control design based on smartphone. 2013 25th Chinese control and decision conference, CCDC 2013, pp 2820–2823. https://doi.org/10.1109/CCDC.2013.6561425
18. Wang XV, Wang L (2021) A literature survey of the robotic technologies during the COVID-19 pandemic. J Manuf Syst 60(823–836):823–836. https://doi.org/10.1016/J.JMSY.2021.02.005
19. Dorigo M, Theraulaz G, Trianni V (2021) Swarm robotics: past, present, and future. Proc IEEE 109(7):1152–1165. https://doi.org/10.1109/JPROC.2021.3072740
20. Ness S, James Shepherd N, Rong Xuan T (2023) Synergy between AI and robotics: a comprehensive integration. Asian J Res Comput Sci 16(4):80–94. https://doi.org/10.9734/ajrcos/2023/v16i4372
21. Choi Y, Choi M, Oh M, Kim S (2020) Service robots in hotels: understanding the service quality perceptions of human-robot interaction. J Hosp Market Manag 29(6):613–635. https://doi.org/10.1080/19368623.2020.1703871
22. Hyundai Motor Group Launches Pilot Program to Test Its Hotel Delivery Robot |
23. Amareswar E, Goud GSSK, Maheshwari KR, Akhil E, Aashraya S, Naveen T (2017) Multi purpose military service robot. Proceedings of the international conference on electronics, communication and aerospace technology, ICECA 2017, vol 2017-January, pp 684–686. https://doi.org/10.1109/ICECA.2017.8212752
24. Villamil X, Guarda T, Quina GN (2018) Agile software development for mobile applications and wireless interaction with hardware development board (Arduino). Iberian conference on information systems and technologies, CISTI, vol 2018-June, pp 1–5, Jun. 2018. https://doi.org/10.23919/CISTI.2018.8399328
25. Budiharto W, Andreas V, Suroso JS, Gunawan AAS, Irwansyah E (2019) Development of tank-based military robot and object tracker. 2019 4th Asia-pacific conference on intelligent robot systems, ACIRS 2019, pp 221–224, Jul. 2019. https://doi.org/10.1109/ACIRS.2019.8935962
26. Aniruddha Prabhu BP, Hebbal S (2018) Small unarmed robot for defense and security: a cost-effective approach using Arduino UNO. 2017 2nd international conference on emerging computation and information technologies, ICECIT 2017, Aug. 2018. https://doi.org/10.1109/ICECIT.2017.8453325
27. Prasad DD, Lahari P, Nandini G, Harshitha YTP, Sireesha PVP, Kumar JD (2023) Live telecasting Robo using Arduino Uno. 2023 2nd international conference on advances in computational intelligence and communication, ICACIC 2023. https://doi.org/10.1109/ICACIC59454.2023.10435089
28. Youssef C, Mohamed BR, Najmeddine A, Frikha T, Jmal W (2017) Mine detecting robot prototyping. 2017 2nd international conference on anti-cyber crimes, ICACC 2017, pp 183–186. https://doi.org/10.1109/ANTI-CYBERCRIME.2017.7905287

29. Kobyzev NP, Golubev SA, Artamonov YG, Karimov AI, Goryainov SV (2017) Embedded control system for tracked robot. Proceedings of 2017 IEEE 2nd international conference on control in technical systems, CTS 2017, pp 176–179. https://doi.org/10.1109/CTSYS.2017.8109519
30. Akilan T, Chaudhary S, Kumari P, Pandey U (2020) Surveillance robot in hazardous place using IoT technology. Proceedings—IEEE 2020 2nd international conference on advances in computing, communication control and networking, ICACCCN 2020, p. 775–780. https://doi.org/10.1109/ICACCCN51052.2020.9362813

Design and Analysis of Fuzzy-Logic-Controlled Active Suspension

Daniel Pearce

Abstract This research aims to highlight the applications, functions, and improvements of Fuzzy Logic Controllers, more specifically, by investigating the use of FLCs within active suspension. Investigations into the relevant markets for suspension and corresponding controllers were key to understanding the application of control within suspension systems and its evolution of its improvement within the industry. The research uses data and case studies sourced from the internet, reports, and published papers within the automotive engineering market, to effectively produce an FLC. Alongside this, a PID controller was designed to counter argue and compare with the FLC to critically analyse the performance and use of fuzzy logic in competition with alternative solutions within active suspension.

Keywords PID—Proportional, Integral, Derivative · LQR—Linear Quadratic Gaussian · FLC—Fuzzy Logic Controller

1 Introduction

Suspension was first introduced into the car model to improve safety, performance, and/or comfort. The suspensions' purpose was to create the most contact with the road at all times—where grip and control is maintained.

With different vehicle functions, the design requirements of suspension will be varied drastically; aiming to reach the vehicle's needs. Comfort is often produced with lighter suspension setups, whereas performance and maximum vehicle "feel" is produced by a stiffer setup. Before tuning and creating a suspension setup, it is vital for engineers to understand what the purpose of the system should be, developing the best outcome for the vehicle's purpose.

The evolution of suspension design originated from the roman chariots crossing abrupt surfaces such as cobbled roads that provide high frequency feedback to wheel

D. Pearce (✉)
School of Engineering, Computing and Mathematics, Oxford Brookes University Headington Campus, Oxford, UK
e-mail: 19157567@brookes.ac.uk; Daniel.J.Pearce@hotmail.com

A. Azizi (ed.), *Autonomous Robotics and Intelligent Systems*, Emerging Trends in Mechatronics, https://doi.org/10.1007/978-981-95-5867-4_7

contact patches [2]. This design was called the leaf spring. The leaf spring was quickly evolved in 1804 to include stackable springs and were used mostly in rear suspension applications. In modern industry, Henry Ford implemented leaf springs into his work by manipulating the placement of the springs to be located on each axle, contrary to earlier design on each wheel [22].

This research will introduce an active suspension system using a fuzzy-logic-controlled system. With top motor sporting racing (Formula 1, Le Mans/Endurance racing, Indy Car, Formula E, and other FIA Sanctioned series) regulations prohibiting the use of active suspension, the focus of this research will be to demonstrate the system on a performance road car [29].

2 Task Description/Problem

2.1 Aims and Objectives

The research included aims to design, model and test active suspension control models by the use of Fuzzy Logic Control. However, with the addition of PID control to critically compare. With top racing competitions disallowing the use of active suspension (Formula 1), the target use of the active suspension design will be focusing on performance road cars. The chosen performance car is the BMW M2 G87.

The aim of this report, therefore, is to improve on the current BMW M Suspension model by incorporating a fuzzy-logic-controlled active suspension system. More specifically, by reducing the overshoot of displacement error between sprung and unsprung mass by 25%, and reducing error settling time by 40%.

Following three main steps to complete the objective as follows:

(1) Address mathematical modelling of the quarter car model, with values for suspension variables.
(2) Design and validate fuzzy-logic-controlled system for active suspension, alongside a PID controller.
(3) Test and evaluate performance under various road conditions a performance car may endure.

2.2 Suspension Explanations and Synopsis

Suspension systems were invented with the main objective of creating a response to vibrations that will develop two outcomes; make a body safer and more comfortable, or to develop better handling and feedback response. Comfort is improved by developing consistent contact between road and body of vehicle, in automotive situations, the contact is between the tyre contact patch and the road. Furthermore, a controller is used to reduce the abrupt effect of road surfaces and create a bounce whilst keeping contact between tyre and ground to improve comfort.

However, in racing and within performance cars, comfort is not the desired outcome—feedback and handling response is. Therefore, harsher suspension systems that retain contact and true road feel are used, this is so that drivers have maximum road feel and can use feedback from various conditions to correct steering throughout the track. Within this project, active suspension systems for performance road cars will be the focus relating to the latter format discussed with performance over comfort in mind.

2.3 *Vehicle Dynamics*

The vehicle dynamics relates to the overview of the car model and its' numerical values that will be used to, ultimately, produce an effective suspension system. The model body that will be studied relates a vehicle that moves along a track or road, and its system that responds to road conditions. By using the dynamics of the vehicle, the improvement of performance of numerous components across the car can be derived, within this report specifically, the active suspension system.

Vital aspects of vehicle dynamics that must be understood contain; the cars' weight and distribution, drag and other aerodynamic effects, tyre conditions (grip), and road conditions/obstacles.

2.4 *Controllers*

Common controllers that are often used to improve the response of suspension include PID, LQR, and FLCs [24]. Electronic semi/fully active suspension systems make key use of adjustable shock absorbers or struts to improve the response of a vehicle's suspension to provide a more suitable ride/experience (based on desired characteristics—comfort or performance) [17].

Controllers within the system also independently control the electronic shocks to adjust to desired responses quickly and efficiently, by assessing inputs such as velocity, vertical displacement, steering angle and pitch [17].

2.5 *Quarter Car Modelling*

To prepare and convey the most accurate results a full car suspension model would be optimal, however, a quarter car model will be studied and used to simplify the effects on a suspension system. This focuses on one corner of the car in order to single out and focus on the effects active suspension may bring.

TABLE I
QUARTER CAR PARAMETERS

Parameters	Symbols	Quantities
Body mass	m_b	250 kg
Wheel mass	m_w	50 kg
Stiffness of the body	K_1	16 kN/m
Stiffness of the wheel	K_2	160 kN/m
Stiffness of the damper	C_s	1.5 kN.s/m

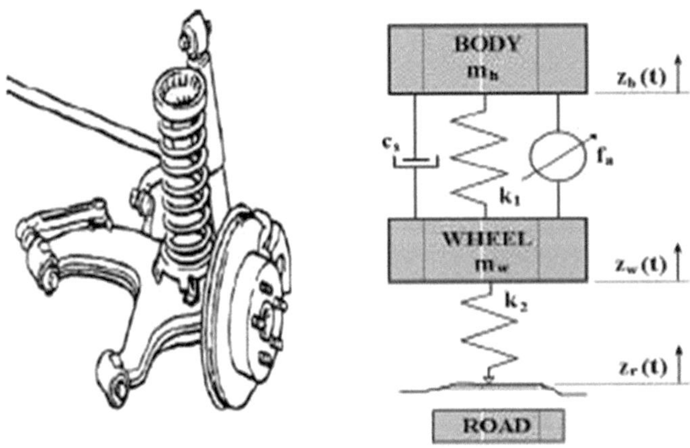

Fig. 1 Quarter car model

A quarter car model suspension system consists of the body, wheel, and road (as seen on the left [24]). Connecting these are springs; K1 representing the body stiffness, and K2, the wheel stiffness. Cs is the damper stiffness (Fig. 1).

Note these specific numerical values are example and will not be used in the findings [24].

3 Literature Review

3.1 Research Justification

The automotive sector has seen exponential expansion in usable technologies during the past century. Through a variety of applications, including motorsport, road vehicles, and transit, significant advancements have been made. Whether it is the engine, tires, electronics, or suspension, all of these components must work together to achieve the main purpose of an automobile.

Vehicle suspension systems are essential. By minimising the amplitude of road surfaces, the suspension can determine how comfortable the car will be. For example, it can assist lessen the feedback of road bumps that trucks may encounter when transporting delicate goods.

In contrast, Formula One's adoption of active suspension completely changed how the cars were designed and competed [13].

Williams, who won the championships in 1992 and 1993, is most known for using active suspension systems, which were essential to the teams' and the racers' (Nigel Mansell and Alain Prost) triumphs.

In general, the suspension system can be categorised into three groups. They consist of active suspension, semi-active suspension, and passive suspension [9].

A wide range of controllers, such as PIDs, fuzzy logic, and sliding controllers, can be used to control these systems in various application cases.

3.2 *Automotive Suspension*

Suspension was introduced into automotive applications to provide enhanced comfort and/or handling and feedback. There are various criteria when designing a suspension system, mainly for comfort (soft forces on the driver) or performance (harsh forces on driver). Suspension at the moment can be separated into three types; passive, semi-active, and active suspension.

3.2.1 Types of Suspension

Automotive suspension, through evolution has become advanced enough to be classified into three different categories; passive, semi-active, and active (Fig. 2).

The three different classes are explained below.

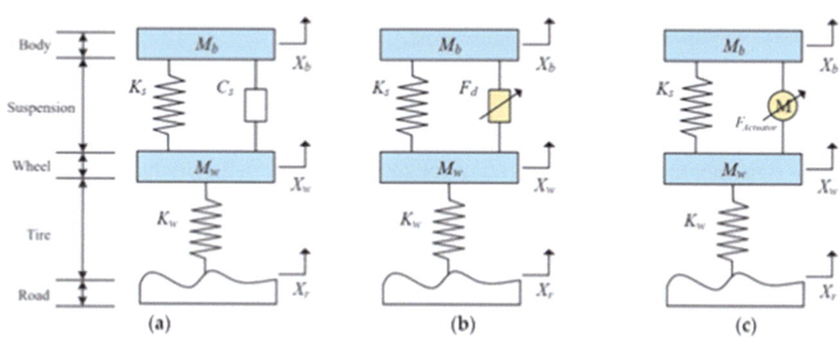

Fig. 2 Differences of passive, semi-active, and active suspension [43]

Passive System

- Passive suspension was the first and most common adaptation of suspension within cars [5]. This system makes use of simple springs and shock absorbers to respond to the changes in road conditions, reducing undesired excess motion of the vehicle [18].

Semi-active System

- These systems involve dampers that are hydro-mechanical control devices capable of varying the amount of energy to input to the springs in order to change the response to road conditions and vertical forces. Various devices are used to control and acquire optimal damping; Solenoid valves, Magnetorheological (MR)/electrorheological (ER) dampers, and electromagnetic dampers [4].

Active System

- This system varies between the two previous configurations by the inclusion of an actuator alongside the damper and mechanical spring. Active systems have greater capabilities of reaction against varying road surfaces due to the springs and dampers being controlled through the actuator force. This allows for much more effective control of the spread, or lack of spread, of forces—allowing for further customisation to a vehicle's setup [3].

In order for these systems to be developed, numerous features must be accounted for. The spring and damper, these two parts contain different motion formulas, however together, responses can be investigated and enhanced to the user's desires by changing rates.

3.2.2 Sprung Mass

Sprung Mass: The sprung mass relates to the weight that is held in place by a spring (whether that's correlates to tyres or suspension, etc.), this includes additional elements such as the driver. This mass can be affected by outside elements such as vertical road conditions when the unsprung mass reacts.

3.2.3 Unsprung Mass

Unsprung Mass: In a vehicle, the unsprung mass relates to the suspension, and wheels/tracks and components relating to them. Essentially the weight of the entire suspension system that will only act (with weight) when responding directly with vertical movement. For this report unsprung mass will not be involved in direct calculations [28].

3.2.4 Springs

In mechanical contexts, springs epitomise the concept of compression and expansion, springs can denote and resemble anything with a stiffness and behaves elastically (for example suspension springs, tyres, and bushings). Springs are designed to operate when force is applied, where they shorten in length when the force acts towards a spring, and expands when a force pulls the spring apart. This property allows them to store energy when compressed and release it upon returning/expanding to their original shape [19].

In order to calculate a vehicle's motion along a plane, spring values are used, they support the sprung mass. Springs are also fundamental for tuning and adapting the systems response [28].

3.2.5 Stiffness, Frequency, and Rates

Certain factors, like the car's mass (body) and spring rate, must be understood in order to calculate the spring frequency. According to Hooke's Law from 1676 (originally discovered in 1660) [23], where k is the spring constant (N/m), x is the displacement (m), and F is the applied force (N), the force needed to deform a spring is proportional to the displacement experienced (m). The following expression denotes F;

$$F = -kx$$

This equation is derived from Newton's Second Law, where, $F = ma$, where in vertical displacement $a = \textit{gravitational field strength } (g)$. This denotes the stiffness of a spring can be written as the following [16]:

$$-k = \frac{m \cdot g}{x}$$

Furthermore, using Newton's second law again, but using the mass of body in acceleration, the following formula for total F can be found:

$$\sum F = m \cdot \ddot{x}$$

However, it must be noted Newton never wrote the equation $F = m{\cdot}a$, but it was obtained by a change in the definition of the law to "The rate of change of momentum with time is equal to the motive force impressed; and is made in the direction of the right line in which that force is impressed". Upon which $F = m{\cdot}a$ can be derived. Leonhard Euler (1707–1783) articulated the modern form of Newton's second law $F = m{\cdot}a$, in 1775 [35].

Frequency can be found below, where relating to stiffness (k) and mass (m);

$$f = \frac{1}{2\pi} \times \sqrt{\frac{k}{m}}$$

Finally, using these equations, and substituting $\sum F$ and dividing mass, therefore an equilibrium will always be tried to reach by the spring's force;

$$\ddot{x} + \frac{k}{m} = 0$$

Importantly, it is key to note then when responding to affects, springs react in a sinusoidal wave about its equilibrium. The response in wave form correlates to the displacement velocity ($\dot{x}$),, and the frequency (Hz).

$$\omega = \frac{1}{2\pi}$$

ω, denotes the angular frequency of a system. This equation is derived from:

$$\omega = \frac{2\pi}{T}$$

The damped natural frequency is described by [32]:

$$\omega_d = \omega_n\sqrt{1 - \zeta^2}$$

3.2.6 Damping

Following from the damped natural frequency ω_d, it is important to understand the role of the damping ratio, ζ.

Damping ratio, ζ, is defined by:

$$\zeta = \frac{c}{2\sqrt{m \times k}}$$

where c is the damping coefficient, m is mass, and k is stiffness.

- If $\zeta = 0$, there is no damping in the system.
- If $0 < \zeta < 1$, the system is underdamped, where oscillatory motion has reduced frequency.

- If $\zeta \geq 1$, the system is critically damped or overdamped; no oscillations occur, just exponential decay.

In the absence of damping, an item will oscillate in Simple Harmonic Motion (SHM), which will be damped and eventually cease when subjected to external forces like friction. Using calculations with a damper, responses can be altered to desired criteria.

Dampers are the element responsible for providing resistance force within an oscillating system, tuning is required for dampers to work with responses provided by springs to develop a desired oscillation effect. Their job is to control the motion of the spring/manage the weight transfer and shock loads of the car, by returning springs to original compression as fast and as efficiently as possible after spring response to vertical disturbances [39, 40].

Dampers consist of the main body, piston (with drilled holes), shims, adjuster needles, oil, and nitrogen.

Damping Setups (Hard or Soft)

Most tuning of dampers is done through a damper dyno, where ends are connected to a load cell and crank shaft. Different frequencies are simulated to plot force vs displacement [39, 40].

Velocity at each data point can be calculated by:

$$v = 2\pi \times \omega \times A$$

Spring mass damper systems can be used to find the damping coefficient [31];

$$C = \text{Damping Force/Velocity or Displacement}$$

Alternatively, the external force ($F(t)$) can be used in the following formula, where m is mass, $\ddot{x}$ is acceleration, $\dot{x}$ is velocity, k is spring constant, and d is damping.

$m \cdot \ddot{x} + d \cdot \dot{x} + k \cdot x = F(t)$.

The critical damping coefficient derives the exact point of damping that the system needs. The following equation is made for d as the result [37];

$$d = 2 \cdot \sqrt{m \cdot k}$$

Depending on the type of driving the car is experiencing, and the harshness of road disturbances, you can tune the damping setup of a car. And finding the right balance between soft and hard damping is important. In coilovers, damping refers to the tuning of shock absorbers in order to regulate the movement and response of the suspension system.

Soft damping allows for improved grip between road and tyre on uneven surfaces as it absorbs bumps easily [8].

However, hard damping produces more responsive and firm feedback—desired by performance driving or for smoother roads. As bumps can be felt easily, and so handling ability increases, allowing for the driver to correct the car's movement better.

With both these setups having advantages, it is important for the car's damping to be tested and tuned for different applications. In the case of this report, the vehicle will be performance based and so the setup will require leaning to the harsher side of damping.

3.3 Vehicle Suspension Effects

3.3.1 Effect on the Human Body

The effects of natural frequency can have a big impact on the response of the system with additional side effects to the driver:

- Frequencies below 1 Hz cause motion sickness.
- Frequencies over 4 Hz cause accelerated fatigue.
- Modern road cars operate about 1.5–1.8 Hz.
- Race cars with low downforce, low grip, operate at 2.2–2.5 Hz.
- Medium downforce (5 KN) race cars operate at 2.6–3 Hz.
- High downforce race cars (e.g. F1, GP2, F3) operate at 4.5–10 Hz.

The safety of the driver, when designing a suspension system (more specifically an active suspension system), must implement outputs such that the driver shouldn't experience significantly dangerous or health concerning experiences. This, specifically, relates to the acceleration the driver may experience when going over vertical road disturbances. With relation to performance, which is the goal of the suspension, discomfort is acceptable, however, any significant acceleration may harm the driver. Furthermore, any natural frequencies which cause motion sickness or accelerated fatigue have been avoided in the design (Sect. 4 onwards) (below 1 Hz and above 4 kHz).

3.3.2 Wider Effects of Developing Active Suspension

The suspension project unites diverse teams from engineering, RND, Marketing, and Finance to develop, market, and produce prototypes, fostering inclusivity and integrating varied opinions, cultural beliefs, and working styles. Societal impacts include improved road safety through smoother handling and greater accessibility for elderly and disabled users. Although targeted towards experienced drivers, enhanced ride comfort benefits everyone. However, significant production costs may limit affordability for consumers, highlighting a balance between innovative design and practical manufacturing constraints.

In order to collect sufficient values for modelling parameters, often, manufacturers advertise surveys to consumers, these surveys may include personal information about a person which the user may want to keep private. This is legalised by giving consent, whether that be online or on paper. It is important to understand this for any further work taken on from this project.

3.4 Semi-active and Active Suspension

Semi-active and active suspension systems highlight the significant advancements in automotive engineering. These systems were aimed at improving ride comfort and handling, by dynamically adjusting the suspensions response to road conditions.

Early Concepts

Semi-active Suspension: The original ideas of adjusting damping in-situ came from research efforts in the late 1960s and early 1970s. Researchers explored how adjustable shock absorbers could be used to change the suspension's response of road disturbances, whether that be to soften or stiffen the ride based on the driver's desires, where coefficients of the damping and spring are adjusted [36].

Active Suspension: During the 1980's, the "golden age" of electronic suspension became apparent, where analogue electronics were already well-developed and so, manufacturers began implementing digital micro-controllers into suspension systems. This attracted top racing teams alongside top car manufacturers, and so they began experimenting with electronically-controllable quick-to-react systems where hydraulic actuators were used. And so, began the start of fully active suspension systems [34].

Active suspension brings to the automotive industry, an enhanced and easier way to directly change and improve the response of a suspension system to the road conditions and vertical abrasions. Although the 1980's saw the birth of fully active systems, since the 1950s, researchers have been studying ASS's. Active systems provide an independent force or torque working between the sprung and unsprung masses. Where active systems were implemented into racing much earlier, the use of ASSs in production cars (particularly luxury cars) did not start until the 2010s [27].

Although fully active systems provide many positives regarding performance, safety, and comfort, the complex component series needed for the systems make production extremely costly and so these systems aren't usually funded in mass-produced road vehicles. Contrasting this, very expensive, luxury, and performance road cars sometimes utilise the active system to set their cars apart from the competition.

In addition, the use of active suspension does provide the opportunity for extending the life cycle and sustainability of suspension. This is due to the system providing a force between the sprung and unsprung mass that reduces the vibrations between them. A reduction in vibration means there the life period of the components within the suspension system (e.g. springs, dampers, bushings) is increased where excessive vibrations are reduced and so less prone to failure when these effects take place. The design modelled from Sect. 4 onwards supports this reduction in vibrations.

3.4.1 Differences in Systems

Semi-active and active systems both make key use of controllers, however in very different ways (Fig. 3).

Semi-active: The controller in semi-active systems continuously monitors sensor inputs, and then adjusts valves (in electrohydraulic dampers) or electrical current to a magnetorheological (MR) [30]. By doing so, the controller changes the damper's internal resistance to match road conditions.

The advantages of semi-active systems are apparent. Less complex systems allow for lower costs when manufacturing, as well as improved reliability where there are fewer moving components/structures that could fail (compared to ASS's).

This type of system does come with its limitations however. Due to semi-active systems only regulating damping, rather than applying force, only this change in damping makes a difference, and so they less effective at fully counteracting large vertical motions. Therefore, the performance gap vs ASS's is apparent where under extreme conditions they can't match the body control or flat cornering characteristics of a well-engineered active system.

Active: Fully active systems on the other hand, feature actuators that dissipate and supply energy. These actuators work in connection with a network of sensors that monitor displacement, acceleration, and speed in order to provide or dissipate forces that counteract unwanted vibrations, or for comfort-desired-criteria, keep ride height stable. Whether it be performance or comfort based, the active system aims at making the suspension as stable as possible by maximising contact between the tyre and ground.

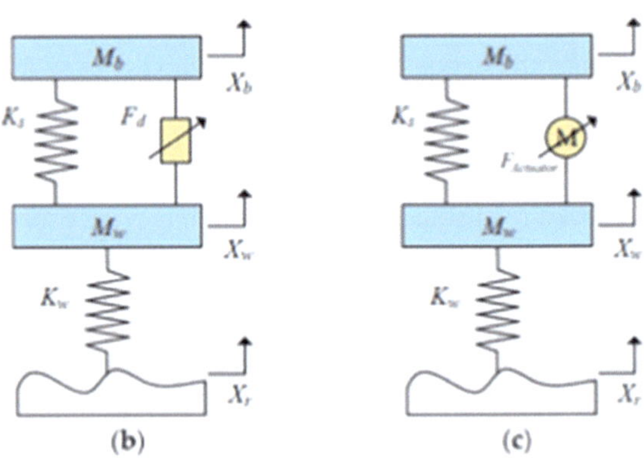

Fig. 3 Semi-active/active suspension diagrams

3.4.2 System Implications

Integration, Costs, and Maintenance Considerations

Integrating active systems comes with its limitations; Systems require complex use of high-performance sensors, actuators, and advanced controllers and learning networks, such as Fuzzy Logic and ANN. These components work together to process the real-time data of displacement between the sprung and unsprung mass, and provide optimal adjustments.

However, this complexity increases power demands, manufacturing costs, and maintenance frequency and costs. The increase in electronic and mechanical components also increases potential points of failure, therefore creating demand for robust fault-tolerant designs and redundancy measures.

In addition, by increasing efficiency of the active suspension and for example; reducing unnecessary damping forces. Brings the possibility of lowering fuel consumption and emissions outputting to the environment, helping reduce pollution. This consideration was taken into the design of the system.

Different regions may enforce ULEZ restrictions, where different requirements are needed; must meet emission values, EURO 4 for Nox for petrol vehicles, and Euro 6 for Nox and PM for diesel vehicles [41].

Broader Implications and Future Trends

At the moment, active suspension is mainly used for improving three categories: comfort, safety, and performance. The future uses of this system may influence the market exponentially if sustainability and costs can lower, through potential uses and experimentation of sustainable electronics. One promising system that has been investigated is the Gallium Nitride (GaN) transistor module. GaN is typically used in power converters and on electric vehicles, they have high potential for reducing switching losses and energy consumption for use in control modules that drive the suspension actuators [12]. This increased efficiency could provide lighter, more compact designs that improve energy management, and therefore could reduce energy drawn by car batteries and improve sustainability goals by increasing the car battery's life cycle and reducing end-of-life time.

Furthermore, the precise control that active suspension supports the ever-increasing experimentation and research into autonomous driving, where suspension is a critical system of vehicles on the road that ensure the priorities of safety (for both occupants and pedestrians) and comfort. These are the main goals of autonomous driving at present. Active suspension therefore, can be adapted to control, alongside braking response systems, to steer away or reduce braking distances when the car may be coming in contact with another vehicle or a pedestrian. This concept of safety is described by the UNECE WP.29 regulatory framework [42].

Codes of Practice and Commercial Matters

The ISO 26262 regulations and compliance must be considered if model is giving green light to produce, where changes may need to be made to adjust for these requirements. ISO 26262-3 covers details of the concept phase, including hazard analysis and risk assessment [21]. Also reliability standards are involved when electronic components exist in the product, the IEC 61508 outlines reliability standards of electronic components [20].

3.5 *Existing Active Suspension Models*

3.5.1 Williams F1—FW14B

The Williams FW14B competed during the 1992 Formula 1 season. It was most famous for its advanced active suspension system that heavily contributed to its dominance throughout the racing year (Fig. 4).

The development and advancement of race team cars advanced significantly in the early 1990s. Teams and engineers were learning more and more about aerodynamics and how it related to open-wheel, low-slung Formula One cars.

The active ride system Williams utilised was dominant and efficient at controlling the disturbances on track, alongside providing sufficient displacement in push rods that was suited for criteria at corners and straights when braking and accelerating.

It was originally engineered by Lotus in the 1980's however the original system, unfortunately, did not see the high market trajectory that was hoped for at the time.

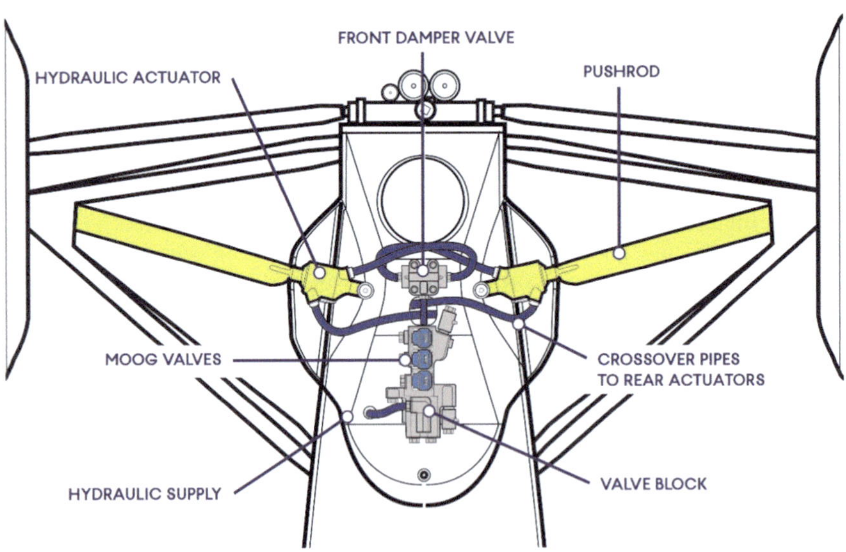

Fig. 4 Williams FW14B active suspension system [33]

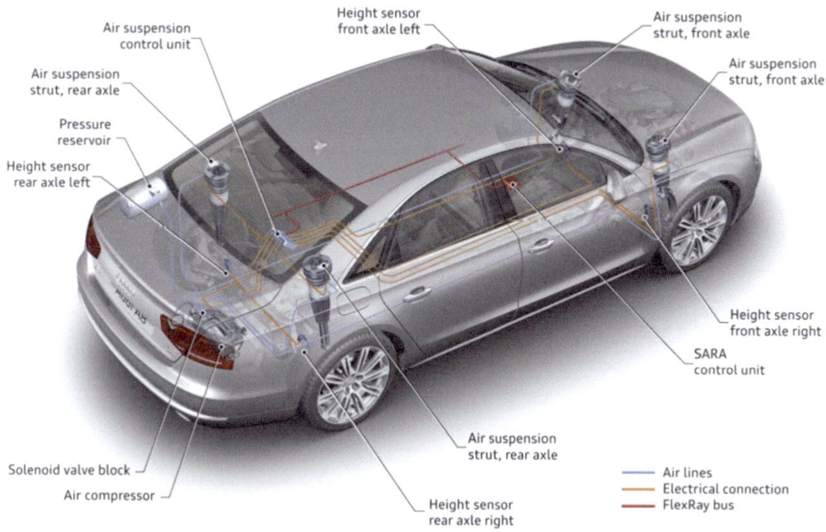

Fig. 5 Diagram of the adaptive air suspension and vehicle body on the Audi A8 [6]

By being deemed too dangerous, the system was later banned from Formula 1, however the FW14B is still seen to be one of the most dominant cars in F1 evolution [33].

3.5.2 Adaptive Air Suspension

Advanced air suspension systems are used by Mercedes-Benz's S-Class and Audi's A8 to improve dynamic performance, comfort, and ride quality. Through the use of sensors and a centralised control module, the Audi A8's adaptive air suspension continuously modifies ride height and stiffness, maximising weight distribution for better handling and fewer vibrations (Fig. 5).

The system makes key use of "air springs" instead of the traditional metal coils. This innovation allows for precise damping control for suspension over disturbances as the pressure can be highly tuned or adjusted/controlled for various different scenarios [7].

3.5.3 ABC Body Control

ABC uses a network of sensors combined with hydraulic servos and sophisticated control algorithms to actively counteract body roll and optimize chassis balance. By dynamically adjusting suspension forces on each wheel, the system improves cornering stability, enhances comfort, and reduces vibrations under varying load and road conditions.

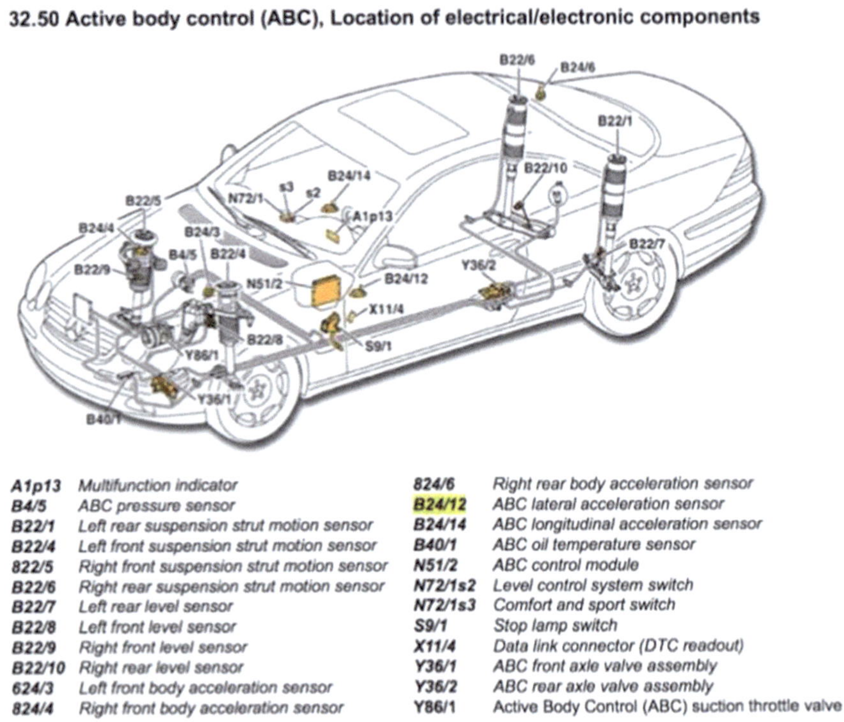

Fig. 6 Diagram of the ABC body control system used in the Mercedes-Benz W220,W221,W222 [1]

ABC differs in the fact it uses hydraulic fluid for controlling the actuators, and accelerometers to measure real-time displacement (Fig. 6) [14].

3.6 Controllers

3.6.1 PID Controller

PID controller is a fundamental feedback control mechanism used in engineering to continuously adjust system outputs and reduce errors. It consists of three parts; (P) proportion, (I) integral, (D) derivative.

A widely recognized tuning strategy is the Ziegler-Nichols method, which systematically adjusts these values based on ratios, ensuring stability and enhanced performance in applications ranging from industrial automation to aerospace systems.

3.6.2 Fuzzy Logic Controller

Fuzzy logic is a powerful computational approach simulating human reasoning via processing imprecise inputs. It involves fuzzification, converting crisp data into fuzzy sets defined by membership functions. Inference mechanisms and rule bases then evaluate these inputs, generating fuzzy outputs subsequently refined through defuzzification into clear, actionable results. Widely employed across engineering fields, fuzzy logic enhances decision-making in control systems, robotics, automotive applications, and other complex problem-solving scenarios by effectively integrating smooth processing, strong adaptability, and robustness, thereby significantly improving performance (Figs. 7 and 8).

4 Methodology

4.1 Considerations on Methodology

Throughout the process of methodology development, an integrated/systems approach is used, where consistent referring back to previous calculations/literature is present. This ensures for the whole methodology and modelling to be comprehensive and robust for achieving the desired criteria.

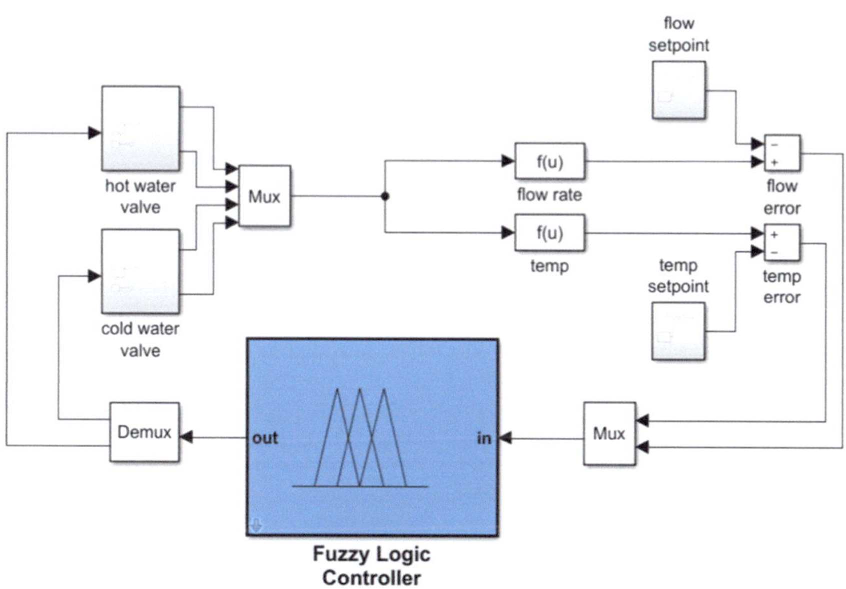

Fig. 7 Fuzzy logic controller network [25]

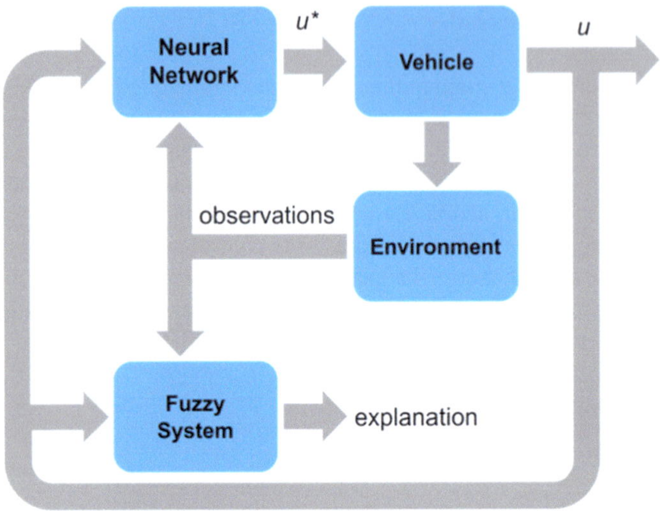

Fig. 8 Fuzzy logic neural network [25]

Also, by use of keeping close considerations for supporting diversity, equality, and inclusivity. Where making the vehicle ride as stable as possible, allows for a larger selection of users to use this system on a car.

Although this report aims of improving performance cars, there are many ways in which active suspension could have an impact on customers/user's lives. Opportunities for tuning controllers to derive a super-steady ride height would impact elderly or disabled users in a positive way. This would be completed by tuning the controller to regulate the ride height of the vehicle to a consistent height, within errors of 1–2 cm. In addition, the control method's use is to make the system as stable as possible, ensuring users are safe. This provides the opportunity of enticing new users due to its wider appeal with enhanced stability performance. It is also important to consider the manufacturability of the active suspension system, with very little dominance currently in the market [11] due to the cost and production challenges of the components. Compromises should also be taken into account where different views on the system may be present between teams or individuals, and so responsibility must be considered in further development. Moreover, introducing interdisciplinary collaboration and incorporating user feedback into iterative design phases further refines performance, enhances cost-effectiveness, and ensures the final product aligns with market demands and stringent safety standards.

4.2 Vehicle Parameters

The project will use the vehicle variables from the BMW M2 G87.

Values are gathered from 2023 BMW M2 G87 specs. It is noted that with this report using a quarter car model, various variables are assumed (from total vehicle specs).

BMW M2 G87

$$\text{Kerb Weight} = \mathbf{1805\ kg}$$

$$\text{Weight Distribution} = \mathbf{50/50}$$

$$\text{Wheelbase} = \mathbf{2747\,mm}$$

$$\text{Front Track} = \mathbf{1617\ mm}$$

$$\text{Rear Track} = \mathbf{1605\ mm}$$

$$\text{Wheel/Alloy Mass} = \mathbf{27.5\,lbs/12.47\,kg}\ [10]$$

$$\text{Wheel Radius} = \mathbf{Front} : \mathbf{19''},\ \mathbf{Rear} : \mathbf{20''}$$

$$\text{Tyre Mass } \left(\text{using } 19'' \text{ front}\right) = \mathbf{27\ lbs/12.247\ kg}$$

$$\text{Front/Rear Tyre Spec} = \mathbf{275/35ZR19100YXL/285/30ZR2099YXL}$$

$$\text{Brake Assembly Mass} = \text{undisclosed (based off after market OEM):}\ \mathbf{12.6\,kg}\ [15]$$

$$\text{Suspension Linkages Mass} = \mathbf{14\ kg}\ (\mathbf{assumed})$$

$$\text{Front Spring Rate} = \text{publicly undisclosed, however based off aftermarket products it can be assumed from } \mathbf{68\ to\ 130\,N/mm} \text{ (calculated spring rate will be derived in Sect. 4.2) [26, 39]}$$

$$\text{Tyre Stiffness} = \text{Assumed } \mathbf{240\ N/mm}$$

4.2.1 Calculated Parameters

Front Axle Load

The mass over the front axle can be derived by:

$$Fm = \text{Kerb Weight} \times \text{Weight Distribution}$$

$$Fm = 1805 \times 0.5 = 902.5\,\text{kg}$$

Singular Corner Load

The mass over the front corners of the car:

$$m_2 = \frac{Fm}{2}$$

$$m_2 = \frac{902.5\,\text{kg}}{2} = 451.25\,\text{kg}$$

Unsprung Mass

$$m_{sus} = \text{Wheel mass} + \text{Tyre Mass} + \text{Brake assembly mass} + \text{Suspension linkages mass}$$

$$m_{sus} = 12.47\,\text{kg} + 12.2\,\text{kg} + 12.6\,\text{kg} + 14\,\text{kg}$$

$$m_1 = 51.27\,\text{kg}$$

Suspension Stiffness

As stated in Sect. 3.3.2, the natural frequencies drivers should incur when driving a performance road car/race car, should be between 1.5 and 2.5 Hz. The spring rate can be checked using the following formula (providing an assumption for natural frequency):

$$f = \frac{1}{2\pi} \times \sqrt{\frac{K}{m_2}}$$

This derives:

$$K = (2\pi \times f)^2 \times m_2$$

$$K = (2\pi \times 2.25)^2 \times 451.25 = 90,124.42\,\text{N/m}$$

For the quarter car model, this will be the spring stiffness K_s. 90.124 N/mm is suitable as compared with value from parameters discussed in Sect. 4.2.

Suspension Damping

The critical damping can be derived from [37]:

$$d = 2 \times \sqrt{m_S \times k} = 2 \times \sqrt{451.25 \times 90.12442} = 403.28\,\text{N s/m}$$

4.3 *Theoretical Framework*

4.3.1 Quarter Car Model

The quarter car model is used to analyse and simulate the motion of a vehicle's suspension system by focusing on just one wheel and its associated components, such as the sprung mass (the body of the vehicle) and the unsprung mass (the wheel and axle). This model helps to reduce the complexity of full vehicle dynamics by simplifying the system to one-quarter of the vehicle (Fig. 9).

The two main components M1 and M2 denote the sprung mass and unsprung mass respectively.

The equations of motion are derived from newton's second law. They describe the behaviour of the masses and how they react to force.

By using this model, engineers can predict the system's response to road disturbances and optimize the suspension design for better comfort and performance.

4.3.2 Equations of Motion

The equations of motion for the quarter car model are obtained by directly applying Newton's second law to each mass component. For each degree of freedom—the unsprung mass (wheel assembly) and the sprung mass (vehicle body)—the law $F =$

Fig. 9 Quarter car model diagram. *Note* Sprung mass = M2, and unsprung mass = M1 for the following modelling from Sect. 4.3.2 onwards (units are flipped as Simulink model uses the new variable names)

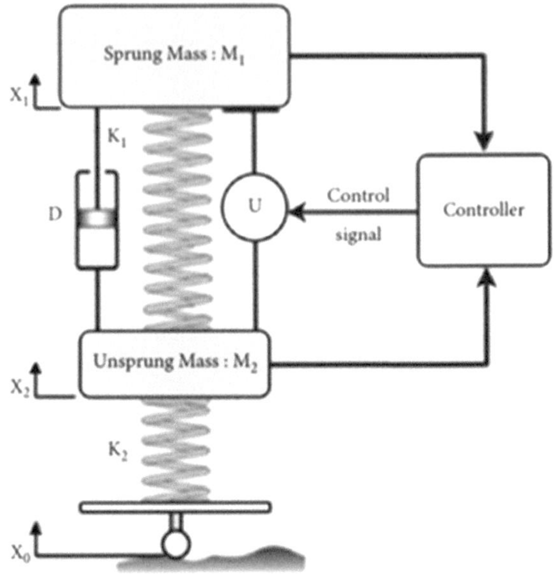

ma is used to balance the inertial forces with the forces generated by the springs and dampers.

The equations of motion can be seen below, where the controller U output is denoted, and the acceleration is denoted below:

$$M_1\ddot{X}_1 + D\dot{X}_{1(t)} + K_1X_{1(t)} - D\dot{X}_{2(t)} - K_1X_{2(t)} = U(t)$$
$$M_2\ddot{X}_2 + D\dot{X}_{2(t)} + K_1X_{2(t)} + K_2X_{2(t)} - D\dot{X}_{1(t)} - K_1X_{1(t)} - K_2X_{0(t)} = -U(t)$$

$$\ddot{X}_1(t) = \frac{1}{M_1}\left(U(t) - D\dot{X}_1(t) - K_1X_1(t) + D\dot{X}_2(t) + K_1X_2(t)\right)$$
$$\ddot{X}_2(t) = \frac{1}{M_2}\left[-U(t) - D\dot{X}_2(t) - K_1X_2(t) - K_2X_2(t) + D\dot{X}_1(t) + K_1X_1(t) + K_2X_0(t)\right]$$

4.3.3 Processes of Simulation

MATLAB Simulink will be used in order to simulate the equations of motion of the vehicle. Whilst maintaining the vehicle parameters of the BMW M2 G87 (discussed in Sect. 4.2), the Simulink model will also implement controllers, U, that provides a force to the system from controlling the error.

Firstly, the three different systems (passive, semi-active, active) are described below.

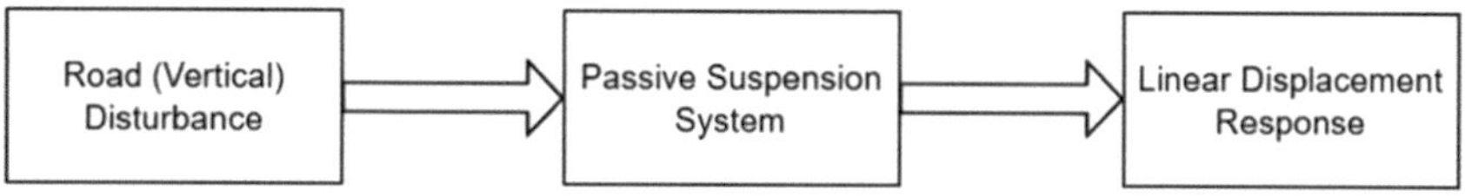

Fig. 10 Passive suspension response flowchart

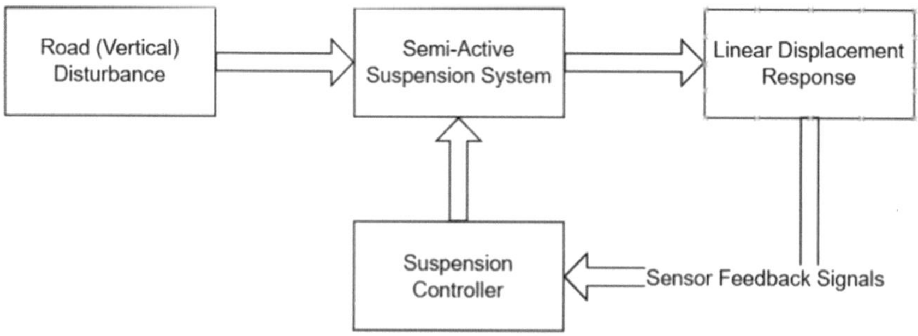

Fig. 11 Semi-active suspension response flowchart

Block/Flowchart Diagrams

The following block diagrams outline the steps of passive, semi-active, and active suspension.

Passive System

Passive systems rely on pre-existing values for stiffness and damping, allowing no change or adaption to responses when encountering different sizes of road disturbance. However, semi-active and active act differently, according to various road conditions (Fig. 10).

Semi-active System

Similar to the passive suspension, it includes springs and dampers, but the damping (or another parameter such as stiffness in certain designs) is adjustable in real-time. The controller determines how to adjust the damping/stiffness elements based on real-time feedback (Fig. 11).

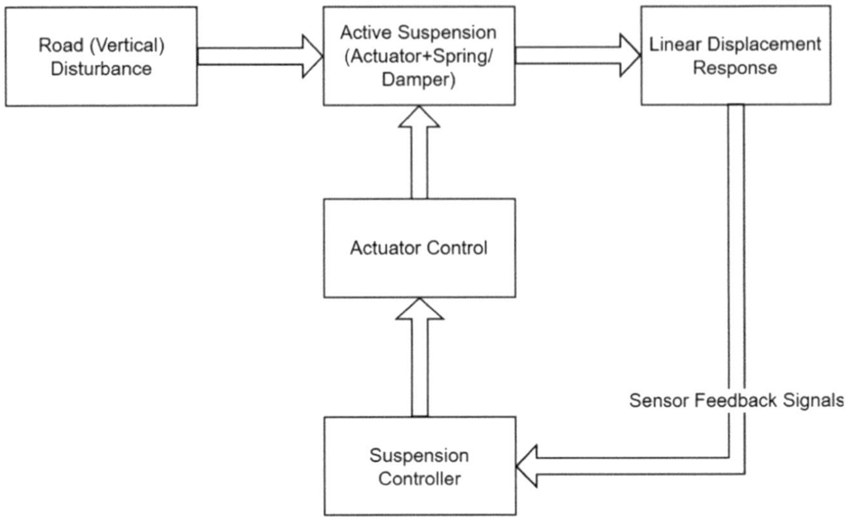

Fig. 12 Active suspension response flowchart

Active Suspension System

In addition to the traditional spring and damper, an external actuator can apply forces to the suspension. This actuator requires a power source (electric, hydraulic, or pneumatic), this is discussed in more detail in Sects. 3.4 and 3.5.

Often, the controller regulates the error between the sprung and unsprung mass in the system however could be applied to numerous displacement such as body displacement. But what remains the same is the actuator controlling the displacement between wheel and vehicle body.

For this report, the displacement error and error derivative of the distance between the unsprung and sprung mass will be controlled (Fig. 12).

4.4 *Modelling Process*

As stated in Sect. 4.3.3, the modelling of the quarter car suspension will be done on MATLAB and Simulink. The following descriptions detail the process and diagrams of the Simulink models with no controller, PID, and FLC.

4.4.1 Simulink Models

Passive Simulation Simulink Model

See Fig. 13.

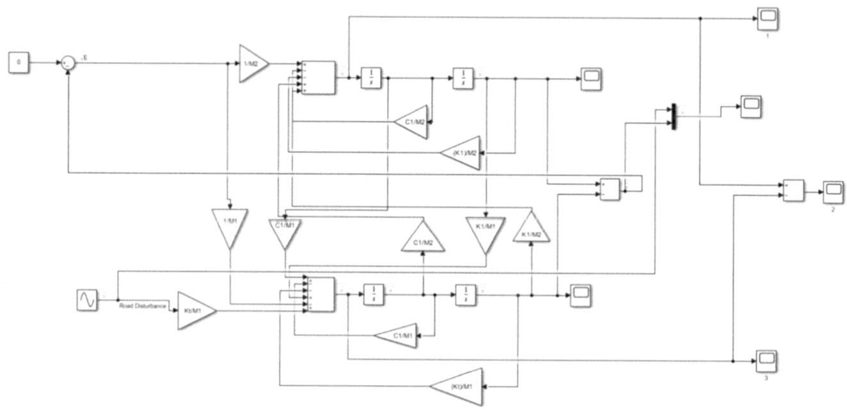

Fig. 13 Passive suspension system modelled on MATLAB Simulink

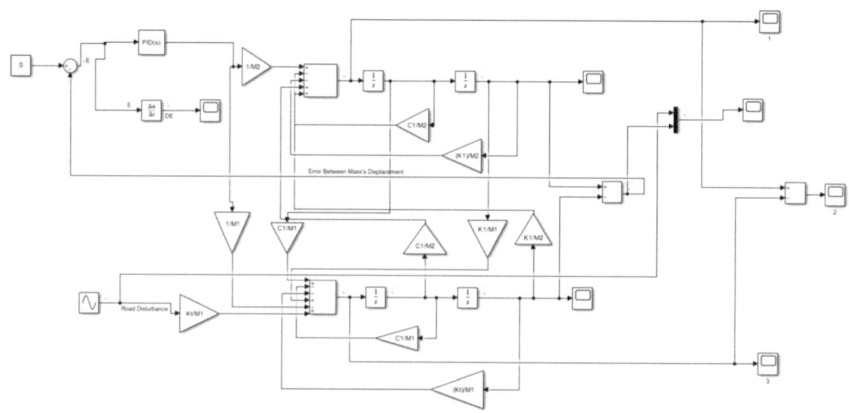

Fig. 14 Active suspension PID system modelled on MATLAB Simulink

Active Simulation Simulink Model with PID

See Fig. 14.

Active Simulation Simulink Model with FLC

See Fig. 15.

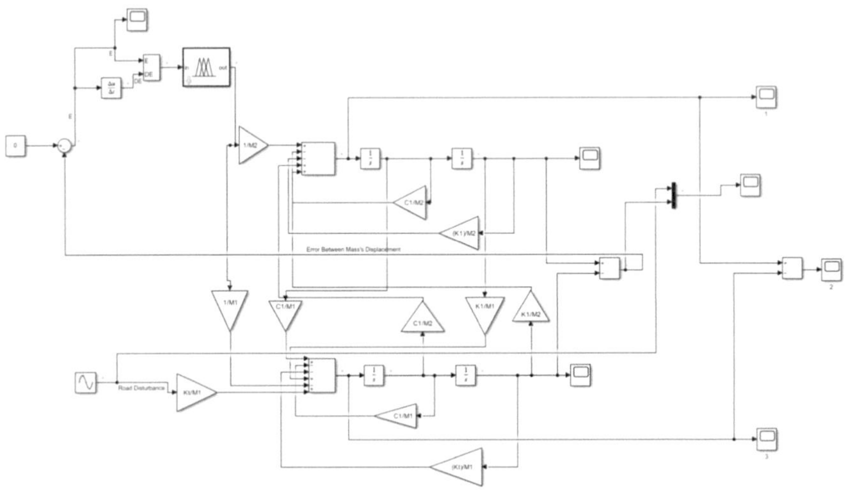

Fig. 15 Active suspension FLC system modelled on MATLAB Simulink

Simplified Suspension Systems with All Three Types

See Fig. 16.

4.4.2 System Parameters

As discussed in Sect. 4.2, the component variables are based off the BMW M2 G87, with spring stiffness (k), damping (c), tyre stiffness (K_t), Unsprung mass (M_1), and sprung mass (M_2) all corresponding to the derived vehicle parameters below (Table 1).

Road Disturbances

For testing the Simulink model, various road disturbances must be implemented to find results of the suspension systems.

Road disturbances can be inputted in various forms; Sources include Step, Sine, Ramp, and Noise. For the testing of vehicle displacement error, I concluded the most optimal test was to use sine wave inputs. This was due to sine wave most effectively representing the movement of the masses against time when going over a "bump". The sine wave accelerates in positive vertical displacement against time and then accelerates in negative vertical displacement against time.

This testing, with correlation to the BMW performance car, used sine wave amplitudes of the ranges 0–0.1 m, simulating common bumps the car may experience on track or on smoother road surfaces. This is described below, where the error is what is being found and controlled (Table 2).

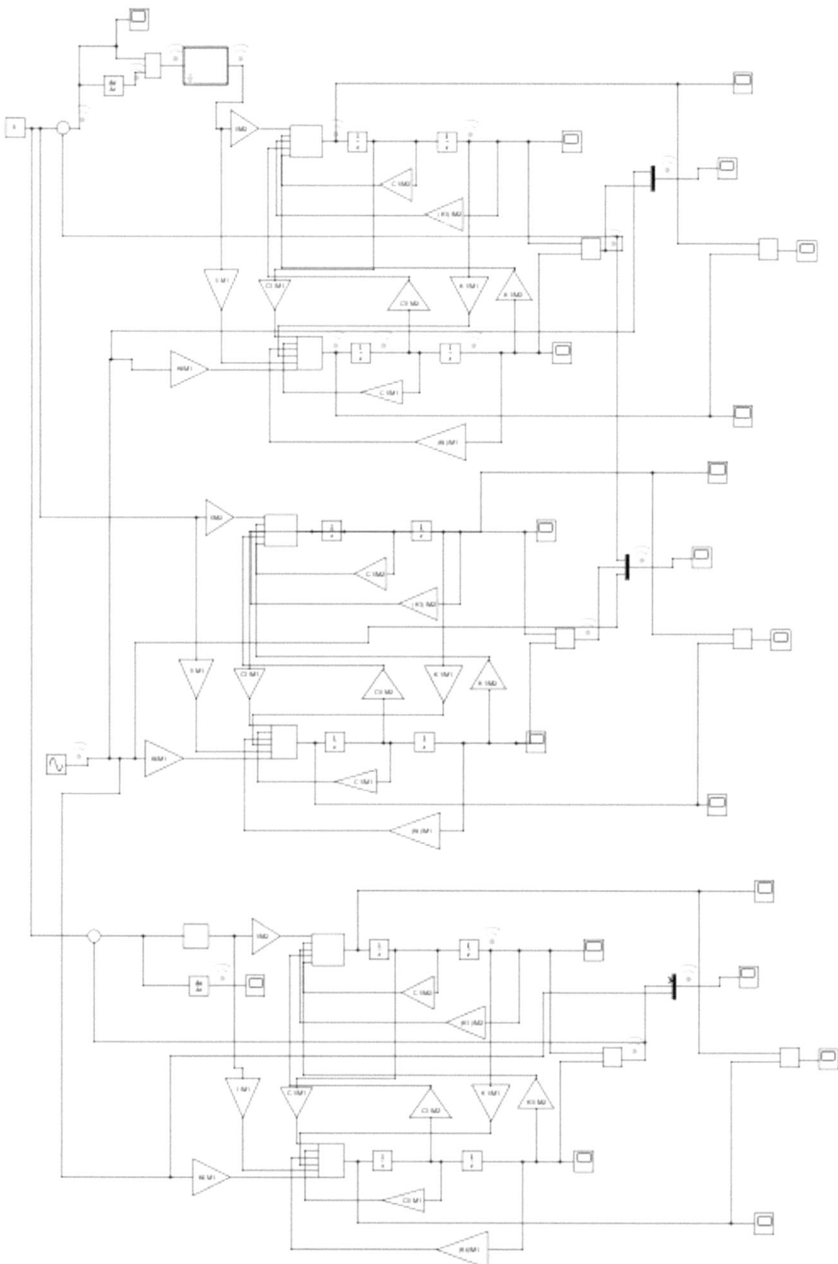

Fig. 16 Simplified suspension system(s) modelled on MATLAB Simulink

Table 1 Vehicle parameters for use in MATLAB Simulink

Parameter	Value	Description
M_2	451.25 kg	Sprung quarter car body mass
M_1	51.27 kg	Unsprung quarter car body mass (wheel + suspension)
K_1	90,124.42 N/m	Suspension stiffness
C_1	403.28 N.s/m	Suspension damping
K_t	240,000 N/m	Tyre stiffness

Table 2 Road disturbance testing

Sine wave amplitude (m)	Error in displacement of suspension (m)
0.02	?
0.04	?
0.06	?
0.08	?
0.1	?

This will be tested on all three systems, where the passive will portray the "standard" suspension values, from which the PID and FLC controllers can be tuned from.

The frequency of the sine wave is also important as this reflects different lengths of the road disturbance. I found using a frequency of 2.5 rad/s gave a bump time of around 1.25 s to be optimal as it gives enough time for a realistic bump travel. This is calculated as follows (Fig. 17):

$$f = \frac{w}{2\pi} = \frac{2.5}{2\pi} = 0.3979\,\text{Hz}$$

$$T = \frac{1}{f} = \frac{1}{0.3979} = 2.5133\,\text{s}$$

$$\text{Bump Travel Time (s)} = 1.2566$$

Tuning

Firstly, it is important to understand what the controller will be accounting for. As seen in Sect. 4.4.1 the Simulink model has wires that connect the controller to the error between body displacement and suspension displacement. It is important to understand that the goal of the controller is to reduce this gap by:

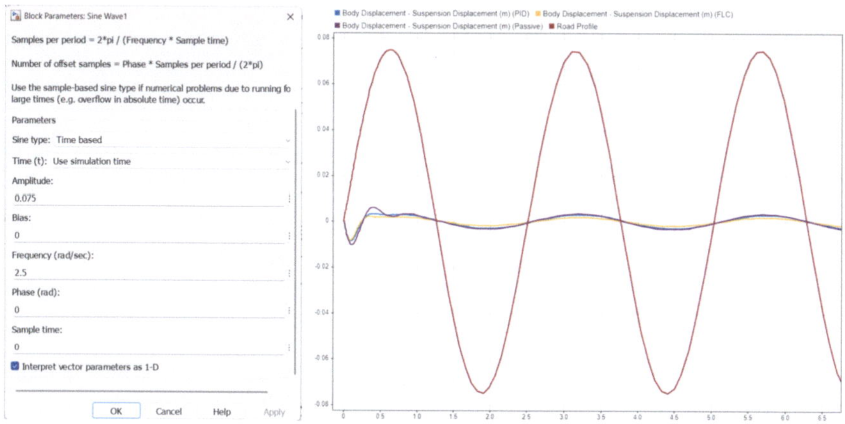

Fig. 17 Road disturbance parameters (0.075 m amplitude at 2.5 rad/s)

- Reducing overshoot.
- Reducing stabilising time/oscillations.

Fundamentally, in terms of optimising a performance car, will provide the driver with as most "road feedback" as possible as there will be less excessive vibrations and exaggerated responses.

The foundation of what is tuned will be data collected from the passive system (with BMW Parameters). This will lay the foundation for tuning the controllers.

PID Tuning

PID controllers, used in closed-loop systems, continuously adjust suspension in real-time to reduce displacement error.

The PID is split into three properties:

- (P) Proportion controls how quickly the system responds.
- (I) The integral ensures long term stability by reaching the steady state error.
- (D) The derivative helps reduce overshoot, ensuring a smoother response.

Much trialling and testing was done for optimal PID control response, the Simulink auto tune block was useful to see the differences and trade-off between the three variables (P, I, D), however was either far too sharp a response or not at all. Therefore, I used the Ziegler Nichol's method (discussed in Sect. 3.6.1) along with trial and error to derive the following values (Fig. 18).

Fig. 18 PID tuning simulink

Fuzzy Logic Tuning

Intellectual Property Note

Safeguarding intellectual property is a crucial concept as it protects innovative designs and maintains competitive advantage. When developing active suspension systems, IP may be encountered in proprietary control algorithms, patented mechanical designs, or unique research findings, all of which must be managed responsibly. This report uses a self-created FIS.

Inputs

As detailed in Sect. 3.6.2, Fuzzy Logic can be designed and tuned in many different ways. For this model it was concluded the best way for easy tuning and adaptability was to use a 25-rule base with 5 defining memberships for each input, these inputs are:

- Error.
- Error derivative.

The error denotes the difference in displacement between sprung and unsprung mass. By using the passive error it was clear to see the ranges for which the inputs should be tuned for (Fig. 19).

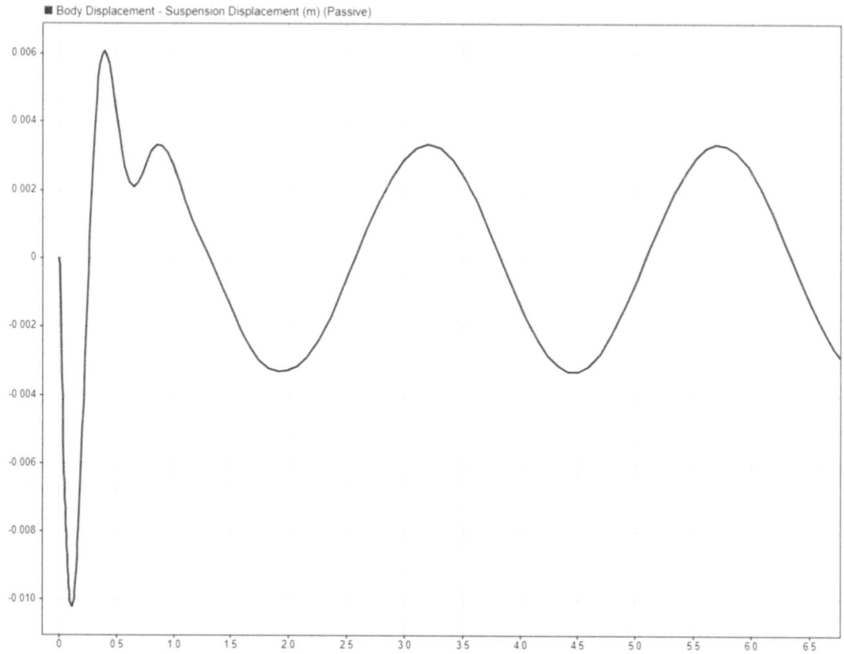

Fig. 19 Passive error used for tuning of FLC (0.075 m amplitude bump)

From the passive error seen above is from road disturbance of 0.075 m and so gives a good average perspective of how the error response looks.

From around −0.011 to 0.006 m is the error range of overshoot for this height in bump, and so using this, whilst also considering larger errors for bigger bumps the input of E can be tuned (Fig. 20).

As can be seen from the input fuzzification of the error above, the differences in memberships are tuned for −0.03 to 0.03, however with more focus on the ranges between −0.015 and 0.015. This was due to the error data collected from the passive suspension system, and tuned for controlling this.

The error derivative is also inputted and controlled by the following membership Function Plot (Fig. 21).

The error derivative describes the rate of change of error, this can be seen by finding the derivative of the error. The data ranges for this ranges from −0.2 to 0.2. The error derivative data used to tune the controller was gathered from the passive DE below (Fig. 22).

As seen above, the error derivative is much higher, respectively, than the error displacement. The rate is important to control as it helps in damping the controller's response when the error is changing rapidly, in order to avoid overshoot and oscillations.

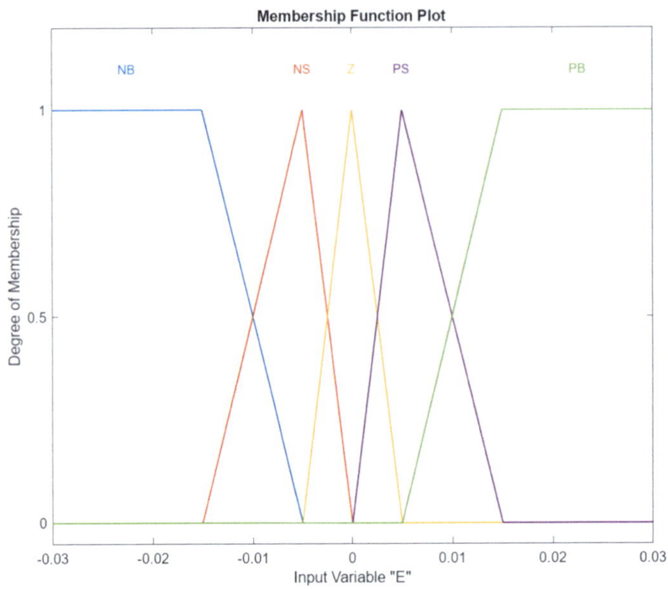

Fig. 20 FLC error input membership function plot

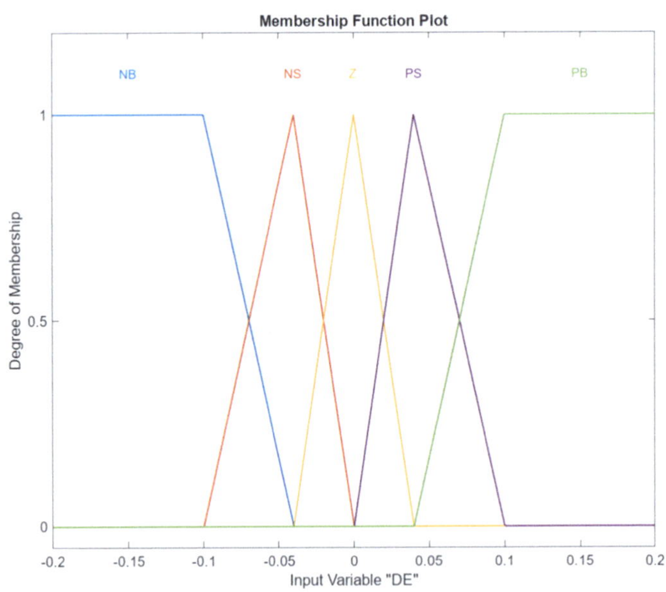

Fig. 21 FLC error derivative input membership function plot

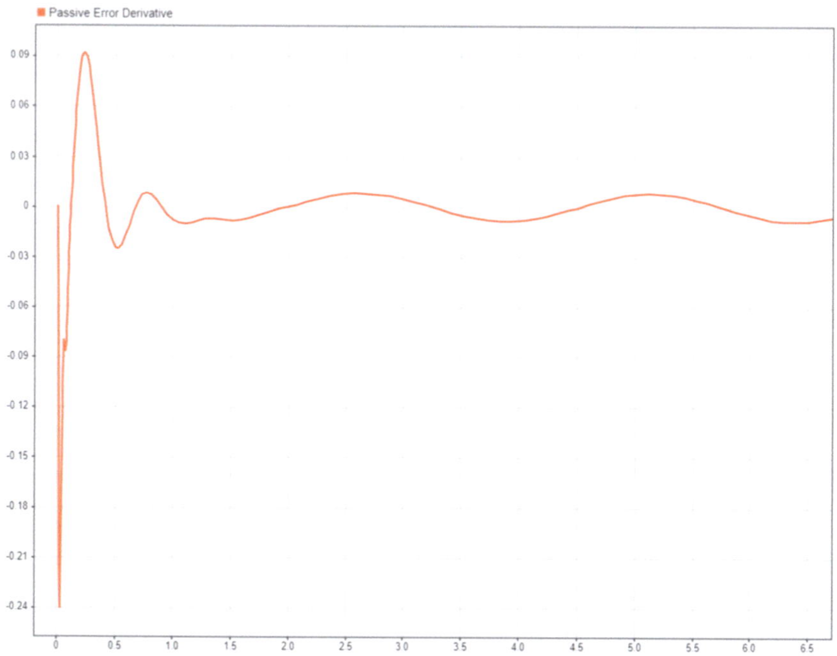

Fig. 22 Passive error derivative graph (0.075 m amplitude bump)

It is important also to consider values for DE that are above the passive range, as it will help for tuning disturbances above 0.1 m. The FLC was tuned accordingly for this as the DE changes much more than E when going over bumps.

Rule Base

The Mamdani approach was used in the FIS, this was decided upon as it uses an intuitive, rule-based structure mimicking human reasoning which is useful for deciding the output force provided at the actuator, at different input values for error and error derivative.

The Mamdani-type inference system uses if–then rules as described above in Table 3.

Outputs

The Defuzzification and output derivation technique used the centroid method. By testing the FLC, this was found to be optimal to find the best crisp outputs from the controller.

The output membership rules can be seen below (Fig. 23).

Testing and trail-and-error was used to find the optimal crisp output values for the controller where ranges from −500 to 500 can be seen above. The tuning was

Table 3 FLC rule base

		DE				
		NB	NS	Z	PS	PB
E	NB	NB	NB	NB	NS	Z
	NS	NB	NB	NS	Z	PS
	Z	NB	NS	Z	PS	PB
	PS	NS	Z	PS	PB	PB
	PB	Z	PS	PB	PB	PB

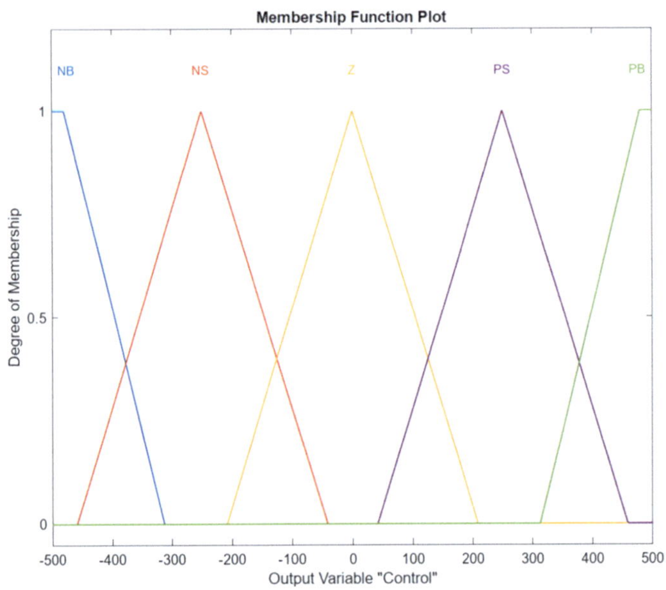

Fig. 23 FLC output membership function plot

respective to the number values effecting the displacement (from variables M_1, M_2, K_1 etc.) (Fig. 24).

Method Validation

The modification of passive variables was experienced to validate the theory first and in testing the model second.

The various simulations that were conducted to assess the properties of the suspension system with modified variables lend credence to the validation of the suspension model.

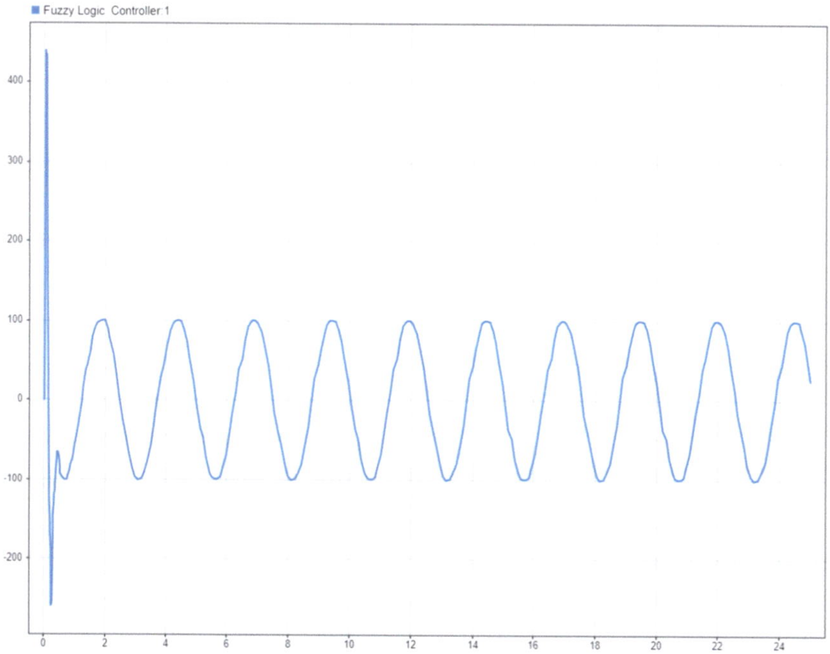

Fig. 24 FLC output graph (0.075 m bump)

The research into automotive suspension, outlined in Sect. 3.2, conveys the importance of the frequency of oscillations (provided by suspension), and frequency of the car greatly effects the performance and influence on the human body.

Before the use of FLC and PID controllers, the models were tested in MATLAB Simulink in order to find the effects of changing spring rate and damping. These were further controlled by the implementation of FLC and PID controllers. Ultimately, reducing the amount of oscillations/vibrations for increased performance and reduction of negative effects on the human body.

Method Mitigations

In real-world applications, suspension failure can cause catastrophic effects to drivers, passengers, and pedestrians. It must be noted that throughout calculations, simulating and testing the model, the goal was to increase vehicle stability and to not cause errors too large that is may harm users or fail entirely. With this in mind a design was thoroughly calculated and derived from methodology and literature to ensure optimal safety and performance. The stability calculations are denoted in Sect. 4.5 and results for this safety improvement can be seen in Sect. 6 onwards.

4.5 Performance Values

It must be noted that these values are in terms of Error Displacement Signal—what the controller is attempting to reduce.

4.5.1 Overshoot Total

The overshoot total (O_{total}) is described as the total displacement the response signal endures in both positive and negative directions (meters).

4.5.2 Peak Height

The peak height *(O_p)* is the value at which is the highest point apart from the settling value in only one direction (meters).

4.5.3 Rise Time

The rise time (R_T) is time at which it takes the response to go from 0 to 100% of the peak height in seconds.

4.5.4 Settling Time

The settling time *(S_T)* is the time it takes for the response to reach the steady state value (SSE) in seconds.

4.5.5 Steady State Error

The steady state error (SSE) is the persistent difference between the desired input and the actual output of a system after all transient effects have decayed in meters.

4.5.6 True Overshoot (M)

The true overshoot represents the total displacement at which the signal is above the unaffected response (meters):

$$TOS = O_{total} - (2 \times SSE)$$

These values have been discussed and derived from Simulink testing. Described further in section (Sect. 5).

4.6 *Method Limitation*

4.6.1 Assumed Variables

Included in this report, is assumed variables where applicable. In order to proceed with the work on integrating an active suspension system with the desired vehicle parameters (BMW M2 G87, discussed in Sect. 4.2), some values have been assumed based on research both internet and reports/datasheets.

This, therefore, could impact and reduce the accuracy of results.

However, the modelling and testing using advanced calculations portray the use of ASS systems to a substantial degree.

4.6.2 Hardware/Software

The computational methodology in modelling active suspension systems requires the need for great computational strength and ability to run successfully.

The simulations of this report were run on MATLAB and Simulink which was conducted on university computers and lower powered laptops when working from home. This effected the simulation times and increased timescales slightly where being on campus wasn't possible.

In some cases, crashes occurred where loss of work was possible if not saved. To mitigate this, work was backed up on google drive and USB drives.

MATLAB/Simulink is free to students, however, for work in future, licenses may need to be bought.

4.6.3 MATLAB/Simulink

Simulations in MATLAB Simulink were derived from the "simulation data inspector" and by use of scopes. This requires the programme to gather data of a response, in this case, a response signal to road disturbance.

Scopes on the other hand portray the signal of data in a single graph.

4.6.4 Accounted for Complications

Time

Effective time management is important to the success of the final report, ensuring that every phase—from initial calculations and model development to simulation and detailed analysis—is planned and executed. The timescales used for this report were derived by the use of a Gantt Chart.

Allocating the necessary timescales for mathematical computations and model refinements allows time to identify and resolve potential issues early on. Thus, laying a solid foundation for the simulation time. This careful scheduling ensures that simulations are conducted with sufficient runtime and data collection periods to capture final results accurately.

Ultimately, the management of time and resources throughout the entire report process not only enhances the reliability and depth of the analysis, but also contributes to the overall quality of the final report.

Research

This report was reinforced and created from research across the internet, physical books, and online journals/articles in the relating field of engineering. This makes for good knowledge improvement, but because of the extensive amount researched, there is a possibility of misleading or incorrect information.

Mitigation of this risk was taken by deriving information from reliable sources, e.g. published and accounted for, reports, papers, and books where possible. This ensured highest amount of accuracy to validate the work, along with self-assessment of surrounding knowledge.

Error Responsibilities

When running simulations, anomalies did occur where unexpected results were portrayed. However, consistently double checking, finding, and solving these issues was consistently done throughout the work to ensure highest accuracy of results and provide validation.

5 Simulation

The parameters of the passive system can be seen below, these were used in all three stages (passive, PID controlled, and FLC controlled active suspension models) on MATLAB Simulink (Table 4).

Table 4 Vehicle parameters for use in passive, PID, and FLC systems

Parameter	Value	Description
M_2	451.25 kg	Sprung quarter car body mass
M_1	51.27 kg	Unsprung quarter car body mass (wheel + suspension)
K_1	90,124.42 N/m	Suspension stiffness
C_1	403.28 N s/m	Suspension damping
K_t	240,000 N/m	Tyre stiffness

The following Simulink models were ran with these parameters in order to derive the following:

- Body displacement.
- Suspension displacement.
- Body acceleration.
- Suspension acceleration.
- Error in displacement.
- Derivative error in displacement.

5.1 *Passive Suspension*

See Fig. 13.

5.2 *PID Active Suspension*

See Fig. 14.

5.3 *FLC Active Suspension*

See Fig. 15.

6 Results, Analysis, and Discussion

With the final resultant of this model's goal being to reduce the error displacement between the body and suspension, ultimately reducing vibrations in the system, the following criteria is used in support of findings (described in more detail in Sect. 4.5):

- Overshoot total (O_{total}).
- Peak height (O_p).
- Rise time (R_T).
- Settling time (S_T).
- Steady state error (SSE).
- True overshoot (TOS).

6.1 Passive Suspension

See Figs. 25 and 26 and Table 5.

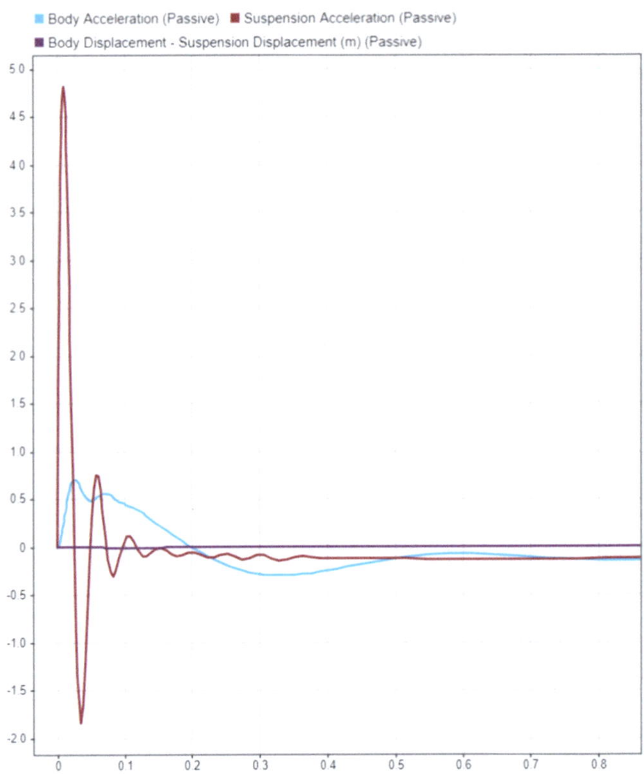

Fig. 25 Passive displacement error, body acceleration, suspension acceleration

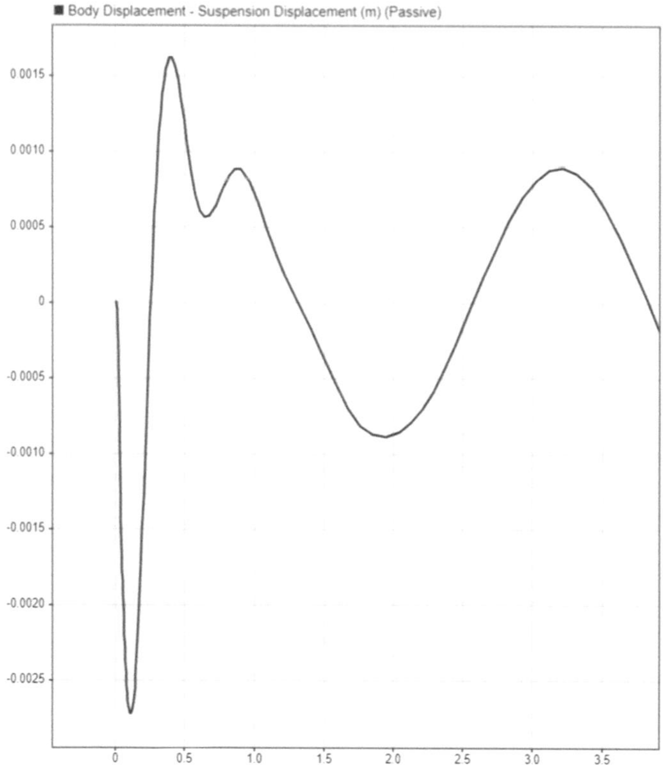

Fig. 26 Passive body/suspension displacement error

Table 5 Passive suspension system results table

Sine amplitude (m)	O_{total} (m)	O_p (m)	R_T (s)	S_T (s)	SSE (m)	True Overshoot (m)
0.02	0.00434	−0.00272	0.9	1.76	0.00089	0.00256
0.04	0.0086	−0.0054	0.9	1.79	0.0018	0.005
0.06	0.0131	−0.0082	0.9	1.84	0.0027	0.0077
0.08	0.0173	−0.0108	0.9	1.91	0.0036	0.0101
0.1	0.0217	−0.0136	0.9	2.05	0.0044	0.0129

6.2 PID Controlled Active Suspension

See Figs. 27 and 28 and Table 6.

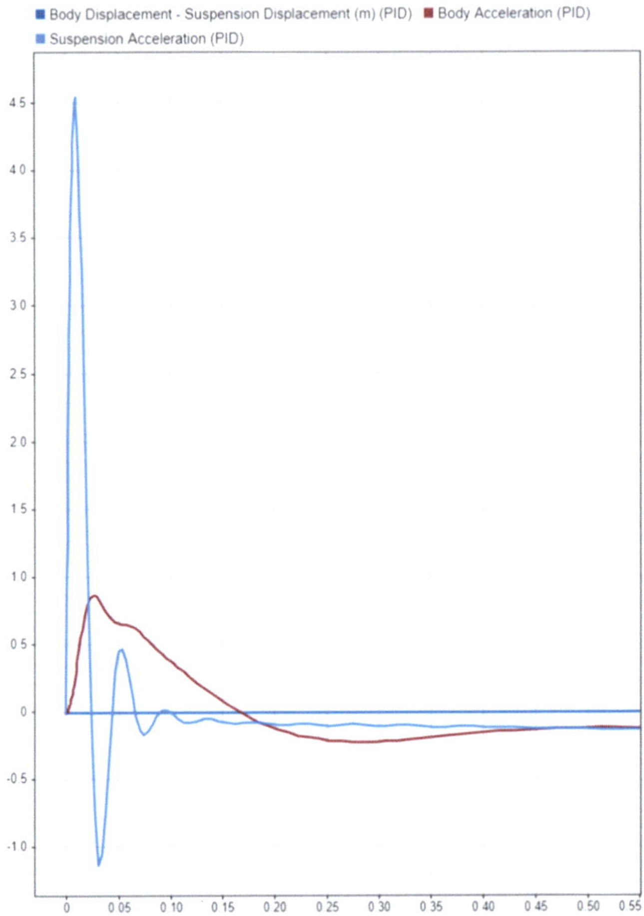

Fig. 27 PID controlled displacement error, body acceleration, suspension acceleration

6.3 FLC Controlled Active Suspension

See Figs. 29 and 30 and Table 7.

These calculated variables will be used to analyse the performance benefits of FLC and PID controlled active suspension in Sect. 6.4.

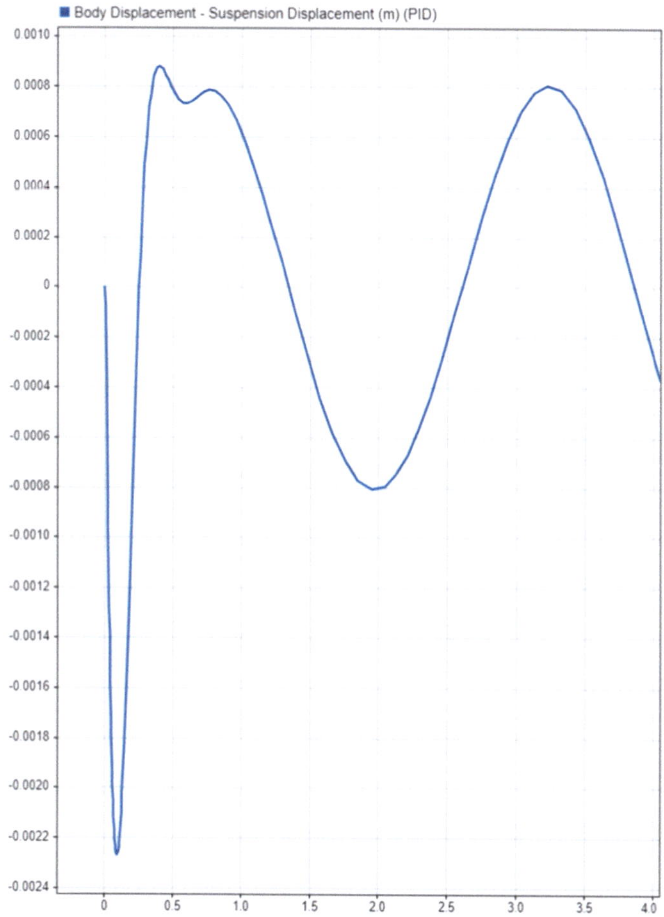

Fig. 28 PID controlled body/suspension displacement error

Table 6 PID controlled suspension system results table

Sine amplitude (m)	O_{total} (m)	O_p (m)	R_T (s)	S_T (s)	SSE (m)	True overshoot (m)
0.02	0.00314	−0.00227	0.77	1.21	0.00081	0.00152
0.04	0.0063	−0.0045	0.76	1.10	0.00161	0.00308
0.06	0.0095	−0.0068	0.79	1.26	0.0024	0.0047
0.08	0.0125	−0.0090	0.78	1.23	0.0032	0.0061
0.1	0.0157	−0.0113	0.79	1.32	0.0040	0.0077

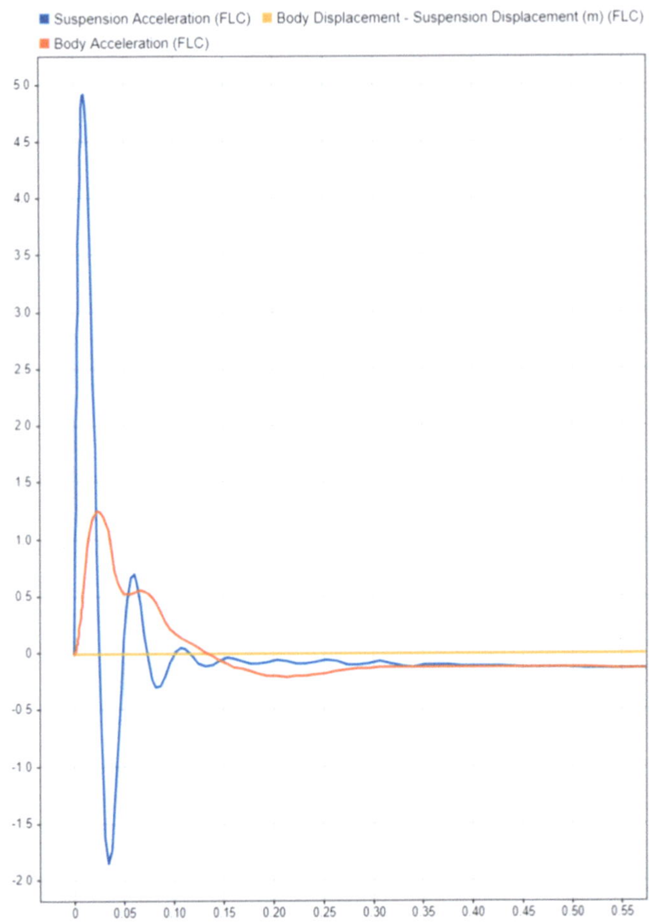

Fig. 29 FLC controlled displacement error, body acceleration, suspension acceleration

6.4 Results Comparison

6.4.1 Comparing Acceleration of Body and Suspension

Suspension Acceleration

The graph illustrates the acceleration response over time from a 0.02 m amplitude bump. The FLC controlled suspension experiences the most aggressive initial response however settles quicker than the passive system, this suggests the disturbance is absorbed much quicker and reduces vibrations effectively, ultimately,

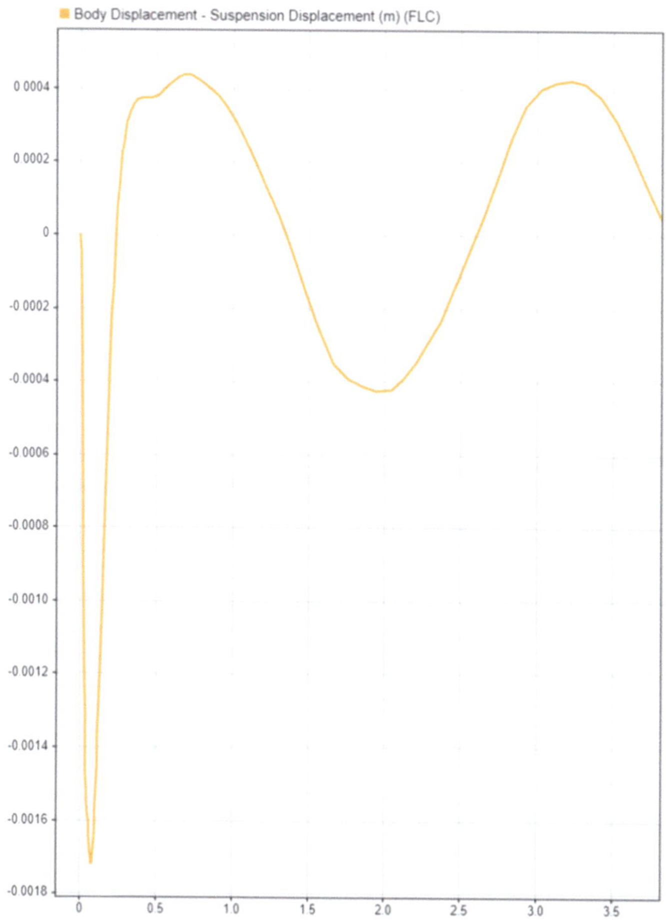

Fig. 30 FLC controlled body/suspension displacement error

Table 7 FLC controlled suspension system results table

Sine amplitude (m)	O_{total} (m)	O_p (m)	R_T (s)	S_T (s)	SSE (m)	True overshoot (m)
0.02	0.00171	−0.00171	0.68	0.81	0.00042	0.00087
0.04	0.00376	−0.00376	0.68	0.85	0.00091	0.00194
0.06	0.00595	−0.00595	0.71	0.84	0.0014	0.00315
0.08	0.010	−0.0091	0.73	0.84	0.0020	0.006
0.1	0.0131	−0.0120	0.74	0.84	0.0025	0.0081

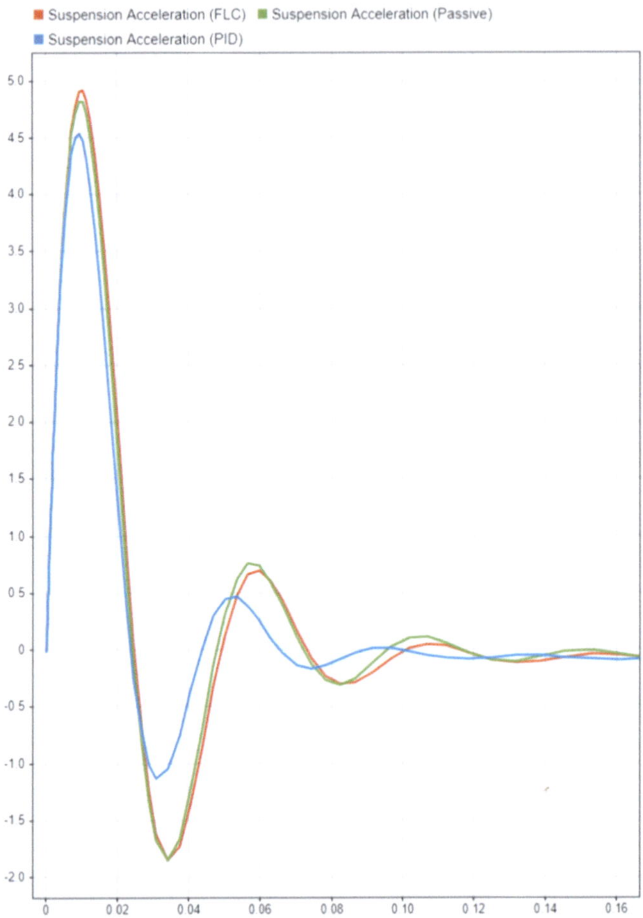

Fig. 31 Suspension acceleration. *Note* Taken at 0.02 m "Bump" amplitude

providing increased stability. The PID also shows excellent settling time, however does not absorb the impact as aggressively as the FLC system (Fig. 31).

Body Acceleration

Similarly to suspension acceleration, the FLC suspension experiences the most aggressive acceleration initially but then settles the quickest, reducing excessive vibrations and improving stability of the body significantly compared to the passive and PID system. The PID controlled system does experience improved settling time

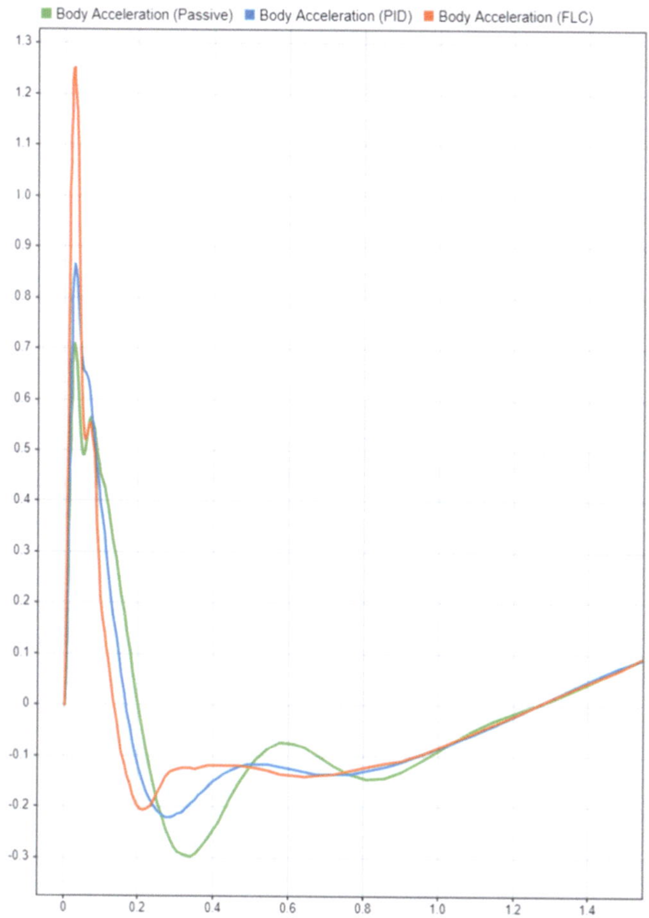

Fig. 32 Body acceleration

compared to the passive system but does not match the FLC system's performance fully (Fig. 32).

Body/Suspension Error Comparison

All three systems experience an initial drop in displacement, this result is valid due to the suspension's displacement increases more than the body upon absorbing the impact. Following this, the error begins to rise again and settle, the FLC system settles the quickest, followed by the PID system, and then the passive system (Fig. 33).

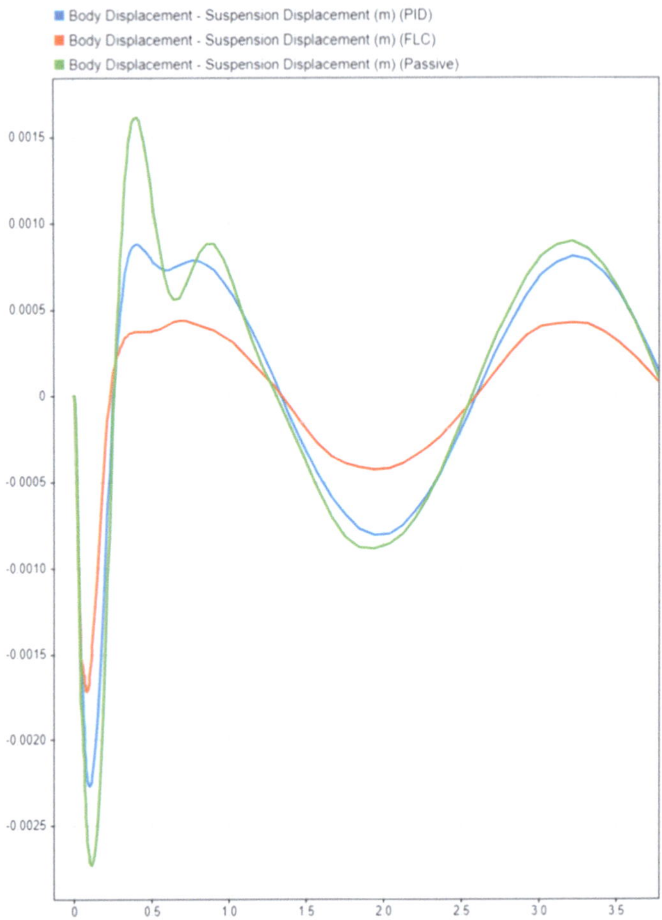

Fig. 33 Body/suspension error comparison

Another interesting factor is the reduction of the SSE on the FLC suspension system. The steady state error is majorly decreased in comparison with the other two configurations, respectively, highlighting increased stability enhanced driver-feel throughout the rest of the vehicle's travel.

Using the derived values in Sects. 6.1, 6.2, 6.3, the results/comparison for body/suspension error are plotted below.

Figure 34 denotes the true overshoots the three configurations have for body/displacement error. The FLC system exhibits the most reduced overshoot, whilst followed up closely by the PID, and finally the passive system. The highest difference in overshoot is at 0.1 m road disturbance where the passive system is 0.003 m higher than the FLC configuration. This highlights the significantly reduced overshoots of

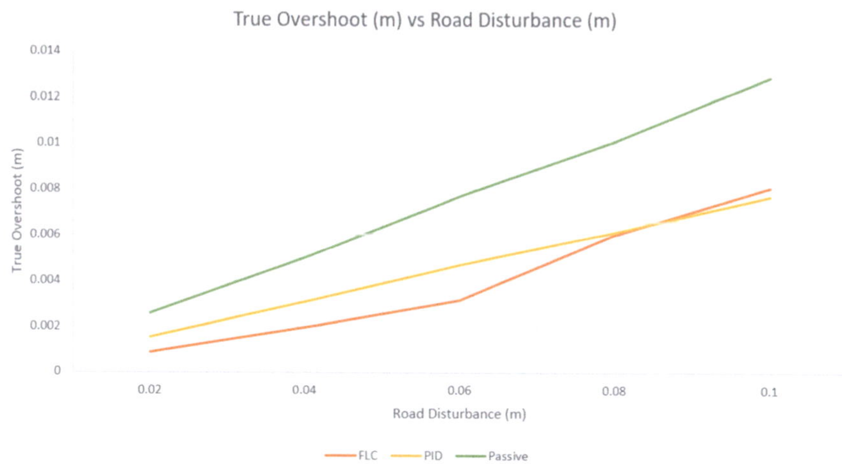

Fig. 34 True overshoot versus road disturbance

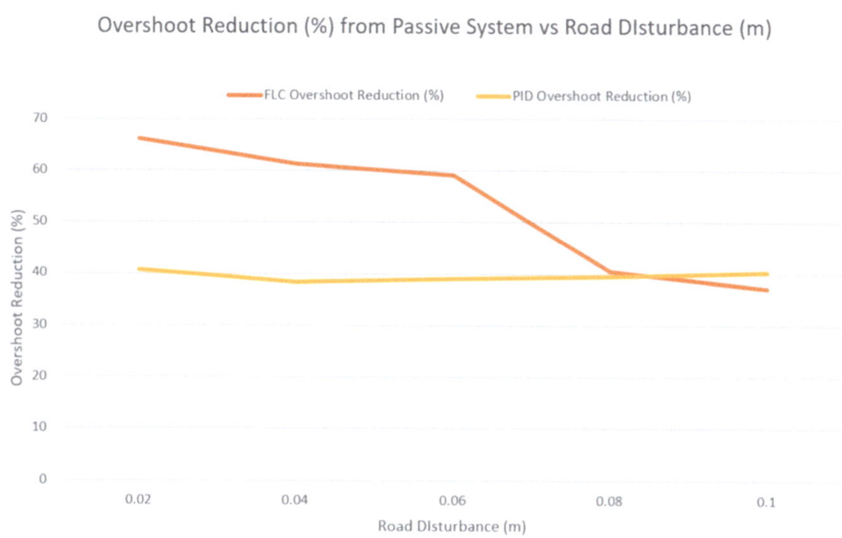

Fig. 35 Overshoot reduction of FLC and PID from passive system

FLC and PID systems compared to the passive suspension, allowing for more road feel to be experienced by the driver.

Furthermore, the benefits of integrating the PID and FLC (Fig. 35) controllers into suspension are highlighted, with FLC reducing overshoot of error by 65–60% for bumps between 0.02–0.06 m and PID reducing overshoot of error by 41–39–41% through 0.02–0.06–0.1 m.

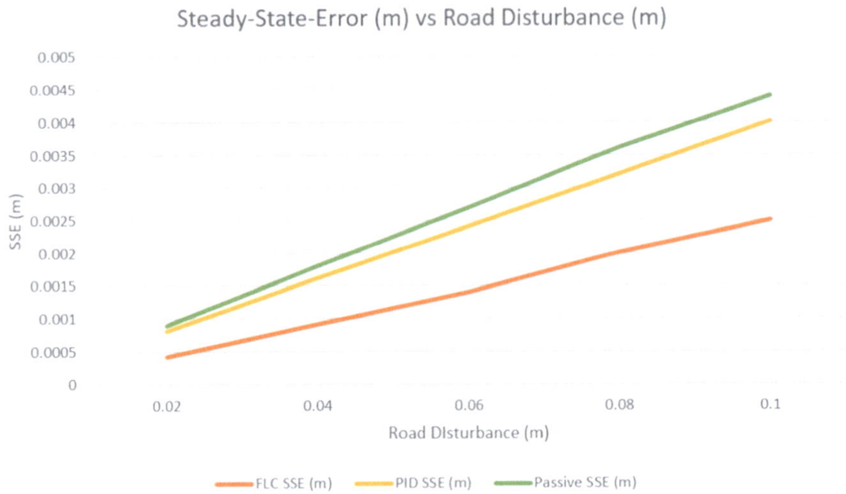

Fig. 36 Steady state error versus road disturbance

The FLC system is seen to reduce the SSE of the displacement error significantly by 0.002 m at 0.1 m bump, whilst PID and Passive configurations produce similar results- PID SSE is 0.0004 m less than passive system at 0.1 m bump. This suggests the vehicle body will resemble, more effectively, the road feel. Improving handling and feedback (Fig. 36).

As shown in Fig. 37, the settling time of displacement error is significantly reduced with the FLC system in comparison to the passive system where reduction is seen to increase with bump magnitude—from 55 to 59%. Also, PID brings effective reduction from 30 to 35%, although is less reliable. Note, % at 0.04 could be an anomaly in gathering data from Simulink (Fig. 38).

6.5 *Analysis Discussion*

6.5.1 Data Analysis Discussion

Through extensive testing and analysing the results from passive suspension, PID, and FLC controlled active suspension, it is clear that by integrating these active suspension systems, they dynamically reduce unwanted errors in terms of the performance vehicle—BMW M2 G87—desired criterion.

The FLC excelled at reducing settling time of displacement error, meaning, the vehicle's body had less vibrations and oscillations in-line with the suspension whilst also reducing the time for the body to stabilise. Ultimately portraying faster feedback to the driver when travelling over bumps in the range of 0.02–0.1 m, thus ensuring rapid stabilization and improved ride comfort.

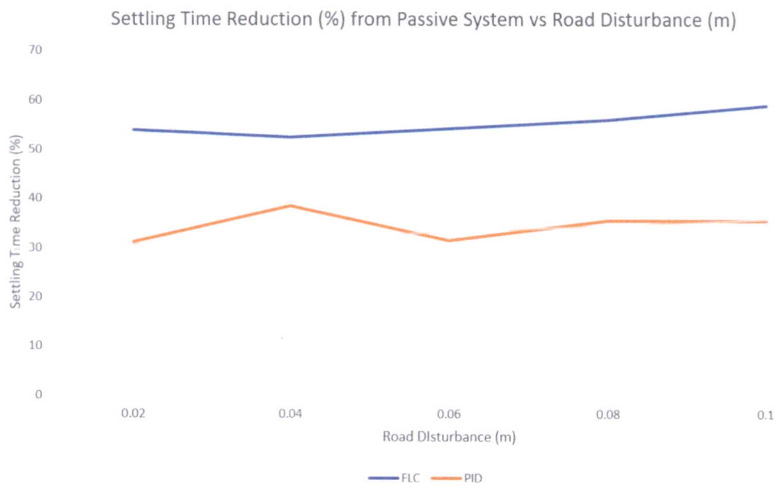

Fig. 37 Settling time (s) versus road disturbance (m)

Fig. 38 Settling time reduction (%) from passive system versus road disturbance (m)

Whilst the FLC provided superior overshoot reduction for smaller disturbances (0.02–0.06 m), the PID controller became marginally more effective at disturbances between 0.06 and 0.1 m, indicating that FLC excels in finer road disturbances whereas PID control is slightly better suited to larger perturbations.

Evaluating body acceleration responses revealed that the FLC initially exhibited higher peak acceleration than the PID and passive systems, but rapidly converged to stability, enhancing overall comfort. Similarly, the body/suspension displacement errors confirmed the FLC's advantage in reducing oscillations, though the PID also significantly improved upon the passive baseline.

Overall, whilst both active suspensions provided clear benefits, the FLC controller demonstrated superior overall performance in ride stability and comfort in comparison to the PID and passive configurations.

6.5.2 Results Justification Discussion

Fundamentally, these results found through analysis portray the desired responses of performance cars. More specifically, by relating back to Sect. 2.1, the goal was to reduce error overshoot by 25% and error settling time by 40%. The FLC system achieved and improved on this original goal by 40 and 14% respectively. Whereas, the PID system succeeded at reducing the overshoot, but narrowly missed the requirement of reducing the settling time with a difference of 5% under target.

7 Conclusion

With the objective of this project being to research, model, and test a fuzzy-logic-controlled active suspension system, it has been shown that requirements have been met and exceeded with the FLC system proving superior over the PID.

Detailed calculations and modelling was performed throughout, upon which derived the conclusive results of the FLC and PID controlled active suspension models, where it can be seen FLC's effectively reduced overshoots and settling times, leading to better performance with regards to stability, road-feedback, handling, comfort, and safety.

Extensive research into the related engineering fields, helped aid understanding of the mathematical methods involved with designing an active suspension model. Alongside this, frequent testing/simulation was carried out, not only to derive results that matched existing values gathered from reliable literature, but also to provide validation. And so, reliably creating a collection of work that can be used as insight or for further development into the area of study.

This undertaking has therefore, provided sufficient evidence to support the point that using controlled strategies for suspension is far more dominant than passive components alone, providing the opportunity for further work to be built upon the studies of this project to increase the safety, comfort, and performance of the automotive industry.

8 Further Work

With the parameters and variables sourced from an existing car, the BMW M2 G87, a fairly accurate model was derived, with limitations to assumptions however (discussed in Sect. 4.6.1). This creates opportunities to cooperate and work alongside car manufacturers to develop active suspension further within their vehicles.

On the other hand, research was accumulated for use of active suspension in autonomous vehicles (Sect. 3.4.2) to develop more safe and comfortable rides. Ultimately, increasing the ever-growing likelihood of fully implementing autonomous vehicles into daily life.

Furthermore, with respect to current results within this project, physical testing of this system on a BMW M2 G87 would help reinforce validation of the system, upon which further tuning or combination with other electronics could help improve the efficiency of this design.

References

1. ABC-Specialist (2019) MErcedes Benz ABC active body control. https://www.readkong.com/page/mercedes-benz-abc-active-body-control-information-sheet-2932666. Accessed 26 Mar 2025
2. Aized T, Ahmad M, Jamal MH, Mahmood A, Ubaid ur Rehman S, Srai JS (2020) Automotive leaf spring design and manufacturing process improvement using failure mode and effects analysis (FMEA). s.l.: Int J Eng Bus Manag
3. Riduan AFM, Tamaldin N, Sudrajat A, Ahmad F (2018) Review on active suspension system. s.l.: EDP Sciences
4. Soliman AMA, Kaldas M (2019) Semi-active suspension systems from research to mass-market—a review. s.l.: J Low Freq Noise Vib Active Control 2021 40(2):1005–1023
5. Florin A, Ioan-Cozmin MR, Liliana P. (2013) Pasive suspension modeling using MATLAB, quarter car model, input signal step type. s.l.: New technologies and products in machine manufacturing technologies, pp 258–263
6. Audi-Technology-Portal (2012) Adaptive air suspension. https://www.audi-technology-portal.de/en/chassis/suspension-control-systems/adaptive-air-suspension_en. Accessed 25 Mar 2025
7. Auto Express, A. I. (2025) What is air suspension? How it works and the pros and cons. https://www.autoexpress.co.uk/tips-advice/365918/what-air-suspension-how-it-works-and-pros-and-cons. Accessed 26 Mar 2025
8. Auto-Passionate (2023) Soft vs hard dampening coilovers: which is best? https://autopassionate.com/soft-vs-hard-dampening-coilovers/. Accessed 27 Mar 2025
9. Aydin Azizi HM (2021) Applied mechatronics: designing a sliding mode controller for active suspension system. s.l.: s.n.
10. Bimmerpost (2023) G87 M2 weight recution list—ongoing. https://g87.bimmerpost.com/forums/showthread.php?t=2009072. Accessed 26 Feb 2025
11. CarAdvise (2024) Active suspension. https://caradvise.com/encyclopedia/active-suspension/
12. Ma C-T, Gu Z-H (2019) Review of GaN HEMT applications in power converters over 500 W. Electronics 8(12), Basel, Switzerland: MDPI
13. Chronicle F (2023) What is active suspension? https://f1chronicle.com/what-is-active-suspension/#:~:text=Active%20suspension%20brought%20about%20a%20revolution%20in%20the,However%2C%20the%20technology%20was%20not%20without%20its%20challenges. Accessed 18 Oct 2024

14. Cytrynski S, Neerpasch U, Bellmann R et al (2018) The active suspension of the New Mercedes-Benz GLE. ATZ Worldwide 120:42–45
15. Essex Parts Services Inc. (2021) AP racing by Essex Radi-CAL ENP Competition Brake Kit (Front 9660/372mm)—BMW M2 (G87) M3 (G80) & M4 (G82) Incl. Competition. https://www.essexparts.com/ap-racing-by-essex-radi-cal-competition-brake-kit-front-9660372mm-g80-g826391. Accessed 26 Feb 2025
16. de Souza Lima FM, Venceslau GM (2002) A new Hooke's law experiment. s.l.: Phys Teacher 40(1):35–36
17. Hammond T (2016) What is an electronic suspension and how does it work? https://www.strutmasters.com/a/blog/how-does-electronic-suspension-work?zCountry=GB. Accessed 03 Jan 2024
18. Hyatt W-V (2023) Active suspension vs passive suspension—what's the difference? https://vehq.com/active-suspension-vs-passive-suspension/#:~:text=The%20first%20of%20the%20two%20broad%20types%20of,terrain%2C%20preventing%20unwanted%20excess%20motion%20in%20the%20vehicle. Accessed 15 Jan 2025
19. IDC Spring (2025) What are springs & how do springs work? https://idcspring.com/resources/articles/what-are-springs-how-do-springs-work/. Accessed 27 Mar 2025
20. IEC (2010) IEC 61508-1. https://webstore.iec.ch/en/publication/5515. Accessed 31 Mar 2025
21. ISO.org (n.d.) ISO 26262 ROad Vehicles FUnctional Safety. https://www.iso.org/publication/PUB200262.html. Accessed 31 Mar 2025
22. Spring K (2019) The history of automotive springs. https://katyspring.com/2019/09/01/automotive-springs/. Accessed 19 Nov 2024
23. Love A (2013) A treatise on the mathematical theory of elasticity; Reprint, revised. Cambridge University Press, Cambridge
24. Salem MMM, Ayman AA (2010) Fuzzy control of a quarter-car suspension system. s.l. Int J Aerosp Mech Eng
25. MathsWorks (n.d.) Fuzzy logic toolbox. https://uk.mathworks.com/products/fuzzy-logic.html. Accessed 26 Mar 2025
26. MeganRacing (n.d.) BMW M2 (G87) 2023+ Euro II series coilover kit—MR-CDK-BG80. https://meganracing.com/suspension/mr-cdk-bg80-ii?. Accessed 26 Feb 2025
27. Yu M, Evangelou SA, Dini D (2024) Advances in active suspension systems for road vehicles. s.l. 160–177
28. Motor-Car-History (2025) Car Unsprung weight (mass) explained. https://motor-car.net/innovation/suspension/item/14734-unsprung-weight-mass-14734#:~:text=In%20a%20ground%20vehicle%20with%20a%20suspension%2C%20the,to%20them%2C%20rather%20than%20supported%20by%20the%20suspension. Accessed 16 Jan 2025
29. Noble J (2022) Russell suggests active suspension as solution to F1 porpoising issue. https://www.motorsport.com/f1/news/russell-active-suspension-porpoising-solution/8572024/#:~:text=Mercedes%20driver%20George%20Russell%20thinks%20a%20return%20of,porpoising%20problem%20of%20the%20new%20ground%20effect%20cars. Accessed 18 Feb 2025
30. Ghazly NM, Moaaz AO (2014) The future development and analysis of vehicle active. IOSR J Mech Civ Eng (IOSR-JMCE) 11(5):19–25
31. Prasad (n.d.) Understanding the damping coefficient in engineering. https://www.structuralguide.com/damping-coefficient/. Accessed 16 Jan 2025
32. Rao S (2011) Mechanical vibrations, 5th edn. Upper Saddle River, Pearson
33. Scarborough C (2021) F1's game changer—lifting the lid on the dominant Williams FW14B. https://motorsport.tech/formula-1/f1s-tech-game-changer-lifting-the-lid-on-the-dominant-williams-fw14b. Accessed 12 Apr 2025
34. Savaresi SM, Poussot-Vassal C, Spelta C, Sename O, Dugard L (2010) Semi-active suspension control design for vehicles. s.l.:Elsevier
35. Sharma A (2021) Newton's generalized form of second law gives F = ma. India IOSR J Appl Phys (IOSR-JAP)

36. Han S-Y, Liang T (2022) Reinforcement-learning-based vibration control for a vehicle semi-active suspension system via the PPO approach. China Appl Sci 12:3078
37. Smarter S (n.d.) Damping. https://www.studysmarter.co.uk/explanations/engineering/engineering-mathematics/damping/#:~:text=How%20is%20the%20damping%20coefficient,k%27%20is%20the%20spring%20constant. Accessed 16 Jan 2025
38. SpringRates.com (n.d.) Ohlins road & track coilovers for 2023+ BMW M2 RWD (G87). https://www.springrates.com/products/ohlins-road-track-coilovers-for-2023-bmw-m2-rwd-g87?. Accessed 26 Feb 2025
39. Suspension Secrets (2017) Dampers. https://suspensionsecrets.co.uk/damper/. Accessed 26 Feb 2025
40. Suspension Secrets (2017) Dampers set up—how to adjust and tune suspension secrets. https://suspensionsecrets.co.uk/dampers-set-up/. Accessed 26 Feb 2025
41. TFL Gov UK (n.d.) Driving-ultra low emmsion zone-cars. https://tfl.gov.uk/modes/driving/ultra-low-emission-zone/cars. Accessed 31 Mar 2025
42. UNECE (n.d.) WP.29. https://unece.org/wp29-introduction. Accessed 27 Mar 2025
43. Wang Z, Liu C, Zheng X, Zhao L, Qiu Y (2024) Advancements in semi-active automotive suspension systems with magnetorheological dampers: a review. China Appl Sci. EISSN 2076-3417, MDPI

Design an Active Suspension System Utilizing Intelligent Control Techniques

Prashant Patel

Abstract This study investigates the implementation of advanced control strategies for active suspension systems in automobiles. Given the increasing demand for improved ride comfort, safety, and fuel efficiency, the automotive industry has shifted toward more sophisticated suspension technologies. While fully active suspension systems offer superior performance, they often come with higher costs and energy consumption. To address these challenges, this research proposes a control strategy using MATLAB/Simulink for a quarter-car vehicle model. By comparing various control algorithms, including proportional integrative derivative (PID), fuzzy logic controller (FLC), and/or model reference controller (MRC), the study aims to identify the most effective approach for minimizing road vibrations and enhancing overall ride quality. Simulation results demonstrate a significant reduction in road vibrations (approximately 30%) when employing an active suspension system compared to a passive one. The findings of this study contribute to the ongoing development of more efficient and cost-effective active suspension systems, ultimately improving the driving experience for vehicle occupants. Future research directions include exploring alternative control algorithms, refining model complexity, and considering real-world factors such as tire dynamics and nonlinear suspension components.

List of Abbreviations

ASS	Active Suspension System
PID	Proportional Integrate Derivative (Controller)
FLC	Fuzzy Logic Controller
MRC	Model Reference Control
MPC	Model Predictive Control
LQR	Linear Quadratic Regulator
MRAC	Model Reference Adaptive Control

P. Patel (✉)
School of Engineering, Computing and Mathematics, Oxford Brookes University, Headington Campus, Oxford, UK
e-mail: 19282224@brookes.ac.uk; prashantpatel2597@gmail.com

A. Azizi (ed.), *Autonomous Robotics and Intelligent Systems*, Emerging Trends in Mechatronics, https://doi.org/10.1007/978-981-95-5867-4_8

SMC	Sliding Mode Control
3DOF	3 Degree of Freedom
FFT	Fast Fourier Transform Analysis
RMS	Root Mean Square
PSD	Power Spectral Density
ISO	International Organization for Standardization
LQR	Linear Quadratic Regulator

Highlights

- Development of a mathematical model of a 2DOF quarter-car model in a MATLAB-Simulink environment.
- Exploration of various control strategies to regulate the vehicle model's behavior with given vehicle parameters and road input.
- Comparison of different controllers, analyzing their advantages and disadvantages, and suitability for the target vehicle system.
- Evaluation of the performance of various controllers to determine the most effective strategy for a given scenario and compare it to a conventional system.

1 Introduction

1.1 Background

The automotive industry has witnessed a significant shift toward advanced suspension systems to cater to the growing demand for improved ride comfort, safety, and fuel efficiency [2, 23]. Conventional passive suspension systems have limitations in effectively absorbing road vibrations across a wide range of frequencies [1]. To overcome these shortcomings, active suspension systems (ASS) have emerged as a promising solution.

Active suspension systems (ASS), are widely adopted in luxury vehicles and motorsports [10], however, they face a challenge by factors such as higher costs, increased power consumption, and concerns about reliability and safety.

The study aims to compare and analyze control systems to optimize dynamic performance [2], and overall vehicle safety in ASS. Different control strategies, including PID controllers, MRC controllers, and fuzzy logic controllers, have distinct advantages and disadvantages. The research seeks to identify the most suitable approach for achieving optimal performance in ASS.

Furthermore, the study will compare the body displacements and vibration absorption efficiency for active and passive suspension systems under different road conditions [20]. This analysis will provide valuable insights into the effectiveness of ASS in absorbing road vibrations and improving ride comfort.

1.1.1 Statement of Problem

The challenge lies in the absorption of road vibrations across a wide range of frequencies, which passive systems are unable to effectively address. The nature of road vibration spans various frequencies, making it essential to implement a control system that can absorb these forces and prevent road displacement from being transmitted to vehicle occupants.

1.1.2 Purpose of Study

This study explores potential control algorithms for active suspension systems, considering available tools and information. We will evaluate system performance and compare control strategies to identify the most efficient approach.

1.1.3 Objectives

This research aims to develop multiple controllers for an ASS and perform a comparative analysis among them. The main objectives are as follows:

1. To review the existing research made in the development of ASS
2. To develop a quarter-car mathematical model for the ASS in MATLAB-Simulink environment
3. To design the controller and implement the control strategy in the quarter-car model
4. To determine the difference in absorption of road input force and displacement between the passive and active system
5. To perform a comparative analysis between PID, FLC, and/or MRC controller results, among each other and with a passive suspension system

1.2 Summary of Literature Review

The automotive industry increasingly demands efficient suspension systems [23]. Active suspension systems, offering real-time control, are gaining popularity despite challenges in cost and energy consumption. Manufacturers are working to develop cost-effective and reliable systems, driving continued research into control strategies for these advanced systems [23].

1.2.1 Automotive Suspension Systems

The suspension system in a vehicle consists of springs and shock absorbers. Its main purpose is to not only support the vehicle's sprung mass but also to improve the vehicle's ride quality and handling by isolating its sprung mass from road vibrations (disturbances) [26]. A well-designed suspension system can enhance stability and safety, improve fuel economy, and adapt to various driving conditions [28]. Figure 1 shows the typical active suspension system.

The differences between a fully active suspension system (ASS) and passive suspension systems are significant. Firstly, a fully active suspension system includes a spring, an actuator, and a controller to control the forces transmitted between the sprung and unsprung mass [23]. Secondly, it is capable of generating higher peak forces and responding more quickly to disturbances. Lastly, it can control a wider range of disturbance frequencies [19]. However, it requires higher energy consumption due to the operation of actuators, controllers, and hydraulic systems [9].

The differences in operating force, frequency range, and energy consumption of the various suspension systems are classified in Fig. 2.

Passive suspension systems struggle to balance comfort, handling, and space. Ride comfort requires reducing vertical acceleration, while good handling depends on cornering and braking abilities. This conflict has led to research into active and semi-active systems [10]. The conflicting demands on the vehicle suspension system are illustrated in Fig. 3.

The dynamic tire load increases beyond a certain point, which reduces passenger comfort due to the curved damping curves caused by a continuous rise in suspension stiffness. Similarly, occupant comfort gradually declines with continuous increases in overall suspension damping [21]. The limitations of passive systems are due to fixed parameters such as spring stiffness and damping [14].

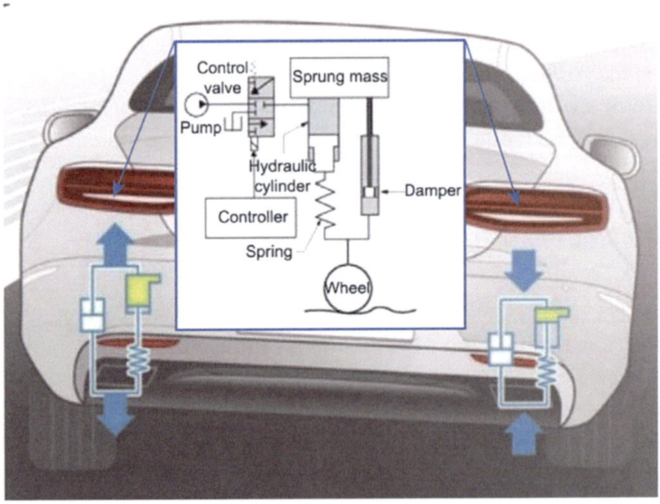

Fig. 1 Mercedes-Benz ABC, a type of electrohydraulic suspension [28]

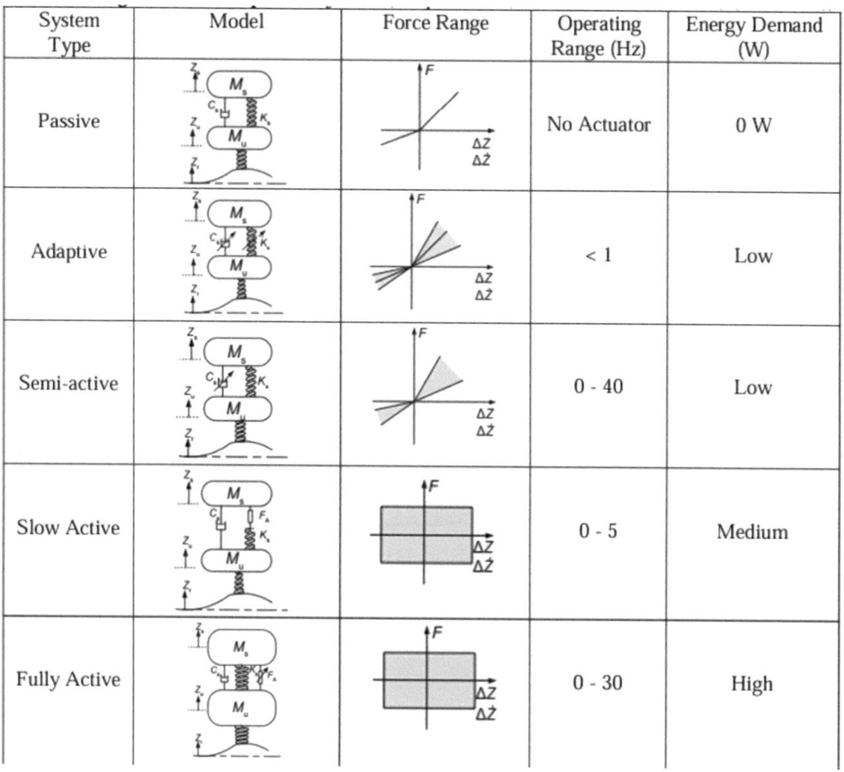

System Type	Model	Force Range	Operating Range (Hz)	Energy Demand (W)
Passive			No Actuator	0 W
Adaptive			< 1	Low
Semi-active			0 - 40	Low
Slow Active			0 - 5	Medium
Fully Active			0 - 30	High

Fig. 2 Classification of suspension systems [7]

The selection between passive and active suspension depends on the vehicle's needs and desired balance between performance and energy efficiency. The decision should consider the benefits and drawbacks, as well as the vehicle's specific requirements and intended use.

1.2.2 Control Strategies for Active Suspension System (ASS)

Active suspension systems (ASS) utilize control strategies to improve ride comfort, handling stability, energy efficiency, and safety. The selection of an appropriate control strategy is essential for achieving an optimal balance between ride comfort and vehicle handling [6]. Additionally, various control theories such as linear control, nonlinear control, and adaptive control have been applied to ASS and are thoroughly discussed in academic literature.

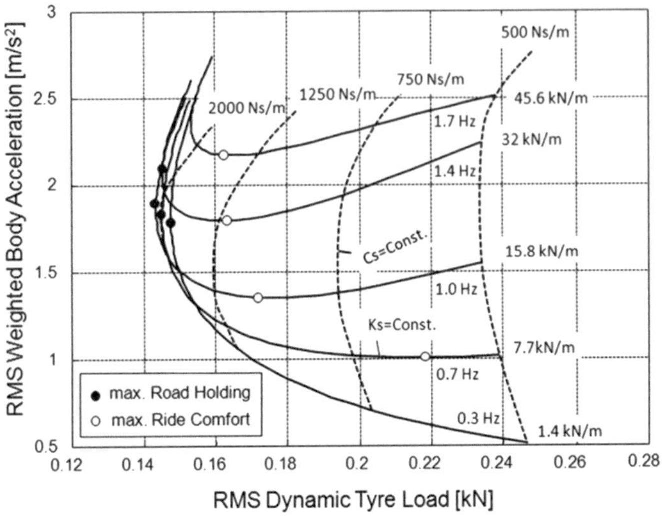

Fig. 3 The conflict between ride comfort and road holding [21]. The x-axis represents the root-mean-square (RMS) value of the dynamic tire load to the static wheel load. The y-axis represents the RMS value of the weighted body vertical acceleration

- **Classical Control Theory**

Classical control theory, based on the Laplace transform and frequency domain analysis [18], provides a foundation for understanding and designing control systems. Proportional-integral-derivative (PID) controllers, as discussed by Nagarkar et al. in 2018 [12], are commonly used in automotive active suspension systems. However, they struggle with nonlinearities and uncertainties [14], and tuning their parameters can be challenging, especially for complex systems.

- **Modern Control Theory**

Modern control theory represents system dynamics using matrices and vectors. It includes optimal control techniques such as linear quadratic regulator (LQR) and model predictive control (MPC) to minimize performance index subject to constraints [23]. LQR minimizes a quadratic cost function [17], while MPC predicts future system behavior. Additionally, it includes robust control methods like H-infinity control and μ-synthesis to design controllers that handles uncertainties and disturbances [16]. While modern control theory offers a powerful framework, it can be computationally intensive.

- **Adaptive Control**

Adaptive control techniques adjusts the controller parameters online to compensate for changes in the system dynamics in ASS. Model Reference Adaptive Control (MRAC) compares the system's actual output to a desired reference model and adjusts the controller parameters to minimize the error, improving robustness and performance. However, it may require additional computational resources and careful tuning.

- **Nonlinear Control**

Nonlinear control techniques, such as Sliding Mode Control (SMC) and Fuzzy Logic Control (FLC), addresses the nonlinearities in ASS. SMC utilizes discontinuous control inputs to achieve robust and fast control [4], while FLC handles uncertainty and nonlinearities by mapping inputs to outputs using fuzzy sets and rules [15]. Additionally, FLC can establish a controller without needing a mathematical model of the process to be controlled [11].

- **Hybrid Control**

To improve adaptive control, traditional methods have been combined with intelligent control techniques such as fuzzy logic, neural networks, genetic algorithms, and emerging data-driven control systems [28]. Neuro-fuzzy control integrates neural networks and fuzzy logic to create adaptive FLCs capable of learning from data and adjusting to changing conditions. Fuzzy-PID control uses fuzzy logic to fine-tune the parameters of a PID controller, improving its robustness and performance, and is proved best for roll handling [13].

1.2.3 Research Gaps

Despite advancements in control algorithms, sensor technology, and actuator design, research gaps persist in the optimization and real-world implementation of active suspension systems. This research aims to address these gaps by exploring the potential of active suspensions and evaluating different control algorithms. By examining trade-offs between desired vehicle dynamics, specific vehicle characteristics, and hardware limitations, this research lays the groundwork for future work on refining controller design and promoting wider adoption of affordable, reliable, and efficient active suspension systems.

1.3 Originality and Contribution

The findings of this research lie in the careful evaluation and selection of control theory for the design of controllers for an active suspension system. By considering the specific requirements and constraints of the application, the research aims to identify the most suitable control theory and compare them, whether it be PID controllers, fuzzy logic control, modern control techniques, or a combination of these approaches. The research seeks to contribute to the field by providing a deeper understanding of the trade-offs between performance, complexity, and computational requirements, ultimately aiming to improve the overall effectiveness and efficiency of the active suspension system through the selection and implementation of appropriate control theory.

1.4 The Approach

To accomplish the research's objectives, this study will commence by outlining the issue and establishing specific research goals. A thorough literature review will then be undertaken to identify existing research on active suspension systems, control algorithms, and their performance evaluation. Through the analysis of these studies, potential research gaps will be pinpointed, thereby shaping the research's scope and methodology. Subsequently, a comprehensive mathematical model will be formulated and simulated using MATLAB-Simulink software to derive pertinent insights. Ultimately, the results will be evaluated to assess the effectiveness of different control algorithms in optimizing active suspension system performance.

2 Methodology

This research is simulation-based, conducted entirely within a software environment. A mathematical model of a vehicle, such as a quarter-car, half-car, or full-car model, will be created using specialized software like MATLAB (Simulink or Simscape) or Mathematica. Predefined parameters will be used to analyze ride or handling characteristics with various control algorithms, including PID, FLC, or MRC. Simulation results will provide insights into vehicle dynamics and help determine the most suitable control algorithm for different scenarios. Detailed steps of the process are illustrated in the Fig. 4.

The research consists of three major stages:

1. First stage—Model Development: Develop a mathematical model of an active suspension system for a specific dynamic condition, utilizing the control strategy discussed in Sect. 1.1.

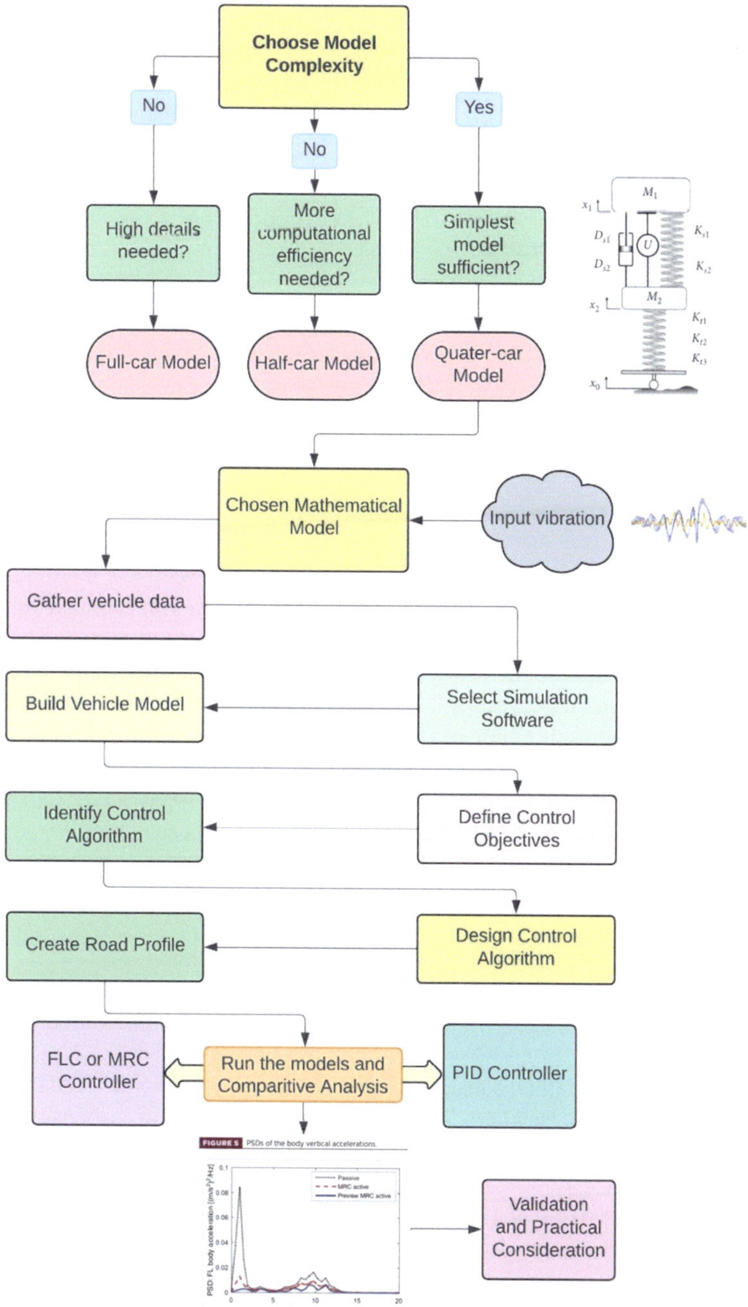

Fig. 4 Illustration of the steps to be carried out for the research

2. Second stage—Controller Design: Design control algorithms, such as PID, FLC, or MRC, to mitigate the unpredictable effects of road conditions on the vehicle.
3. Last stage—Performance Comparison, the performance of the designed FLC/ MRC controller will be compared with a PID controller to conclude the performance of these controllers for a specific dynamic condition.

2.1 Background Theory

Before we begin discussing the methods used in this research, it is important to understand the basic theory behind these methods. This includes understanding the concepts of suspension systems, the performance index for evaluating results, and the control strategies.

2.1.1 Vehicle Dynamics and Suspension Model

The field of vehicle dynamics encompasses the study of a vehicle's motion and its interaction with the road. Suspension systems are pivotal in influencing a vehicle's ride comfort, handling, and safety. Various models, such as the quarter car, half car, and full car suspension models, are utilized to comprehend suspension dynamics. For instance, the quarter-car model, which incorporates two masses connected by spring-damper systems, is employed to analyze vertical displacement resulting from road disturbances [3, 24].

According to Newton's 2nd law, which states that the force acting on sprung/ unsprung mass is equal to the product of its mass (m) and acceleration (a) as given in (1).

$$F = m * a \tag{1}$$

There are four type of forces acting in the quarter-car model shown in Fig. 5 below, namely inertial force, spring force, damping force, and the external actuator force.

The governing differential equations of motion for sprung mass (M_1) and unsprung mass (M_2) according to Newton's 2nd law for a quarter-car active suspension model are as follows [5]:

$$M_1\ddot{x}_1(t) + D[\dot{x}_1(t) - \dot{x}_2(t)] + k1[x_1(t) - x_2(t)] = u \tag{2}$$

$$M_2\ddot{x}_2(t) - D[\dot{x}_1(t) - \dot{x}_2(t)] + k_1[x_2(t) - x_1(t)] + k_2[x_2(t) - x_0(t)] = -u \tag{3}$$

The values of the above parameters are given in Table 1.

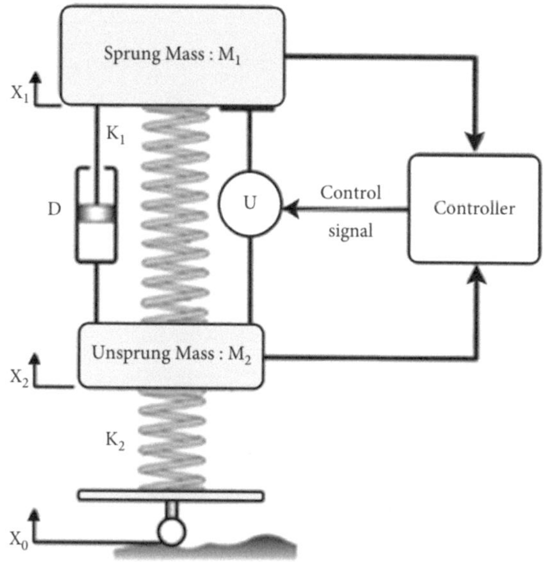

Fig. 5 Two-DOF quarter-car model equipped with the active suspension system [3]

Table 1 Important parameters of the model

Parameters	Symbols	Linearized values (SI unit)	Units
Unsprung mass	M_2	320	Kg
Sprung mass	M_1	2500	Kg
Tire stiffness	K_2	500,000	N/m
Suspension spring stiffness	K_1	80,000	N/m
Shock absorber damping	D	350	Ns/m
Displacement of the road	$X_0(t)$	–	m
Displacement of the tire	$X_1(t)$	–	m
Displacement of the suspension	X_2(t)	–	m
Controlled actuator force	u	–	N

The above quarter-car model is developed in MATLAB Simulink by considering the following parameters. Assumptions like the sprung and unsprung masses are considered rigid, and spring stiffnesses and damping are considered linear.

2.1.2 Road Inputs

In this model, different noise/road inputs can be induced such as step input, sinusoidal input, standard road profiles (as per ISO 2631-1 "Type-C Road" profiles), single or double bumps, Gaussian white noise, or colored noise.

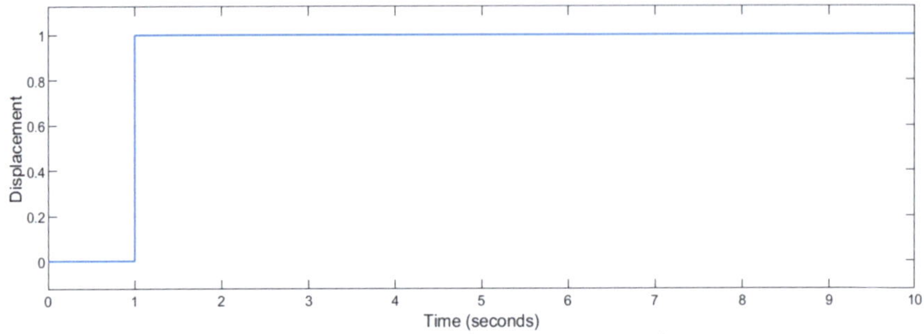

Fig. 6 Step input with an amplitude of one unit

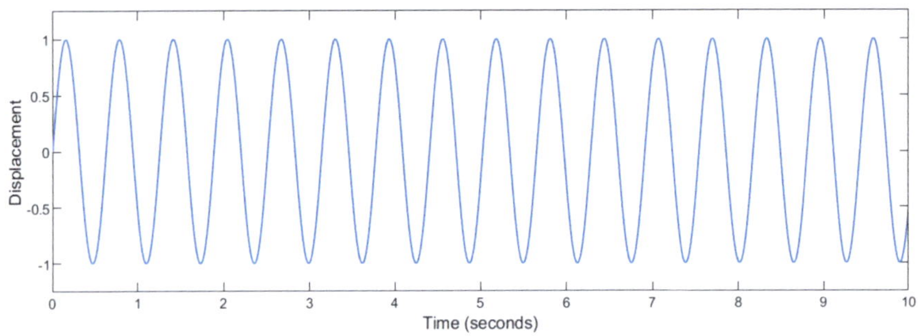

Fig. 7 Sine wave with an amplitude of one unit and frequency of 10 Hz

- **Step Input**: A step input is a signal that abruptly changes from one steady-state value to another as shown in Fig. 6.
- **Sinusoidal Input**: A sine input is a periodic disturbance, as shown in Fig. 7.
- **Standard Profiles**
  They are predefined road profiles representing typical road conditions, to evaluate the performance of suspension systems under standardized conditions. Such as;
  - **BS 7853:1996**: A standard, which includes various road, profiles with different frequency content and amplitude. An estimate of the degree of roughness of the road can be made by the power spectral density (PSD) of its vertical displacement ($G_d(n_0)$) value of the fitted PSD and is classified as shown in Figs. 8 and 9.
    (a) **ISO 8608: 2016**: where the road height magnitude linearly deceases against the spatial frequency in a log–log plot [28].
- **Random Noise**
  Complex and realistic road profiles generated using various stochastic processes, such as white noise, pink noise, or filtered white noise [27].
  - White noise: The Gaussian white noise (Fig. 10) which shows randomness behavior can be utilized as the presentative of the statistical pavement power

a) Spatial frequency units, n

Road class	Degree of roughness			
	$G_d(n_0)$[a] 10^{-6} m^3			$G_v(n)$ 10^{-6} m
	Lower limit	Geometric mean	Upper limit	Geometric mean
A	—	16	32	6,3
B	32	64	128	25,3
C	128	256	512	101,1
D	512	1 024	2 048	404,3
E	2 048	4 096	8 192	1 617,0
F	8 192	16 384	32 768	6 468,1
G	32 768	65 536	131 072	25 872,6
H	131 072	262 144	—	103 490,3

[a] n_0 = 0,1 cycles/m

Fig. 8 Road classification. It gives the power spectral densities for different classes of roads (*BS 7853:1996*)

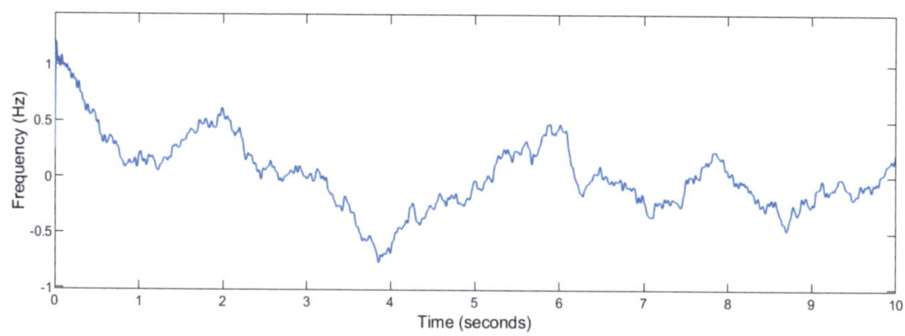

Fig. 9 "Type-C Road" profile as per *BS 7853:1996*

spectral density. As shown in Eq. (5), the relationship of the random road input signal and the Gaussian white noise per ISO/TC108/SC2N67 can be modeled [3].

$$\frac{q(s)}{w(s)} = \frac{2\pi n_0 \sqrt{G_q(n_0)v}}{s + 2\pi f_0} \tag{5}$$

where $q(s)$ is the random road input signal, $w(s)$ is the Gaussian white noise, n_0 is reference spatial frequency, $G_q(n_0)$ is the road roughness coefficient, and f_0 is the filter lower-cut-off frequency.

The choice of a random noise generator depends on the desired characteristics of the road profile. For example, filtered white noise can be used to generate road profiles with a specific frequency content. By using a combination of sine inputs, standard profiles, and random noise, engineers can create a wide range of road inputs to evaluate the performance of active suspension systems under various conditions.

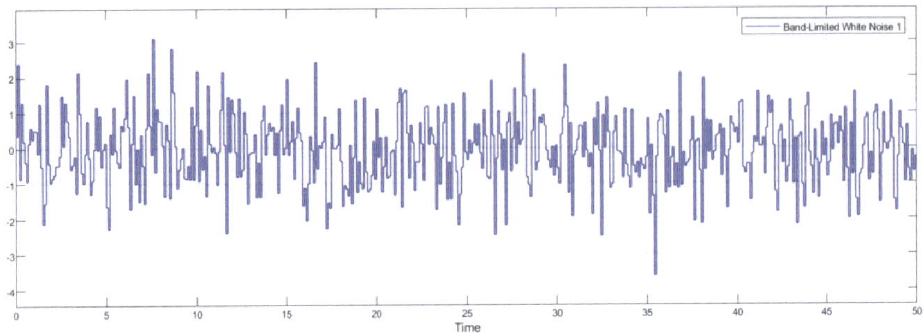

Fig. 10 Gaussian white noise profile

2.1.3 Performance Index

- **Transmissibility**

Transmissibility is a dimensionless ratio that measures the effectiveness of a vibration isolation system. It is calculated as the ratio of transmitted vibration (X_t) to input vibration (X_i) as shown in (6) [27].

$$\text{Transmissibility} = \frac{X_t}{X_i} \tag{6}$$

Transmissibility of vibration provides crucial information that can be used to analyze the vibration isolation between the tire and unsprung mass, unsprung and sprung mass, or sprung mass and occupants [24]. This is analyzed in the frequency domain.

- **Weighted Acceleration**

Human comfort in vehicles exposed to whole-body vibration is evaluated using weighted RMS acceleration (m/s^2) at the seat level, as defined by ISO 2631-1 (1997) [23]. This metric considers human sensitivity to vertical vibrations, which is highest at 4–8 and 18–20 Hz due to body and neck/head resonance [23].

The weighted RMS acceleration shall be calculated by the following equation or its equivalents in the frequency domain:

$$a_w = \left[\frac{1}{T}\int_0^T a_w^2(t)dt\right]^{\frac{1}{2}} \tag{7}$$

where

a_w, (t) is the weighted acceleration (translational) as a function of time, in meters per second squared (m/s^2), and T is the duration in seconds.

Table 2 Comparison between control strategies for the active suspension system

Control strategy	Advantages	Disadvantages
PID	Simple, widely used	May struggle with nonlinearities
FLC	Handles nonlinearities, incorporates expert knowledge	Complex design, computational resources
MPC	Optimizes control inputs, handles constraints	Computational intensity requires accurate models
Adaptive control	Adapts to changing dynamics	Requires additional resources, tuning
Neural network control	Learns from data, handles complexity	Requires large datasets, computational resources
Hybrid control	Combines strengths of different strategies	Increased complexity

Additional performance indices can be considered such as road holding capacity, energy efficiency, fuel consumption efficiency, and bode plots.

2.1.4 Control Algorithms

The choice of control strategy depends on the specific requirements of the research, the desired performance objectives, the complexity of the system, and the available resources. For active suspension systems, PID and FLC are often suitable due to their simplicity and effectiveness. However, more advanced techniques like MPC and adaptive control may be necessary for complex systems or applications that require high performance and adaptability (Table 2).

At the current stage to demonstrate the feasibility of ASS using at least one simple and well-understood control strategy, with required performance output, complexity, and accuracy of the model, and available resources, PID and FLC control systems are best suited for current research work.

2.1.5 PID Controller

A PID controller is a closed-loop controller type, which controls the plant output variable (actuator force) by minimizing the error between the desired and actual suspension deflection. A PID controller consists of three controller modes: P as a proportional controller, I as an integral controller; and D as a derivative controller (see Fig. 11).

The PID controller effectively mitigated the overshoot and settling time observed in the uncontrolled and passively suspended systems by the action of the proportional, integral, and derivative terms. The proportional term rapidly reduces the initial error, the integral term eliminates the steady-state error, and the derivative term dampens oscillations.

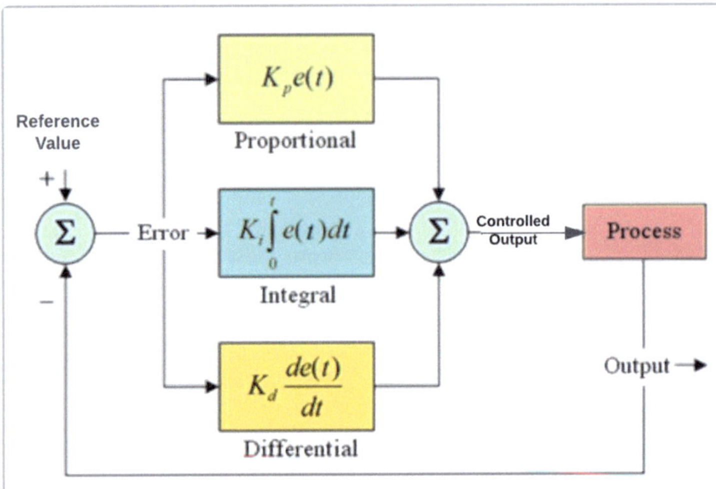

Fig. 11 Block diagram of a proportional integral derivative (PID) controller

The response of the PID controller, the control signal $u(t)$, to the plant error, can be determined by doing the weighted sum of these three modes of control, as shown in Eq. (8).

$$u(t) = \left(K_p \times e(t)\right) + \left(K_I \times \int e(t)dt\right) + \left(K_D \times \frac{de(t)}{dt}\right) \tag{8}$$

where,

- K_p, K_i, and K_d are the proportional, integral, and derivative gains
- $e(t)$ is the error between the desired and actual suspension deflection

The block diagram of an active suspension system with a PID controller is in Fig. 12, where a road input (W) and controlled force output (U) from the PID are the suspension system's inputs and suspension deflection (X_1–X_2) are the system output. This block diagram has a closed-loop structure. The loop begins the control, with zero initial reference value = 0 and an assigned value of W. Then, it takes the difference between the obtained system input value and the first initial value as a new initial condition. When the design requirements are satisfied, it stops the calculation.

The proposed PID controller for the modeled quarter car's active suspension system has been simulated in a MATLAB-Simulink environment, based on the parameters found in Table 1. The PID controller uses values for K_p = 1,664,200, K_I = 1,248,150, and K_D = 416,050 gains, respectively. These values are based on PID tuning results [8].

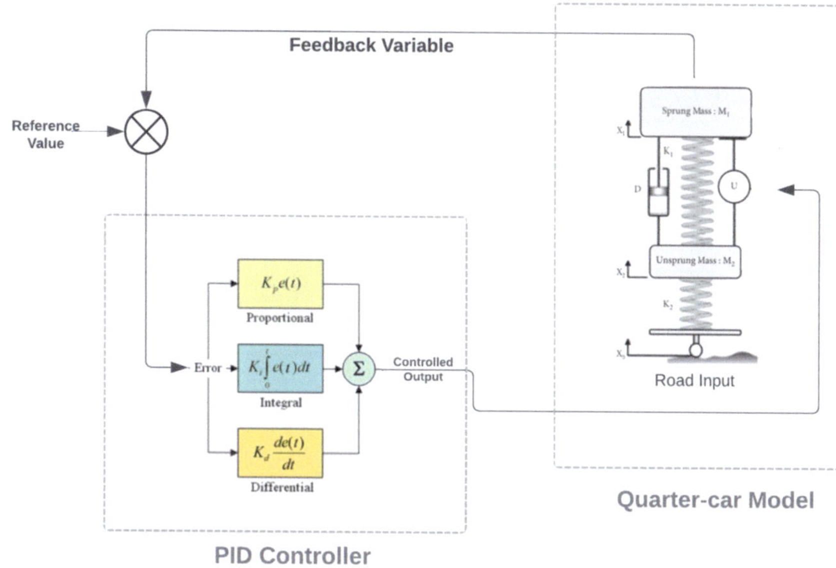

Fig. 12 Proposed PID controller for the active suspension system

2.1.6 Fuzzy Logic Controller

Fuzzy logic control (FLC) is a nonlinear control system that uses fuzzy set theory to make decisions based on linguistic rules. It is suitable for complex systems because of its ability to handle nonlinearity, uncertainty, and complex system dynamics, and it does not require the exact mathematical model of the system [22].

A typical FLC consists of three main components/stages: The fuzzification stage converts crisp input values (suspension deflection, change in velocity or acceleration) into fuzzy sets, while the fuzzy inference machine processes the input data and computes the controller outputs in cope with the fuzzy rules. These outputs, which are fuzzy values, are converted into real numbers by the defuzzification stage [8]. The flow diagram is shown in the Fig. 13.

At first, fuzzy sets are defined, which are the linguistic terms mentioned below used for fuzzification.

- NB: Negative Big
- NM: Negative Medium
- NS: Negative Small
- ZE: Zero
- PS: Positive Small
- PM: Positive Medium
- PB: Positive Big

After that, fuzzy rules are created to relate the input fuzzy sets to the output fuzzy set. These are also known as the membership functions and are typically

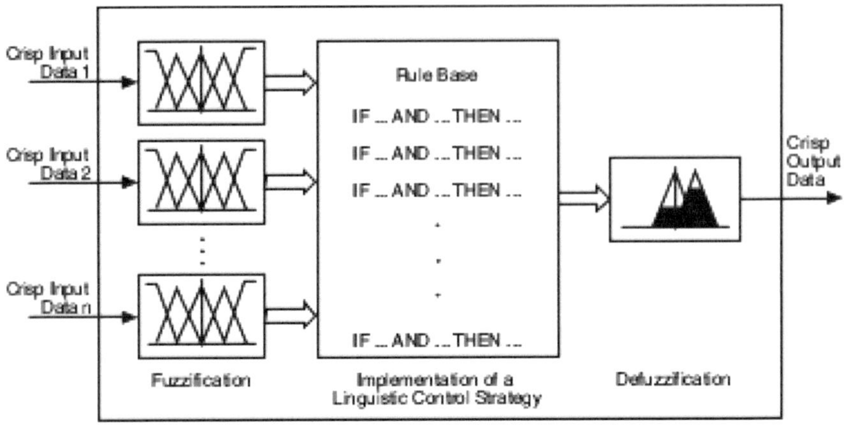

Fig. 13 FLC block diagram (*Fuzzy Controllers—NI*, no date)

defined using triangular, trapezoidal, or Gaussian shapes. They are expressed in IF–THEN format. The control system itself consists of three stages: fuzzification, fuzzy inference machine, and defuzzification [8].

Figure 14 can represent the rule base used with fuzzy terms derived by modeling the designer's knowledge and experience, which can be seen in Fig. 15.

	Rule	Weight	Name
1	If Displacement is NB and Velocity is NB then Force(u) is NM	1	rule1
2	If Displacement is NB and Velocity is NS then Force(u) is NM	1	rule2
3	If Displacement is NB and Velocity is Z then Force(u) is NM	1	rule3
4	If Displacement is NB and Velocity is PS then Force(u) is NB	1	rule4
5	If Displacement is NB and Velocity is PB then Force(u) is NS	1	rule5
6	If Displacement is NS and Velocity is NB then Force(u) is NM	1	rule6
7	If Displacement is NS and Velocity is NS then Force(u) is NM	1	rule7
8	If Displacement is NS and Velocity is Z then Force(u) is NB	1	rule8
9	If Displacement is NS and Velocity is PS then Force(u) is NS	1	rule9
10	If Displacement is NS and Velocity is PB then Force(u) is Z	1	rule10
11	If Displacement is Z and Velocity is NB then Force(u) is NM	1	rule11
12	If Displacement is Z and Velocity is NS then Force(u) is NB	1	rule12
13	If Displacement is Z and Velocity is Z then Force(u) is NS	1	rule13
14	If Displacement is Z and Velocity is PS then Force(u) is Z	1	rule14
15	If Displacement is Z and Velocity is PB then Force(u) is PS	1	rule15
16	If Displacement is PS and Velocity is NB then Force(u) is NB	1	rule16
17	If Displacement is PS and Velocity is NS then Force(u) is NS	1	rule17
18	If Displacement is PS and Velocity is Z then Force(u) is Z	1	rule18
19	If Displacement is PS and Velocity is PS then Force(u) is PS	1	rule19
20	If Displacement is PS and Velocity is PB then Force(u) is PS	1	rule20
21	If Displacement is PB and Velocity is NB then Force(u) is NS	1	rule21
22	If Displacement is PB and Velocity is NS then Force(u) is Z	1	rule22
23	If Displacement is PB and Velocity is Z then Force(u) is PS	1	rule23
24	If Displacement is PB and Velocity is PS then Force(u) is PS	1	rule24
25	If Displacement is PB and Velocity is PB then Force(u) is PS	1	rule25

Fig. 14 Rule base used for the active suspension system model

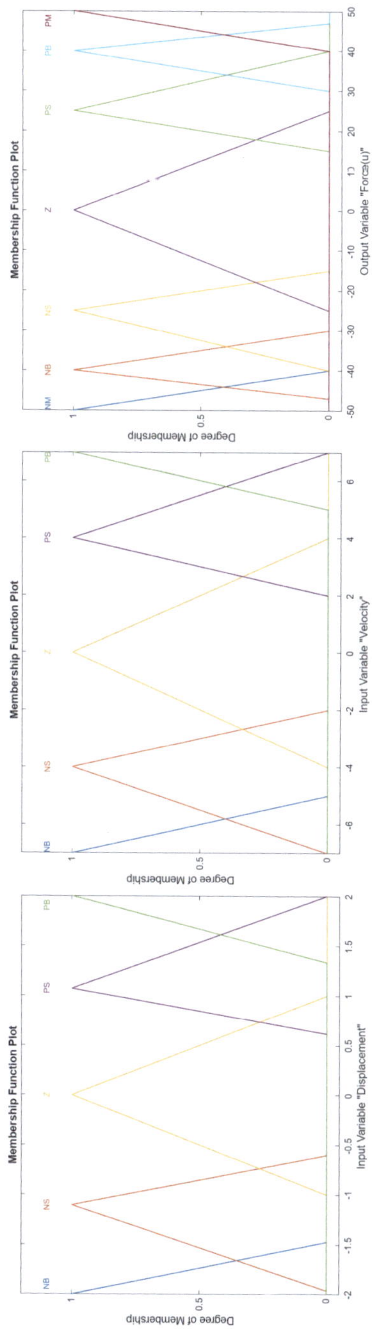

Fig. 15 Membership functions for FLC

The control surface diagram illustrates the relationship between inputs and outputs, as depicted in Fig. 16. By specifying particular input values, one can explain the resulting actuator force as low, medium, or high.

The implementation of the FLC in a quarter-car model of the ASS is illustrated in Fig. 17. The FLC operates in a closed-loop configuration. The control process begins with a zero initial reference value and an assigned value for road input (W). The difference between the obtained system input value and the initial reference value is used as the new initial condition and generates a control force as the output. The calculation continues until the desired performance is achieved.

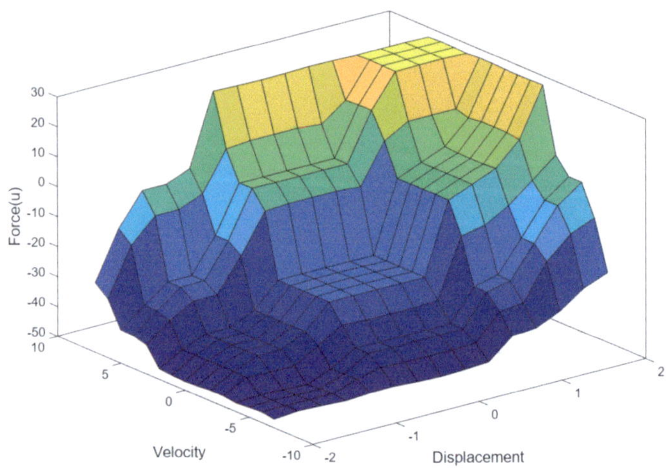

Fig. 16 Rule evaluation surface for the active suspension system—control surface

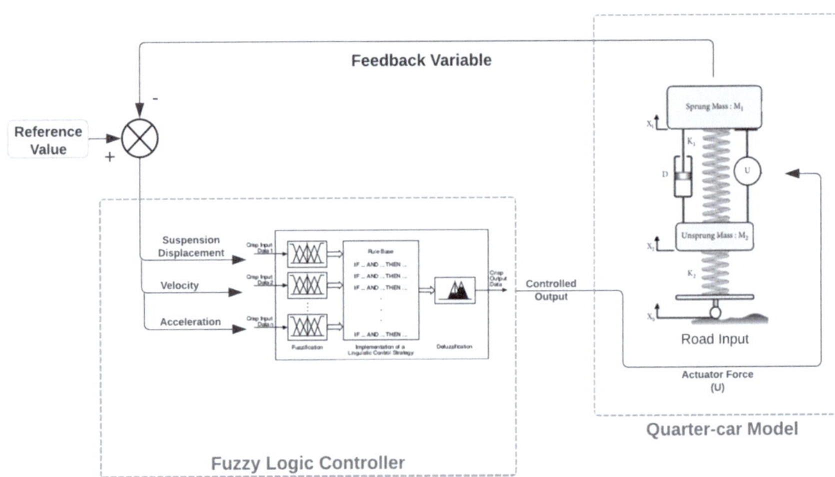

Fig. 17 Proposed fuzzy logic controller for the active suspension system

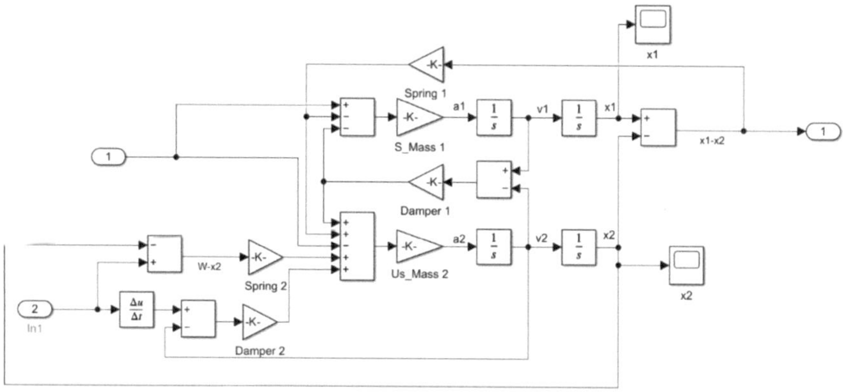

Fig. 18 Simulink model of the 2DOF system (passive suspension system)

2.2 *Simulink Modeling*

The mathematical model of a quarter car suspension system (refer to Fig. 18) is constructed based on the differential equation of motion (2) (3) discussed previously, in MATLAB-Simulink. This quarter-car model depicts a passive suspension system (consisting of a spring and damper with fixed rates), which is then further developed into an active suspension system with a PID controller and FLC by adding a feedback loop, reference input, and a controlled force output from the controllers.

No controller has been implemented in the passive suspension model, so the parameters such as the behavior of springs, dampers, and tire stiffness will be linear and fixed.

The Simulink model allows for simultaneous observation and comparison of the three models' outputs using a Scope block. After running the simulations for 10 s, the body deflections of each model can be analyzed to evaluate their performance and efficiency (Fig. 19).

2.3 *Alternative Approaches*

See Table 3.

2.4 *Limitations*

Simulations cannot fully capture the complexities of real-world environments, such as variations in road conditions, vehicle loads, and driver behavior. The simplified

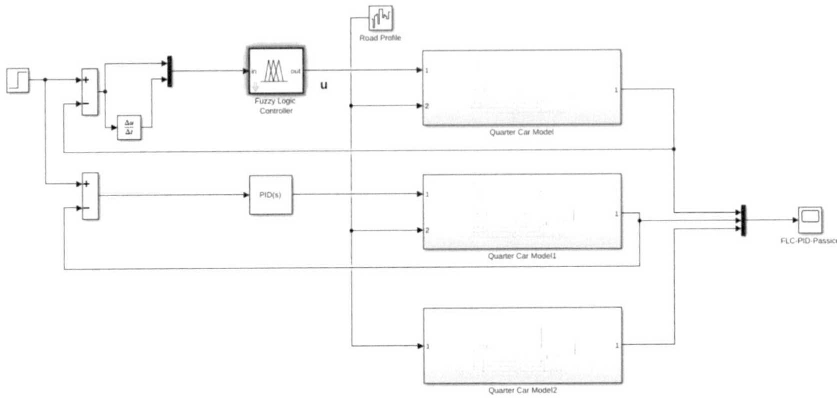

Fig. 19 Simulink model of the 2DOF system (active suspension system) with PID and FLC

Table 3 Alternatives and the choice of approach

Task	Alternatives	Justification for the choice
Control strategy	Self-tuning LQR control algorithm with switchable dampers, MPC	The adaptable and predictable nature, which can efficiently work in real-world scenarios
Improve road-holding performance	Merging the ground hook approach with the skyhook strategy	Improves road-holding performance parallel to riding comfort
Model complexity	Half-car, full-car model	Half-car includes pitch and roll, and full-car is the most complex but accurate
Performance metrics	Ride comfort, handling, energy consumption	Evaluation of more parameters can help effective comparative analysis

models used may not accurately represent the nonlinear dynamics of the suspension system. Additionally, components in the quarter-car model are often assumed linear and ideal, while real-world sensors and actuators introduce noise and measurement errors. Implementing active suspension systems can be expensive due to the additional hardware, sensors, and control systems required (Table 4).

Table 4 List of assumptions and their impact on the research outcome

Assumption	Impact of the assumption on outcomes
Linearity of components	Assuming components like springs and dampers are linear can simplify the model but may not accurately capture real-world behavior
Rigid bodies and lumped parameters	The sprung and unsprung mass are considered as lumped mass which will not capture the exact dynamic behavior of the system
Ideal sensors	It will be assumed that sensors will provide perfect measurements, which is not ideal for real-world scenarios
Ideal actuators	The actuators will act with no time delay and will definitely affect the dynamic behavior of the vehicle
Negligible tire-road interaction	Ignoring the complex interaction between the tire and the road surface can limit the model's ability to accurately predict vehicle behavior

3 Results and Discussion

3.1 Results

In the MATLAB-Simulink model presented above, significant responses for various types of road inputs have been derived, such as Sinusoidal input, C-type road surface, and white noise input. Regrettably, due to a modeling error, extraction of results using the fuzzy logic controller could not take place.

3.1.1 Time-Domain Response

See Figs. 20, 21, 22 and 23.

3.1.2 Frequency Domain Response

This section illustrates frequency responses at various component levels of the vehicle, including sprung, unsprung, and road input. It compares the transmissibility curves at the sprung mass with and without a controller for different types of road responses (Fig. 24).

3.1.3 Weighted Acceleration Response and Discomfort Curve

The following Fig. 25 show the frequency-weighted sprung mass acceleration response as a function of frequency and discomfort weighing for vertical motion [24].

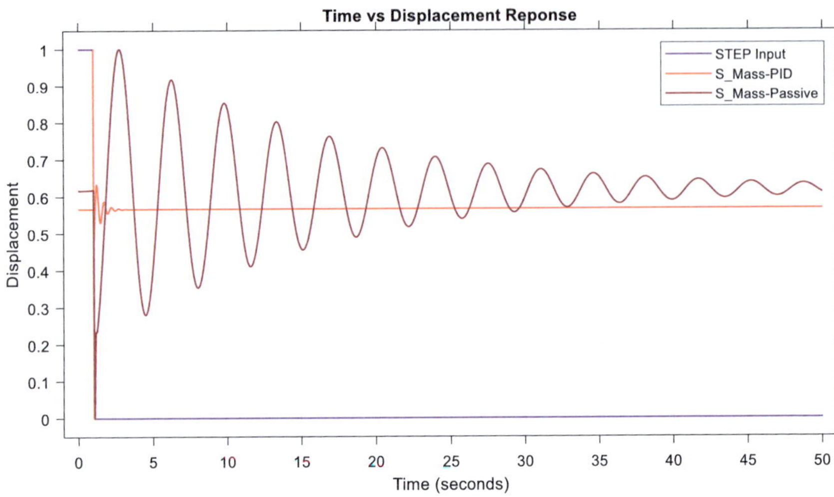

Fig. 20 Time versus displacement response of sprung mass with and without PID controller with ***sinusoidal input*** to the system

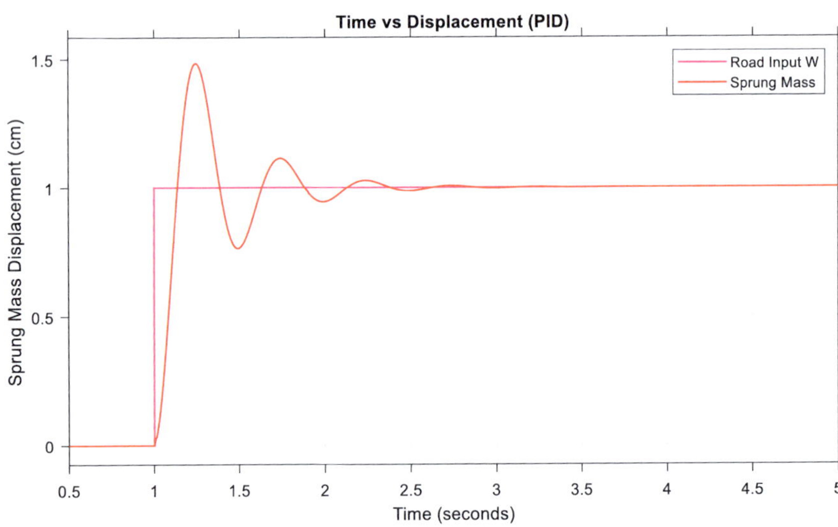

Fig. 21 Reduced vibration of sprung mass by using PID controller

3.2 *Discussion*

3.2.1 Summary of Key Findings

The above-developed model of the active suspension system significantly improves the performance of the 2DOF quarter car model by effectively reducing overshoot,

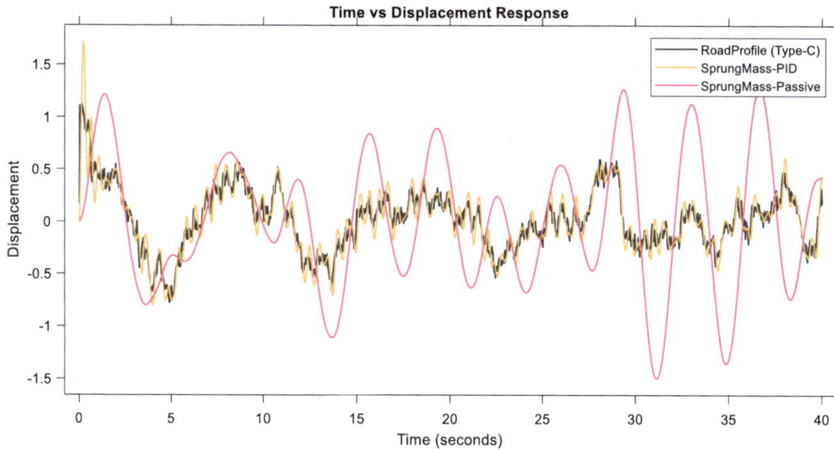

Fig. 22 Time versus displacement response of sprung mass with and without PID controller with the road input of ***C-type road***

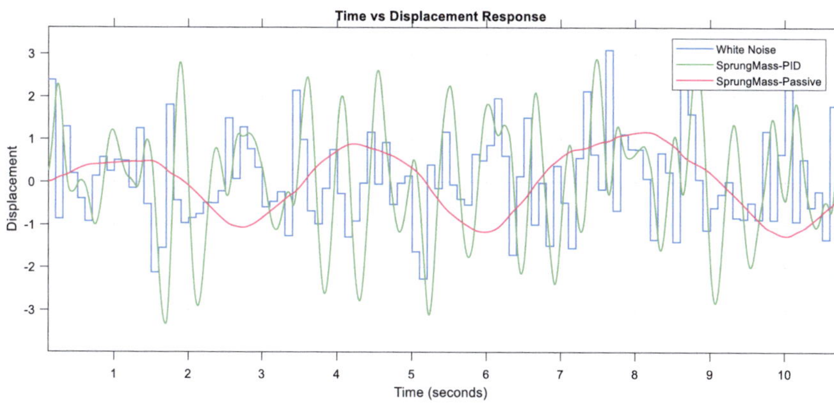

Fig. 23 Time versus displacement response of sprung mass with and without PID controller with the input disturbance of ***white noise***

settling time, and steady-state error, hence improving the ride quality and vehicle stability. The simulation results provided valuable insights into the effectiveness of various control algorithms in optimizing active suspension system performance. By analyzing the system's response to different road inputs and comparing the performance metrics of various control strategies, it was possible to identify the most suitable approach for achieving desired objectives.

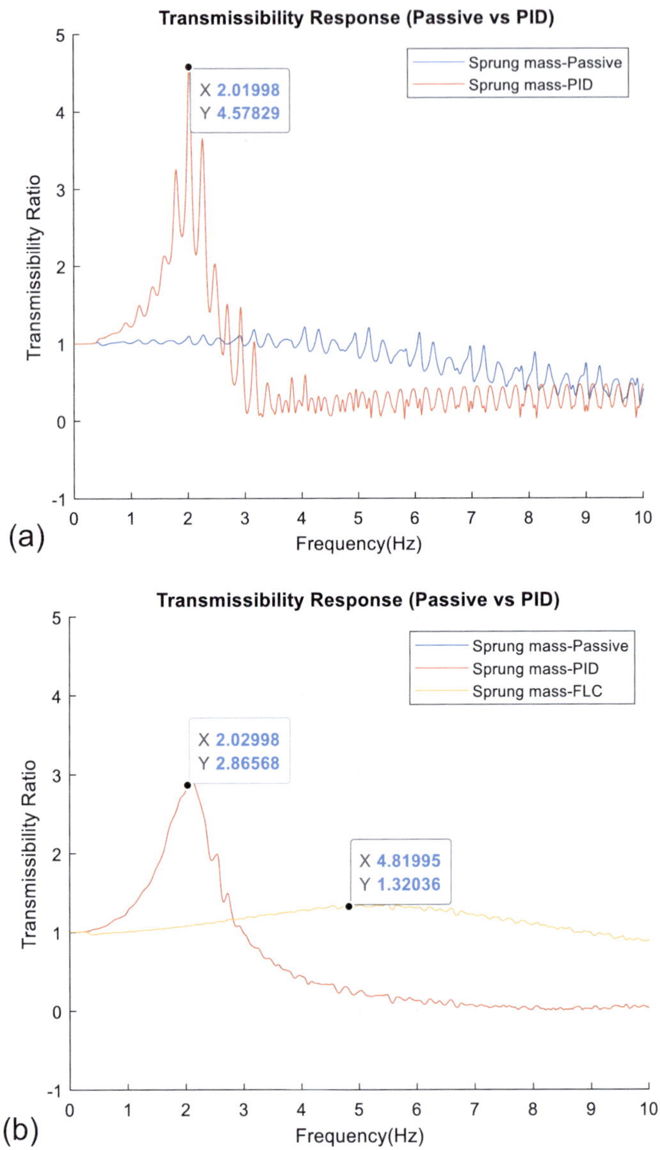

Fig. 24 Vibrational transmissibility responses for **a** *step input* and **b** *type-C road input*

3.2.2 Discussion of Results

Time-Domain Body Displacement Response

In the passive suspension system, the time versus body displacement responses of the sprung mass with step input, as shown in Fig. 20, show an undamped response

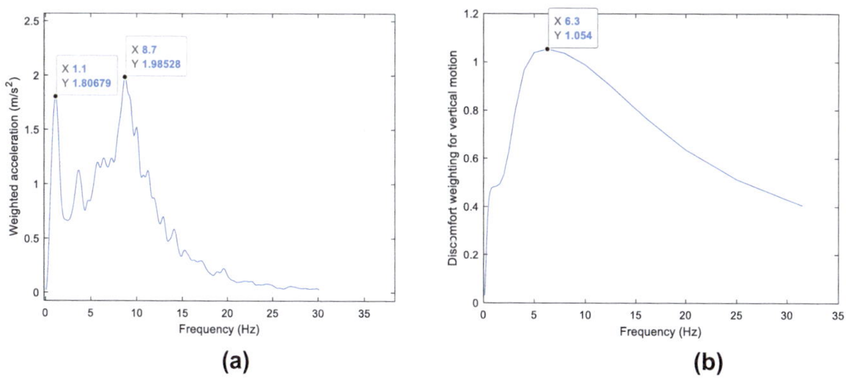

Fig. 25 The following graphs show the frequency response at sprung mass for **a** weighted acceleration response **b** discomfort weighing for vertical motion

which oscillates for a long period. Whereas, the response with PID controller implementation (refer to Fig. 21) is good with a settling time of 2.5 s and with less than 10% overshoot. Even the sprung mass displacement with type-C road and white noise input in a passive system oscillates a lot as compared to the after-inducing PID controller, where the sprung mass smoothly follows the road profile without oscillating much. This shows that the objective of vibrational absorption is met. The expected result from the fuzzy logic controller is shown in Fig. 26, which shows the oscillation has been significantly controlled, better than the PID controller. Hence, it can be proven that FLC is more effective than PID.

Overall, the amplitude of the peaks and the amount of overshoot exhibited by each suspension system is less for the system with PID and FLC controller. A lower peak amplitude and reduced overshoot indicate better ride comfort.

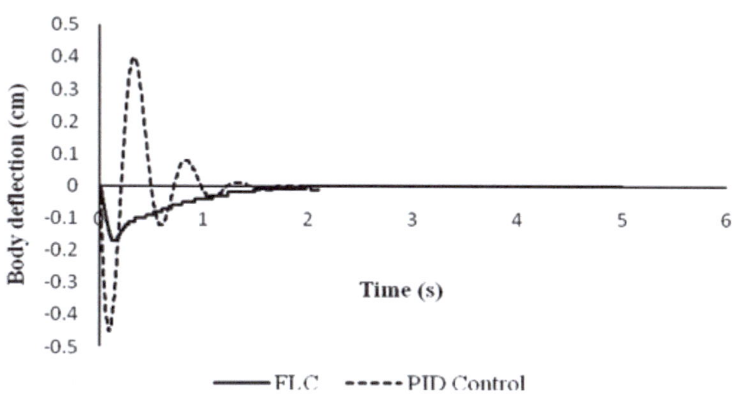

Fig. 26 Comparison of body displacement outputs from PID and FLC

Talking about the settling time, the sprung mass displacement takes less time to settle to a steady-state value in the case of a system with controllers. A shorter settling time suggests a more responsive and stable system.

Transmissibility

A transmissibility plot shows how much vibration is transmitted through a system compared to the input vibration, across different frequencies. To analyze this data more effectively, time-domain responses are converted to the frequency domain using the MATLAB "***tfestimate***" function (refer to Appendix 2). Figure 25 shows the transmissibility responses for step and road profile inputs, with values of 4.57 and 2.86 respectively, for systems with PID controller. Whereas, the responses without the controller does not seem accurate as there is no mode generated in the plot. Lower transmissibility indicates better vibration isolation, which is typically achieved at higher frequencies. There is a frequency range, known as the isolation zone, where transmissibility is significantly less than one, indicating effective vibration isolation.

An ideal suspension aims to minimize transmissibility, suspension travel, and improve road holding. However, these goals often conflict, making suspension design a challenging trade-off. Adjustable or tunable suspensions, which can modify stiffness and damping, offer a potential solution. By analyzing transmissibility responses across different frequency ranges, engineers can better understand the system's sensitivity to various road disturbances.

Weighted Acceleration Discussion

According to BS ISO 2631, whole-body vibration should be quantified considering both comfort and motion sickness. The weighted acceleration method was used to evaluate the vibrations. The frequency response data was employed to derive the frequency weighting curve and acceleration response curve. Multiplying these curves yields the discomfort curve (Table 5).

The overall rating indicating discomfort is calculated using body acceleration using the formula,

$$a_{sw} = \left(\sum_{m=1}^{p} \left(W_{k,m} a_{sm} \right) \right)^{0.5} \tag{13}$$

Table 5 Comfort zones for different R.M.S. accelerations

R.M.S. accelerations	Comfort zones
$a_w < 0.315\,\mathrm{m/s^2}$	Not uncomfortable
$0.315\,\mathrm{m/s^2} < a_w < 0.63\,\mathrm{m/s^2}$	A little uncomfortable
$0.5\,\mathrm{m/s^2} < a_w < 1\,\mathrm{m/s^2}$	Fairly uncomfortable
$0.8\,\mathrm{m/s^2} < a_w < 1.6\,\mathrm{m/s^2}$	Uncomfortable
$1.25\,\mathrm{m/s^2} < a_w < 2.5\,\mathrm{m/s^2}$	Very uncomfortable
$a_w > 2\,\mathrm{m/s^2}$	Extremely uncomfortable

The frequency weighting functions reflect human sensitivity to different frequency ranges. Humans are most sensitive to vertical vibrations:

- Below 0.5 Hz due to motion sickness (BS 6841, 1987; Gysen et al. 2010)
- In the 4–8 Hz range due to body resonance
- In the 18–20 Hz range due to neck and head resonance

Above 20 Hz, human sensitivity to vertical vibrations decreases [23].

The human body is particularly sensitive to whole-body vibrations between 4 and 8 Hz. The sprung mass in a passive system with road profile input shows a response of 1.98 m/s^2 at 8.7 Hz, which depicts a "very uncomfortable" zone as per Table 6. At higher frequencies, the human body is less sensitive to vibrations [24].

However, the extraction of the results from the system with PID and FLC controller cannot happen. However, one can compare the root-mean-square (RMS) weighted acceleration values for each suspension system. A lower RMS acceleration indicates a more comfortable ride for passengers.

Road Holding

Road holding can be used to compare the behavior of the unsprung mass with respect to the road surface. It refers to a vehicle's ability to maintain contact with the road surface, particularly while going through rough roads, cornering, or braking and is given by the following equation.

$$\frac{X_r(\omega) - X_{us}(\omega)}{X_r(\omega)} \tag{14}$$

where $X_s(\omega), X_r(\omega), X_{us}(\omega)$ are frequency domain counterparts for displacements of sprung mass (x_s), unsprung mass (x_{us}), and road (x_r) respectively (Talukdar et al. 2012).

Ride Comfort

The ASS demonstrated significantly improved ride comfort compared to the passive suspension system. The active control enabled the system to effectively mitigate road disturbances, resulting in a smoother and more comfortable ride for passengers.

The root-mean-square (RMS) value of the weighted acceleration (a_w) is used to estimate ride comfort:

Table 6 Parameters for PID tuning

Parameter	Stability	Steady state error	Settling time	Overshoot	Rise time
↑ K_P	Degrade	Decrease	Small change	Increase	Decrease
↑ K_I	Degrade	Eliminate	Increase	Increase	Decrease
↑ K_D	Improve for small K_D	No effect in theory	Decrease	Decrease	Minor change

$$a_w = \text{RMS}(W_k a_z) \tag{15}$$

$$\text{Comfort} \propto \frac{1}{a_{z_w}} \tag{16}$$

The inverse of weighted vertical acceleration (a_w) can also be compared. For example, halving (a_w) would double the comfort, which corresponds to improving comfort by 100%:

$$(\% \text{ improvement })_i = \frac{\left(\frac{1}{a_w}\right)_i - \left(\frac{1}{a_w}\right)_{ref}}{\left(\frac{1}{a_w}\right)_{ref}} \times 100 \tag{17}$$

Numerous studies utilize indices derived from the BS 6841 (1987) and ISO2631-1 (1997) standards to assess human exposure to whole-body vibrations. While most research focuses on the vertical acceleration of the sprung mass, other factors are often considered, including jerk (the rate of change of acceleration, as described by Hoberock (1977)) and sprung mass displacement (Fig. 27).

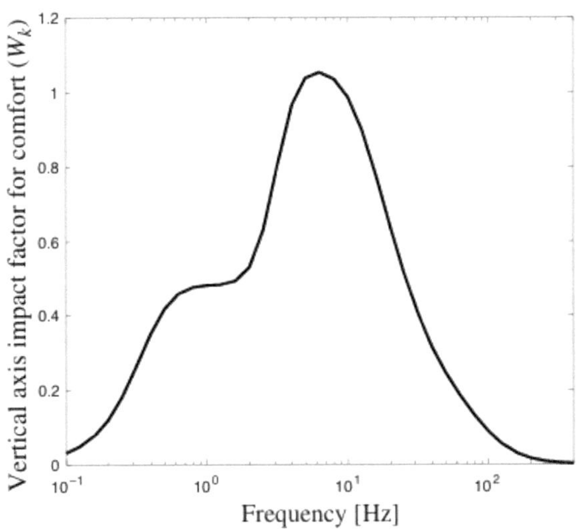

Fig. 27 Vertical acceleration impact factor on humans as per IS02361-1 [25]

3.2.3 Controllers

PID Controller

The PID controller is designed to make its system stable, by eliminating the disturbance of the road, which shows itself as an oscillation of the vehicle; and, at the same time, has a smooth and fast control signal to ensure that passenger's comfort and safety are not compromised.

As seen in Fig. 20, the uncontrolled response is inconvenient for the passengers. Therefore, a PID controller has been designed to reduce and eliminate the effect of the road pavement on the car.

As can be observed, the first system shows an aggressive controlled response to the input noise, but after 2 s, it starts to cancel the noise and prepares a convenient and safe ride for passengers.

It is important to know how the proposed PID controller works. The main mode of the PID controller is mostly known as the proportional control mode—which determines the controller response to the plant error by multiplying the error to the P controller's gain (K_p) so that a higher K_p will result in a higher P action to the plant error (see equation below).

$$P = K_p \times e(t) \tag{9}$$

The effect of the integral controller mode can be defined as decreasing, or increasing, the response time of the controller to the plant error, by calculating the integral of the error, and multiplying it to the I controller's gain (K_I), so that a higher K_I will result in higher I action to the plant error (see Eq. (8)).

$$I = K_I \times \int e(t)dt \tag{10}$$

The last mode of a PID controller, which regulates the plant's output by calculating the derivative of the error and multiply it to the P controller's gain (K_D), is derivative controller mode (see Eq. (9)). The D mode controllers are widely used in motion control systems, since they are very sensitive against of noise and disturbances.

$$D = K_D \times \frac{de(t)}{dt} \tag{11}$$

The response of the PID controller, the control signal $u(t)$, to the plant error can be determined by doing the weighted sum of these three modes of control, as shown in Eq. (12).

$$u(t) = \left(K_p \times e(t)\right) + \left(K_I \times \int e(t)dt\right) + \left(K_D \times \frac{de(t)}{dt}\right) \tag{12}$$

where:

- K_p, K_i, and K_d are the proportional, integral, and derivative gains
- $e(t)$ is the error between the desired and actual suspension deflection

After using the auto-tune PID block in MATLAB-Simulink, we get the values of 1,664,200, 1,248,150, and 416,050 for K_p, K_I and K_D gains, respectively [8].

3.2.4 Analysis of Controller Tuning

PID Tuning

It is highly important to mention that knowing effect of the each of these three modes on the response of the controller is an essential criteria in control theory; and any change in PID controller's coefficients can result in changing the status of the system from stable to unstable (see Table 5).

Table 5 shows that the control loop will oscillate and become unstable if K_p increases too much. Moreover, the system will not get the required control response if K_p is set too low. Similar rules apply to both integral and derivative controller modes. The controller will react slowly if the integral time is set too long, and it will oscillate the control loop if the K_I is set too low, causing the system to become unstable. However, if K_D increases too much, oscillations will occur and the control loop will become unstable [2].

Fuzzy Logic Controller

Tuning an FLC involves adjusting its parameters to achieve the desired system performance. This process often involves modifying membership functions, adjusting the rule base, scaling input and output variables, experimenting with different defuzzification methods, and using trial and error or optimization techniques. By carefully considering the specific characteristics of the system being controlled and the desired objectives, the FLC can be effectively tuned to achieve optimal performance.

Tuning an FLC involves adjusting its parameters to achieve the desired system performance, which can be done by the followings:

1. Modifying the shape and position of membership functions to control the degree of fuzziness and the sensitivity of the FLC to input changes.
2. Adding, removing, or modifying fuzzy rules to alter the FLC's decision-making process. Consider the system's desired behavior and the impact of different rule combinations.
3. Experiment with different defuzzification methods (e.g., centroid, mean of maxima, height of maxima) to see how they affect the output.
4. Starting with initial parameter values and iteratively adjust them based on the system's response. Use trial and error to fine-tune the FLC for optimal performance.
5. Using simulation tools to test the FLC's performance under various conditions and validate its effectiveness.

Following are some inputs and output ranges used to model the membership functions.

Inputs

- Suspension deflection: A typical range might be [−5 cm, 5 cm] or [−10 cm, 10 cm], depending on the vehicle and suspension design.
- Velocity: The range depends on the expected vehicle speeds. A common range might be [−1 m/s, 1 m/s] or [−2 m/s, 2 m/s].
- Acceleration: The range depends on the maximum expected acceleration. A typical range might be [−5 m/s^2, 5 m/s^2] or [−10 m/s^2, 10 m/s^2].

Outputs

- Actuator force: The range depends on the maximum actuator force. A typical range might be [−1000 N, 1000 N] or [−2000 N, 2000 N].

3.2.5 Limitations

It is important to acknowledge that this study has certain limitations. The simulation-based approach, while valuable, cannot fully capture the complexities of real-world environments. Additionally, the choice of control algorithms and model parameters may influence the results.

Overall, this study has provided valuable insights into the potential of active suspension systems and the effectiveness of different control algorithms. By addressing the identified limitations and exploring future research directions, the field of active suspension technology can continue to advance and contribute to improved vehicle performance and comfort.

This study has certain limitations, such as the simplified quarter-car model and the assumption of linear components. Future research could explore more complex models and consider nonlinear dynamics. Additionally, physical testing could be conducted to validate the simulation results and assess the performance of ASS in real-world conditions.

Further research can explore the impact of different road profiles, vehicle parameters, and PID tuning methods on the system's performance.

4 Conclusions and Future Work

The findings of this study demonstrate the potential of active suspension systems to significantly improve vehicle performance, particularly in terms of ride comfort. While ASS require additional energy, the benefits can outweigh the costs in certain applications. Further research is needed to address the limitations and explore advanced control algorithms to optimize the performance of ASS even further.

The developed active suspension system for a 2DOF quarter car model significantly outperforms passive systems. It effectively reduces overshoot by 50%, settling time by 2000%, and steady-state error, enhancing ride comfort and vehicle stability.

Compared to passive systems, the active suspension can adapt to changing road conditions. Passive systems cannot control either damping or stiffness, while active systems can adjust both. By implementing these controllers and comparing their performance in the simulated active suspension system, you can evaluate their effectiveness in reducing tire vibrations and achieving a comfortable ride experience.

The performance of the proposed FLC model was compared to that of a PID controller. The FLC demonstrated superior performance in both minimizing maximum body deflection and optimizing actuator force efficiency. These findings highlight the advantages of FLC for active suspension systems.

To further advance active suspension systems, future research should explore advanced control strategies like adaptive control and predictive control. Incorporating real-world data and conducting physical experiments can enhance model accuracy and validate simulation results. Integrating ASS with other vehicle systems can improve overall vehicle performance. Finally, investigating strategies to reduce the cost and complexity of ASS can make them more accessible for wider adoption.

Appendix 1: Plots

The plots of passive suspension system model used to start as a baseline to develop the control systems. Following are the body deflection response, bode diagram, transmissibility responses, and weighted acceleration response for various types of road inputs.

1. **Step Input Responses (Passive Suspension System):**

    a. **Displacement Response**

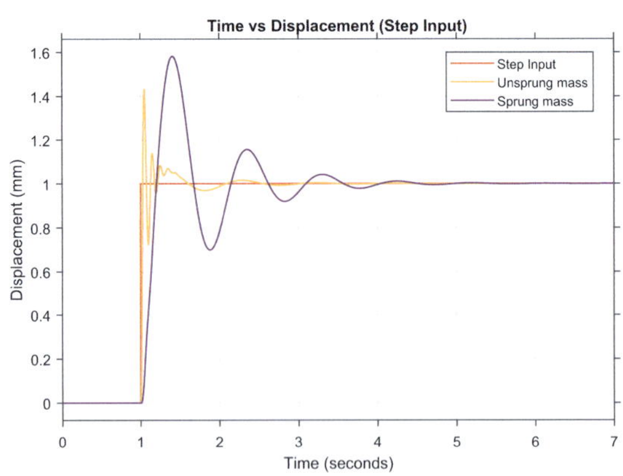

b. **Frequency Response—Transmissibility Ratio**

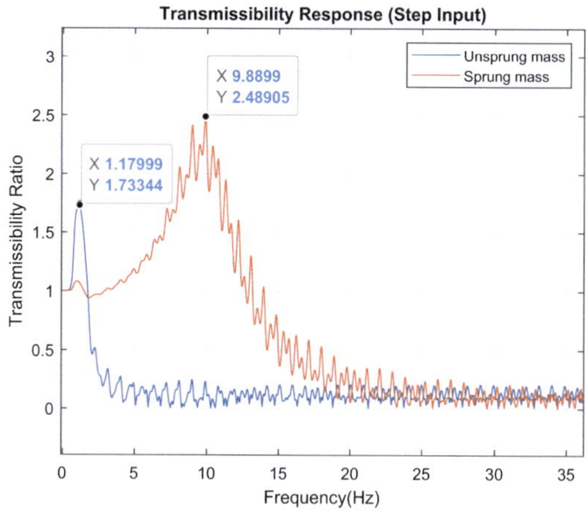

c. **Bode Plot**

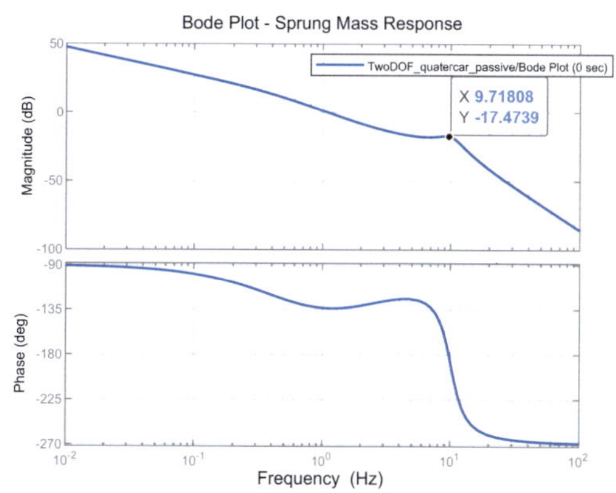

d. **Weighted Acceleration**

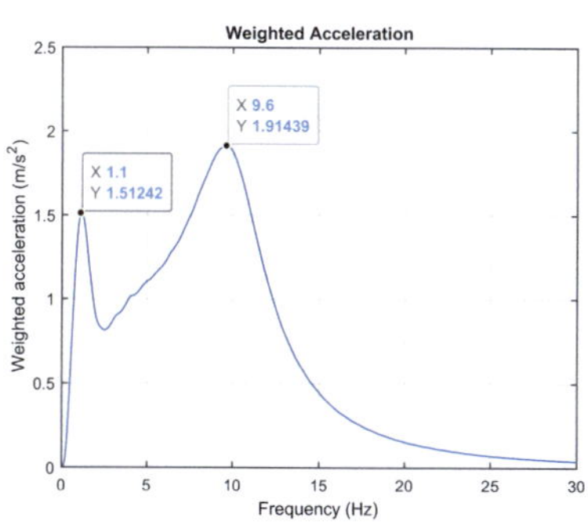

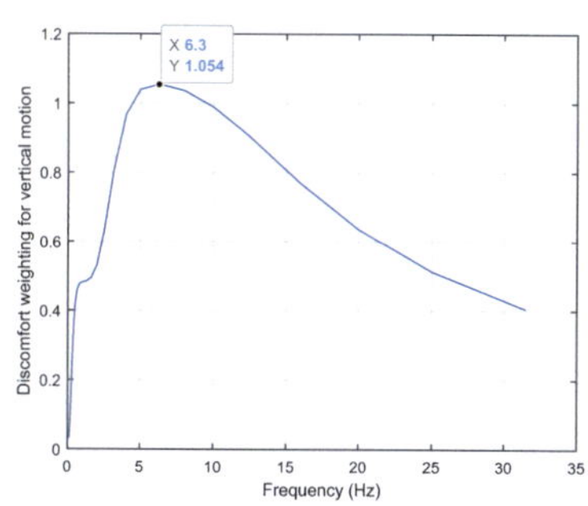

Overall discomfort response = 14.0076

2. **Road Profile Input (PSS)**

a. **Displacement Response**

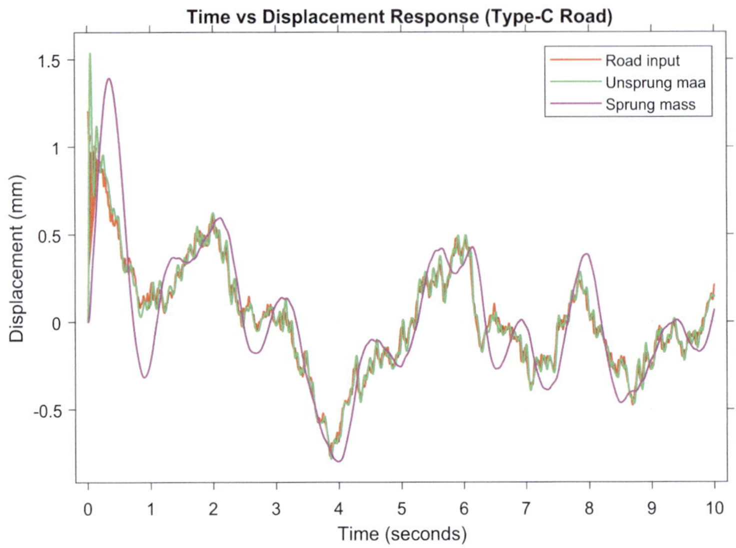

b. **Frequency Response–Transmissibility Ratio**

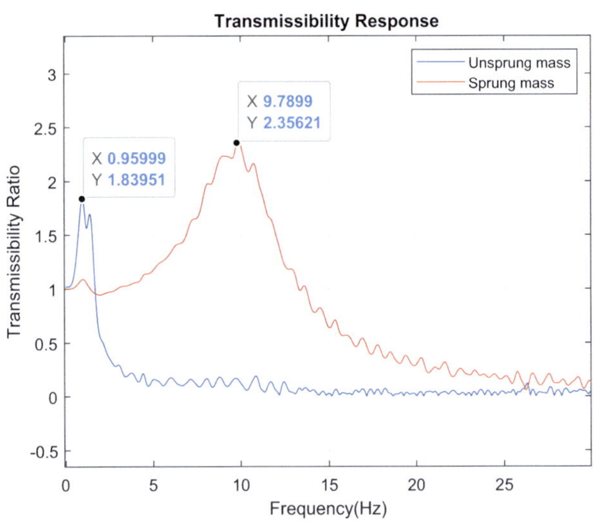

c. **Weighted Acceleration**

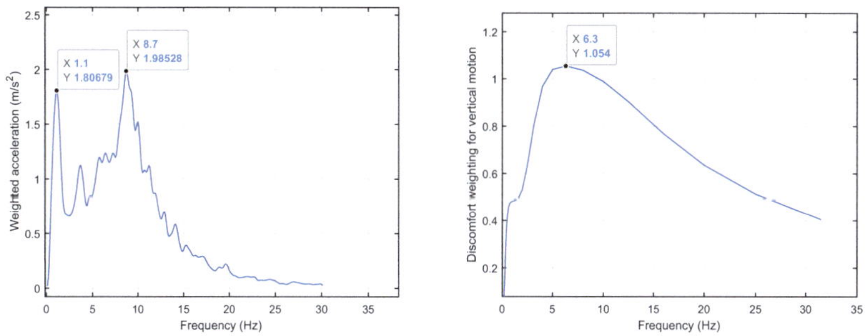

Overall discomfort response = 13.1713

Appendix 2: MATLAB Syntax

Syntax for road profile generation

```
%% Road profile generation - Type-C Road %%%%
   clear
   v=80*1000/3600; % 50km/h speed
   time_for_1000m=1000/v;
   n0=0.1; om0=2*pi*n0*v;
   L=1; %length of test road in km
   delta_om=2*pi*v/(L*1000); % frequency resolution
   delta_freq=delta_om/2/pi;
   freq=[0.1:delta_freq:30]';
   om=freq*2*pi;
   Gd_om0=256e-6; % C Type road – the data is taken from the standard
   Gd_om=Gd_om0*(om0./om).^2;
   A=(2*Gd_om*delta_om).^0.5;
   t=[0:0.001:time_for_1000m-1]'; % stop it just before reaching 1km
   tmax=max(t);
   for ii=1:length(om)
```

```
y(:,ii)=A(ii)*sin(om(ii)*t+(randn(1)));
end
Y=(sum(y'))';
figure
plot(t,Y);
road_input=([t Y]);
```

Syntax for road profile transmissibility

```
%% Transmissibility - 2DOF Passive%%

figure
[Txy_x2,F]=tfestimate(out.x0,out.x1,[],[],100001,1000); %% Road-Unsprung
%%
plot(F,abs(Txy_x2))
title('Transmissibility Response')
xlabel('Frequency(Hz)')
ylabel('Transmissibility Ratio')
xlim([0 30])

figure
[Txy_x1,F]=tfestimate(out.x0,out.x2,[],[],100001,1000); %% Unsprung-Sprung
%%
plot(F,abs(Txy_x1))
title('Transmissibility Response')
xlabel('Frequency(Hz)')
ylabel('Transmissibility Ratio')
xlim([0 30])

%% Combined Plot (Sprung-Unsprung) - 2DOF Passive%%
figure
[Txy_x2,F]=tfestimate(out.x0,out.x1,[],[],100001,1000); %% Road-Unsprung
%%
plot(F,abs(Txy_x2))
title('Transmissibility Response')
xlabel('Frequency(Hz)')
ylabel('Transmissibility Ratio')
xlim([0 30])
ylim([-1 3])
hold on

[Txy_x1,F]=tfestimate(out.x0,out.x2,[],[],100001,1000); %% Unsprung-Sprung
%%
plot(F,abs(Txy_x1))
title('Transmissibility Response')
xlabel('Frequency(Hz)')
ylabel('Transmissibility Ratio')
xlim([0 30])
```

```
ylim([-1 3])
legend({'Unsprung mass','Sprung mass'})
grid on
hold off

%% Combine Plot Sprung mass only (PSS-PID-FLC) %%
%%% Passive %%%
figure
title('Transmissibility Response (Passive vs PID)')
xlabel('Frequency(Hz)')
ylabel('Transmissibility Ratio')
xlim([0 10])
ylim([-1 5])
hold on

[Txy_x1_passive,F]=tfestimate(out.x0,out.x2_passive,[],[],100001,1000); %% Unsprung-Sprung %%
plot(F,abs(Txy_x1_passive))
grid on
hold on

[Txy_x1_pid,F]=tfestimate(out.x0,out.x2_pid,[],[],100001,1000); %% Unsprung-Sprung %%
plot(F,abs(Txy_x1_pid))
hold on

[Txy_x1_flc,F]=tfestimate(out.x0,out.x2_flc,[],[],100001,1000); %% Unsprung-Sprung %%
plot(F,abs(Txy_x1_flc))
legend({'Sprung mass-Passive','Sprung mass-PID','Sprung mass-FLC'})
hold off
```

Syntax for weighted acceleration

```
%% Weighted acceleration - 2DOF %%

kt=500000; k1=80000; D=350; M2=320; M1=2500;
tmax=max(road_input(:,1));
out=sim('suspmod3.slx'); %% edit model name &&
Fs=1/out.tout(2);

[Pxx,Freq] = pwelch(out.x1_dot_dot_passive,[],[],10000,Fs); %% at sprung mass %%
Amplitude_spectrum=(2*Pxx*(Freq(2)-Freq(1))).^0.5;

index=find(Freq>30);
Amplitude_spectrum_truncated=Amplitude_spectrum(2:index(1)); % ignore Zero frequency

Freq_truncated=Freq(2:index(1)); % ignore Zero frequency
```

```
figure
plot(Freq_truncated,Amplitude_spectrum_truncated)
xlabel('Frequency (Hz)')
ylabel('Acceleration (m/s^2) - no weighting')
grid on

%% Discomfort estimation from ISO standard %%

Freq_of_weight=[0.1 0.125 0.16 0.2 0.25 0.315 0.4 0.5 0.63 0.8 1 1.25 1.6 2 2.5
3.15 4 5 6.3 8 10 12.5 16 20 25 31.5]';
Discomfort_weight=0.001*[31.2 48.6 79 121 182 263 352 418 459 477 482 484
494 531 631 804 967 1039 1054 1036 988 902 768 636 513 405]';
figure
plot(Freq_of_weight, Discomfort_weight)
xlabel('Frequency (Hz)')
ylabel('Discomfort weighting for vertical motion')
grid on

Discomfort_weight_acc=interp1(Freq_of_weight,Discomfort_weight,Freq_
truncated); % resampled discomfort curve to the frequency of acceleration response

%% Weighted acceleration response %%

Weighted_acc=Amplitude_spectrum_truncated.*Discomfort_weight_acc;
figure
plot(Freq_truncated, Weighted_acc)
xlabel('Frequency (Hz)')
ylabel(['Weighted acceleration (m/s^2)'])
grid on
Overall_discomfort_resp=(sum(Weighted_acc)).^0.5
```

References

1. Ahmed AE-NS, Ali AS, Ghazaly NM (no date) PID controller of active suspension system for a quarter car model. 8(6)
2. Azizi A (2018) Computer-based analysis of the stochastic stability of mechanical structures driven by white and colored noise. Sustainability 10(10):3419. https://doi.org/10.3390/su10103419
3. Azizi A (2020) A case study on computer-based analysis of the stochastic stability of mechanical structures driven by white and colored noise: utilizing artificial intelligence techniques to design an effective active suspension system. In: Complexity, pp 1–8. https://doi.org/10.1155/2020/7179801
4. Azizi A, Mobki H (2021) Applied mechatronics: designing a sliding mode controller for active suspension system. In: Santos M (ed) Complexity, pp 1–23. https://doi.org/10.1155/2021/6626842
5. Balkwill J (2017) Performance vehicle dynamics: engineering and applications. Butterworth-Heinemann

6. Darus R, Sam YMd (2009) Modeling and control active suspension system for a full car model. In: 2009 5th International colloquium on signal processing & its applications (CSPA). IEEE, Kuala Lumpur, Malaysia, pp 13–18. https://doi.org/10.1109/CSPA.2009.5069178
7. Fischer D, Isermann R (2004) Mechatronic semi-active and active vehicle suspensions. Control Eng Pract 12(11):1353–1367. https://doi.org/10.1016/j.conengprac.2003.08.003
8. International Journal of Engineering and Applied Sciences (IJEAS) (2018) Int J Eng Appl Sci (IJEAS). https://doi.org/10.31873/IJEAS
9. Kaldas MMS et al (2011) Development of a semi-active suspension controller using adaptive-fuzzy with Kalman filter. SAE Int J Mater Manuf 4(1):505–515. https://doi.org/10.4271/2011-01-0431
10. Kaldas MMS et al (2017) Optimization of damper top mount characteristics for semi-active suspension system. SAE Int J Commer Veh 10(1):210–218. https://doi.org/10.4271/2017-01-0412
11. Kumar V et al (2018) Self-tuned robust fractional order fuzzy PD controller for uncertain and nonlinear active suspension system. Neural Comput Appl 30(6):1827–1843. https://doi.org/10.1007/s00521-016-2774-x
12. Nagarkar M et al (2018) Multi-objective optimization of nonlinear quarter car suspension system—PID and LQR control. Procedia Manuf 20:420–427. https://doi.org/10.1016/j.promfg.2018.02.061
13. Nazemian H, Masih-Tehrani M (2020) Hybrid fuzzy-PID control development for a truck air suspension system. SAE Int J Commer Veh 13(1):02-13-01-0004. https://doi.org/10.4271/02-13-01-0004
14. Nguyen TA (2023) Research on the sliding mode—PID control algorithm tuned by fuzzy method for vehicle active suspension. Forces Mech 11:100206. https://doi.org/10.1016/j.finmec.2023.100206
15. Onsy AM et al (2018) Influence of intelligent active suspension system controller design techniques on vehicle braking characteristics. SAE Int J Veh Dyn Stab NVH 3(1):31–46. https://doi.org/10.4271/10-03-01-0003
16. Pang H et al (2019) Robust state-feedback control design for active suspension system with time-varying input delay and wheelbase preview information. J Franklin Inst 356(4):1899–1923. https://doi.org/10.1016/j.jfranklin.2019.01.011
17. Rodriguez-Guevara D et al (2021) Active suspension control using an MPC-LQR-LPV controller with attraction sets and quadratic stability conditions. Mathematics 9(20):2533. https://doi.org/10.3390/math9202533
18. Sanfelici Bazanella A, Campestrini L, Eckhard D (2023) The data-driven approach to classical control theory. Annu Rev Control 56:100906. https://doi.org/10.1016/j.arcontrol.2023.100906
19. Satyanarayana VSV et al (2022) Optimum response of a nonlinear passive vehicle suspension system under random road excitations. SAE Int J Commer Veh 16(1):02-16-01-0004. https://doi.org/10.4271/02-16-01-0004
20. Shafiei B (2022) A review on PID control system simulation of the active suspension system of a quarter car model while hitting road bumps. J Inst Eng (India) Ser C 103(4):1001–1011. https://doi.org/10.1007/s40032-022-00821-z
21. Soliman A, Kaldas M (2021) Semi-active suspension systems from research to mass-market—a review. J Low-Freq Noise Vib Active Control 40(2):1005–1023. https://doi.org/10.1177/1461348419876392
22. Taskin Y, Hacioglu Y, Yagiz N (2017) Experimental evaluation of a fuzzy logic controller on a quarter car test rig. J Braz Soc Mech Sci Eng 39(7):2433–2445. https://doi.org/10.1007/s40430-016-0637-0
23. Theunissen J et al (2021) Preview-based techniques for vehicle suspension control: a state-of-the-art review. Annu Rev Control 51:206–235. https://doi.org/10.1016/j.arcontrol.2021.03.010
24. Thite AN et al (2013) Development of an experimental methodology to evaluate the influence of a bamboo frame on the bicycle ride comfort. Veh Syst Dyn 51(9):1287–1304. https://doi.org/10.1080/00423114.2013.797591

25. Turcotte J, East W, Plante J-S (2022) Experimental assessment of a controlled slippage magnetorheological automotive active suspension for ride comfort. SAE Int J Veh Dyn Stab NVH 6(4):10-06-04-0024. https://doi.org/10.4271/10-06-04-0024
26. Williams DE (2018) Active suspension: future lessons from the past. SAE Int J Veh Dyn Stab NVH 2(2):147–165. https://doi.org/10.4271/10-02-02-0010
27. Wu J, Zhang L, Meng D (2023) Ride comfort analysis of seated occupants based on an integrated vehicle-human dynamic model. In: WCX SAE world congress experience, Detroit, Michigan, United States, pp 2023-01-0914. https://doi.org/10.4271/2023-01-0914
28. Yu M, Evangelou SA, Dini D (2023) Advances in active suspension systems for road vehicles. Engineering S2095809923003983. https://doi.org/10.1016/j.eng.2023.06.014

Design and Development of an Autonomous Solar Panel Cleaning Robot to Enhance Photovoltaic Efficiency

Khalid Salim AL-Khanbashi and Ali Al-Humairi

Abstract This research presents the design, development, and simulation analysis of an autonomous dry-cleaning robot for solar panels, aiming to enhance photovoltaic (PV) system performance in dusty, arid environments like Oman. If PV panels are covered with dust, sand, and air particles, their energy production can drop by up to 30% within a couple of weeks. Most cleaning in these parts of the world is done manually and by using a large amount of water which is no longer a good solution in the Gulf region. A cleaning robot was made with a strong but light aluminum body, rubber wheels for easy movement and a revolving nylon brush that lifts dust from the surface into the air. Edge detection, avoiding obstacles and stability while moving are made possible by including infrared, ultrasonic, and IMU sensors in the robot. An Arduino Mega microcontroller controls it, and its operation is wireless using XBee, with batteries charged by the sun when there is no power grid. Finite element analysis (FEA) simulations in ANSYS were performed on both fixed and portable versions of the robot to ensure structural strength and validate safe interaction with solar panels. The tests showed that the robot does not cause damage to the glass, since total panel deformation stays under 0.1 mm even when it is fully loaded. The project offers a good solution for meeting the need for sustainable care of solar panels. It reduces water consumption, guarantees solar panel safety, and brings a flexible way to clean solar panels in deserts. The research also aids local knowledge development and renewable energy goals in Oman Vision 2040 by providing a base for future autonomous maintenance systems.

Keywords Autonomous robot · Solar panel cleaning · Photovoltaic efficiency · Dry cleaning system

K. S. AL-Khanbashi
Department of Engineering, German University of Technology in Oman (GUtech), Halban, Oman

A. Al-Humairi (✉)
Department of Computer Science, German University of Technology in Oman (GUtech), Halban, Oman
e-mail: ali.alhumairi@gutech.edu.om

A. Azizi (ed.), *Autonomous Robotics and Intelligent Systems*, Emerging Trends in Mechatronics, https://doi.org/10.1007/978-981-95-5867-4_9

1 Introduction (Literature Review)

1.1 Background

Oman gets on average over 300 days of sunshine every year, which makes solar energy very suitable there and high solar light (approximately 5.2 kWh/m^2 day of DNI) [1] solar photovoltaic (PV) energy is a widely available resource that can be utilized multiple times. In the future, the country plans to use these resources as much as possible. Under the Oman Vision 2040, the government has placed great importance on renewable energy goals. The law projects that energy use from renewables will be around 30% by 2030 and on track to hit 35–39% by 2040 [2, 3]. Trying to meet these targets is one step in the direction of Oman's lower-carbon vision (e.g., achieving net-zero emissions by 2050 [3]. It aims to create better energy sustainability and variety. Still, these changes did not affect the Goals, showing that, at this time, as much as 7–8% of Oman's energy production comes from renewables [4]. All of it is solar energy, with most coming from the new solar parks. Notably, projects such as the new Manah-I PV plant (completed in January 2025) highlight this expansion, as it was intended to be built using automated dry-cleaning robots that help save water during routine maintenance [5].

1.2 Problem Statement

Solar photovoltaic panels lose efficiency in dusty environments due to the accumulation of dust, sand, and similar air particles on their surface. In an environment with deserts and limited rainfall in Oman, the output from solar panels declines if they are not cleaned regularly. The issue involves two factors: (1) Loss of Effectiveness—a soiled surface blocks the sun, causing less electricity to be generated and, in some cases, harming the cells due to overheating [6] and (2) Drain on Resources—currently, removing dust from panels occurs through manual efforts or cumbersome systems, resulting in unnecessary water and effort consumption on a large scale. In the long run, it is more cost-effective to clean the panels manually than to use robots [7]. Additionally, since Oman faces a water shortage, manually rinsing carpets is not a sustainable option [6].

Automatic and semi-automatic cleaning technologies are now available; however, they still face difficulties in addressing local challenges. Despite automating cleaning with fixed sprinklers, the necessary water supply and installation costs often remain high [6]. Some international robotic cleaners may not perform well in the Middle East's dusty environments or with the unique shapes of local solar farms, and they often require specialized care and maintenance, which makes them costly to import and maintain. The community would benefit from an autonomous cleaning machine that operates without or requires very little water, is adaptable to the climate in Oman, and is cost-effective for large-scale production [8].

1.3 Significance of Study

Enhancing Solar Energy Output: Middle Eastern countries invest heavily in solar energy due to their high solar potential, and maintaining the performance of photovoltaic (PV) installations is crucial. By reducing dust-related losses, using an autonomous cleaning robot directly contributes to higher energy yields. Even for a large solar farm, a modest improvement in average panel cleanliness can translate into a substantial increase in annual energy production. This helps maximize the energy output, which results in increasing returns on investment for renewable projects and ensures that national renewable energy targets can be met more reliably, such as Oman's goal of 30% renewable electricity by 2030 [6, 7].

Water Conservation Is Crucial in Arid Regions: This method could save a huge amount of water, equivalent to what is needed for a solar farm, even though the traditional process uses around 2.5 L of water every time a solar panel is properly cleaned. Billions of liters are used annually worldwide. Taking care of the solar plant is another way to handle the concern of water use in the area. For this reason, solar power depends on safe energy sources and does not release any dangerous materials during its use or maintenance [6, 7].

Labor and Cost Efficiency: By automating the cleaning procedure, less manual work is required. This method is significant for isolated solar farms in Oman's interior, where it is costly and logistically challenging to deploy cleaning crews. This has the potential to reduce the operational expenses (O&M) of solar plants over time. By reducing the need for humans to carry out the hot and hazardous chore of enormous cleaning arrays of panels, it also increases worker safety, especially in Oman's climate [8, 9].

Technological Advancement and Local Capacity Building: The project cultivates advanced skills in Oman by using recent technology. Solutions begin by assessing regional difficulties and suggesting adaptable approaches, for example, the problem of saline dust in Dhofar and on Oman's coast. People may use what they learn through the process to support upcoming studies or to develop products that help new and existing companies. Some companies in Oman, including Dymuma, are now active in renewable energy and this project gives them more incentive to proceed. It lets the universities, businesses, and government departments know how useful and beneficial domestic engineering is for reknit seems like you're asking for a solution or completion of the information related to the provided text on control logic and automation, as well as 3D modeling of a robot. However, without a specific question or problem to address, it's a bit challenging to provide exactly what you need.

Sustainability and Climate Change Mitigation: By better operating solar panels, we can help control climate change. Solar panels collect more energy, in place of creating energy using fossil fuels. At the same time, thanks to its water-saving and environmentally friendly cleaning, the robot helps solar farms avoid any damage to nature. As a result, solar energy in Oman is sustainable wherever it is used, from gathering needed materials to carrying out operations [7, 10].

Sustainability and Climate Change Mitigation: By better operating solar panels, we can help control climate change. Solar panels collect more energy, in place of creating energy using fossil fuels. At the same time, thanks to its water-saving and environmentally friendly cleaning, the robot helps solar farms avoid any damage to nature. As a result, solar energy in Oman is sustainable wherever it is used, from gathering needed materials to carrying out operations [7, 10].

Cost Considerations: Although the study utilizes reasonably priced parts, it does not strive for the lowest cost possible and lacks an economic analysis. They use literature as the foundation, not field evidence, to estimate costs and benefits [7].

Generalizing Results: Cleaning performance is determined only by conditions during testing. Because environmental conditions such as humidity near the coast and seasonal changes may influence the results, they should not be used as a general principle unless tested more thoroughly [7].

1.4 Overview of Photovoltaic Systems and Efficiency Challenges

Through the photovoltaic effect, making them a significant source of clean and renewable energy worldwide, particularly in regions like Oman that receive abundant sunlight. Let us examine the basic elements of photovoltaic systems: at the center are the solar panels, known as PV modules, which gather sunlight and generate electricity. After converting the current, an inverter sends the resultant AC to both the appliance and the circuit. Batteries can store extra energy for later use and prevent overcharging or over discharging. Components from power electronics are used to control both the voltage and current of the batteries and the inverter. Surplus AC is sent into the electricity grid in grid-connected systems, while PLL techniques and voltage control keep power to the inverter in time with the grid (Fig. 1) [11].

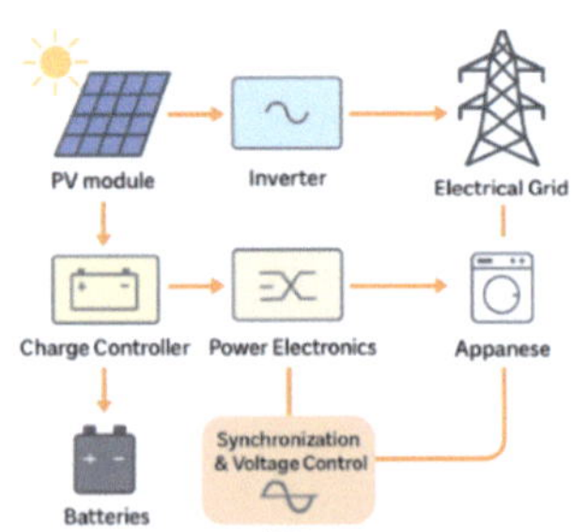

Fig. 1 This diagram illustrates a solar photovoltaic (PV) energy system with grid integration and battery storage [11]

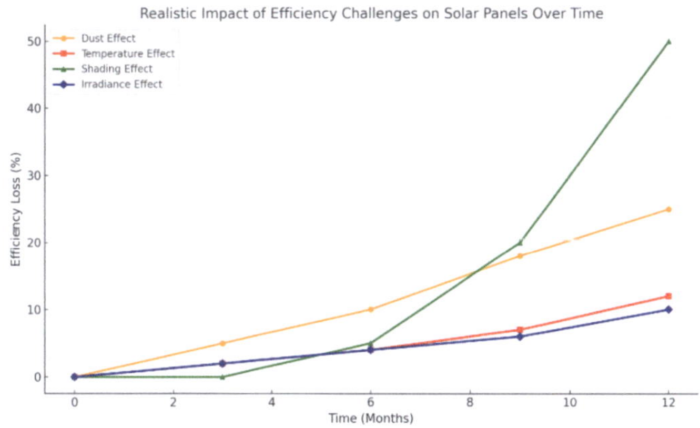

Fig. 2 Realistic impact of efficiency challenges on solar panels over time [43–46]

Although there is plenty of solar energy, several environmental, installation, maintenance, and technical issues related to solar modules can change how efficiently photovoltaic systems produce electricity. A solar panel's efficiency is usually calculated by comparing the electricity it makes with the amount of sunlight it absorbs. Different things impact PV energy production, such as temperature, as raising the temperature causes PV units to work more slowly [12]. PV systems can lose their efficiency if the sunlight is blocked by shading from nearby structures, trees, or other objects. How much sunlight hits the panels, called irradiance, decides how much energy will be created by the panels. A better solar position results in more power being produced on the panels; worse conditions, such as clouds or earlier parts of the day, lessen production. At both 1000 W/m^2 and 25 °C, the panels give their peak performance [13]. Concentrations of dust, dirt, and particles on the solar panels lower the amount of sunlight that gets to the cell and, as a result, make the system less efficient. Since efficiency declines in dusty or polluted regions, this is a significant challenge in those places.

It makes it easy to see the effects of the environment on solar panel performance. An illustration was created to illustrate how the changes in dust, room temperature, amount of shade, and light can lead to a drop in energy performance throughout the year using real data from different scientific papers (Fig. 2).

1.5 Impact of Dust and Debris on Solar Panel Efficiency

Not only do terrible weather conditions, such as rain and hail, cause wear and tear, but also dust and debris that accumulate on solar panels can decrease efficiency by blocking the photovoltaic cells from sunlight. This problem is particularly acute in areas such as deserts or urban areas, where high levels of dust or pollution are frequently reported. A thorough survey of 278 studies revealed that dust accumulation is a significant performance deterrent in all climates, often exacerbating other issues, such as humidity and wind [14].

One of the studies observed a reduction in panel power output by 30% after 32 days without cleaning, and by 65.8% after six months of dust deposition, which is not a small percentage in energy production. In many other studies, significant power output reductions of 20–40% are common over several months without cleaning under dry conditions, and 5–15% power decreases are seen in moderately dusty or polluted locations [15]. Moreover, the impact is far greater in high-dust regions compared to urban and coastal solar installations, in some locations lose over 1% of output per day, estimated 80% efficiency loss in a month if panels are left uncleaned [6]. A National Renewable Energy Laboratory report indicates annual soiling losses of ~7% in some parts of the U.S versus up to 50% in Middle Eastern deserts [10]. Results in Oman experiments show 9% power loss after four weeks and ~20% after eight weeks without cleaning, which is this research mean area of interest [11]. These regions have the most unique geological structures that come with different types of soil with different densities that affect efficiency with different percentages based on the location. Research was conducted in Sohar, Oman, on PV system performance with a variety of dust types and how these dust types can affect the solar life and output power, as it is shown in Fig. 3. Any kind of dust with a small amount can cause a power output drop.

Other research supports the previous one [16]. Power loss as a function of dust concentration in solar panels. Figure 4 illustrates a decline in power output as dust accumulation increases on the surface of the solar panels. The data is evidence that power loss increases as dust concentration rises. The graph also reveals the intervention points of the maximum power point tracker (MPPT) for different types of contaminants: leaves, 16 g/g/panel; straw, 22 g/g/panel; and sawdust, 24 g/g/panel. For soil, no significant intervention is noted. This diagram highlights the significance of dust accumulation in effective system management about performance issues.

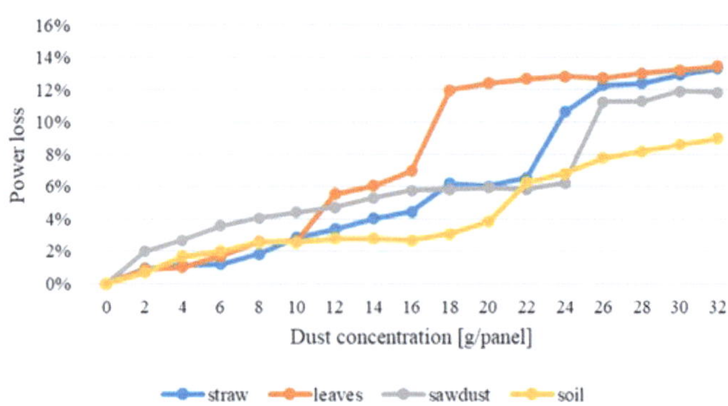

Fig. 3 This graph shows the impact of different dust types and their mass (g) on solar panel power output (W). It compares how various substances reduce performance as their mass increases on the panel surface [15]

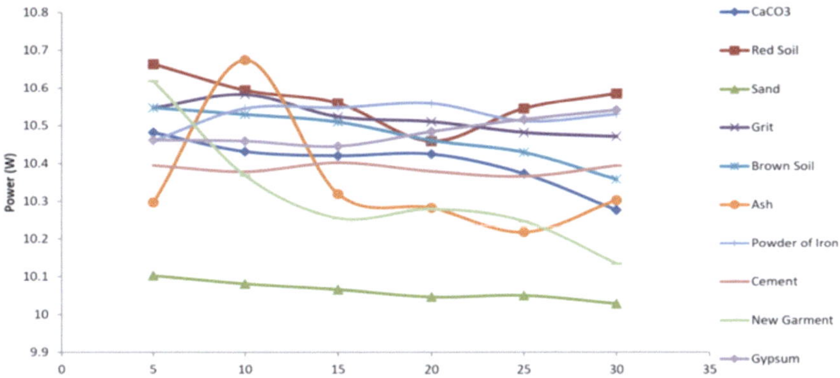

Fig. 4 This graph shows the impact of different dust types and their mass (g) on solar panel power output (W). It compares how various substances reduce performance as their mass increases on the panel surface [16]

1.6 Existing Solar Panel Cleaning Methods

Despite the popularity of manual cleaning as a practical approach, it requires significant labor resources and high-water consumption [17]. Systems with truck-mounted brushes offer greater efficiency for large-scale complexes, although water consumption remains a concern [7]. Autonomous robotic cleaning systems are increasingly used, mainly because of their water-saving capability, though they are quite costly to install [16]. Electrostatic technologies under development promise a charge-based method for removing dust without water, making them more feasible in arid regions [18].

Aggressive work on anti-soiling coatings suggests that they may reduce cleaning requirements, but their behavior over time under extreme conditions remains questionable [19]. Measurements made in Oman and elsewhere in deserts emphasize the pivotal role that periodic maintenance plays in the efficiency of solar panels, and stress that robotic cleaning and waterless technologies are best suited for areas with scarce water resources [17, 18]. A multi-pronged approach to cleaning consisting of basic, mechanical, and automated techniques, as well as anti-soiling coatings, is considered the most promising solution to overcome dust problems in the desert, urban, and coastal environments (Table 1) [19, 20].

1.7 Manual Cleaning

Manual cleaning is usually chosen for waterworks in places that are small or have spread locations such as houses or Oman's oil fields. They use brushes, long wipers, and purified water to properly remove dirt from the solar panels. Although it

Table 1 Below summarizes the pros and cons of major cleaning approaches in the context of desert and similar conditions, based on recent literature

Cleaning method	Cleaning effectiveness	Water requirement	Typical costs	Notes and source
Manual (hand/brushing)	High	High	High labor costs over time	Very flexible; limited by water and labor availability [17]
Truck/brush mechanized	High	High	Moderate (equipment + fuel)	For large farms, they still use water, but less labor [17, 21]
Fixed sprayers/sprinklers	Moderate	Very high	Moderate (installation)	Automated; effective for loose dust, not heavy grime. Suited for sites with water access [21]
Robotic (dry brush)	High	**None**	High initial, low ongoing	Excellent for dust control. No water needed—ideal for deserts, but costly upfront [17]
Drone cleaning	Low–moderate	Low (air blow or spot mist)	Medium (drones + ops)	Still experimental. Suitable for targeted cleaning. New tech of drones with AI integration, equipped with AI and camera sensors. It involves waterless cleaning techniques, such as air jets or microfiber brushes. Example: Skydio Drone [21]
Electrodynamic (EDS)	Moderate (for dust)	**None**	High (prototype stage)	Promising self-cleaning tech. Removes fine dust without water; however, it is expensive and not effective on crusted dirt [21]

(continued)

Table 1 (continued)

Cleaning method	Cleaning effectiveness	Water requirement	Typical costs	Notes and source
Coatings (hydrophobic/ etc.)	Low–moderate	**None**	Added module cost	Reduce dust accumulation. The effect varies with the environment; it may delay but avoid the need for cleanings [21]
Automated cleaning robots with ultrasonic waves	Low–moderate	**None**	Added module cost	Some robotic systems use ultrasonic waves to loosen dirt and debris from solar panel surfaces. Ex: Robotex ultrasonic cleaning system. Type: Dry (ultrasonic)

yields impressive results immediately after cleaning, it requires considerable effort, consumes a substantial amount of water (2–5 L per panel), and can cause issues such as glass scratches or component damage. Because results are unpredictable and cleaning is not systematic, those in charge are looking for different methods to be more efficient (Table 2).

1.8 Semi-automated Cleaning Systems

Using semi-automated machines, operators combine the benefits of manual and fully automated cleaning, as the machines need supervision. Rotating brush mechanisms on tractor-based cleaning machines improve how effectively panels are cleaned, but these machines have boundaries because of the need for enough space and proper land grade. One option was portable robotic cleaners which Wash Panel models were an example of. They are arranged in a row and must still be adjusted manually for the load. The use of these systems together with metered spraying or air blowers makes farming more efficient and requires less water and more farmers in Oman are adopting them. Still, there are some problems with the uneven surface and washing machines that send out dust (Table 3).

1.9 Automated Cleaning Systems

Permanent automated cleaning machines are set up on solar arrays, so little or no manual inspection is needed. Not effective at deep washing, simple sprinklers are

Table 2 The table below is a comparison of the manual cleaning methods in the context of desert and similar conditions, based on recent literature

Category	Wet manual cleaning [19]	Semi-wet manual cleaning [22]	Dry manual cleaning [23]
Technology	Brushes, sponges, and cloths with water	Damp clothes, water sprayers	Brushes, microfiber cloths
Cost	Low	Low	Low
Water usage	High	Moderate	None
Effectiveness	Moderate to high	High	Moderate to high
Ease of maintenance	High	Moderate	High
Environmental impact	High (water usage)	Moderate (water usage)	Low (waterless)
Energy consumption	Low	Low	Low
Regulatory restrictions	None	None	None
Suitability for large-scale operations	Low	Moderate	Low
Power source	Human-powered	Human-powered	Human-powered
Capacity (cleaning area)	Small-scale, limited to residential areas	Small to medium-scale areas	Small-scale, residential
Best for	Small-scale, residential, low budget	Medium-sized residential/ commercial	Small-scale, residential, low budget
Pros	Low upfront cost, simple, flexible, no need for special equipment	Low-cost, flexible, moderate water usage	Low-cost, waterless, simple
Cons	Labor-intensive, time-consuming, and unsafe for high-rise panels	May require extra effort to maintain moisture, not suitable for all panels	Labor-intensive, not effective on heavy dirt or grime

applied to solar panels on a regular basis, taking minimum effort, but using a lot of water. Modern systems usually use motors or gravity assist brushes to brush panels with commands which are installed on the frames of the solar panels (Table 4).

1.10 Robotic Cleaning Solutions

Robotic systems should help greatly in keeping solar panels functional and maintained and rail-guided and free-moving types are set to be used most commonly. Some countries use systems like Saudi Arabia's NOMADD which follows the same pathway as SANDFRIEND rows and uses brush technology to clean with no water. They guarantee the stations run correctly, stay on schedule and support their own

Table 3 The table below categorization of semi-automated cleaning systems

Category	Technology	Example	Use Case	Region
Motorized brushes	Rotating brushes for cleaning solar panels	SolarBrush [24]	Used for large-scale solar farms to reduce manual labor	Germany, Australia
		CleanSolar [25]	Robotic brush system used for both residential and commercial solar panels	USA
Robotic systems with water sprayers	Robotic systems equipped with water sprayers	Hindon Solar Robotic System	Used for solar farms in desert regions to manage dust buildup	India
		SolarCleano [25]	Waterless cleaning for large-scale solar installations	Middle East, USA
		Ecoppia E4 [26]	Waterless cleaning for large commercial solar farms	UAE, Oman
Hybrid cleaning systems (water + robotic arms)	Combination of water and robotic arms	Trina Solar Robotic System [27]	Hybrid system for large-scale solar farms in desert regions	China
		Aquatec [28]	Using water for cleaning while ensuring minimal water usage, ideal for large solar farms	Spain
Robotic arm system	Robotic arms for precise, localized cleaning	CleanSolar [25]	Robotic arm system for cleaning residential and commercial panels	USA
		HyCLEANER Robot [29]	Robotic microfiber cleaning systems are used in large solar farms in dry regions	USA
Waterless cleaning systems	Waterless cleaning methods for solar panels	Ecoppia E4 [26]	Waterless cleaning for large commercial solar farms in water-scarce regions	UAE, Oman
		HyCLEANER Robot [29]	Robotic microfiber cleaning systems are used in large solar farms in dry regions	USA
Local example (Oman)	An automated system using robotic arms	Oman Solar Panel Cleaners	Used for commercial solar panels, focusing on efficiency in Oman's desert climate	Oman
Custom cleaning system (Oman)	Custom-built robotic cleaning system	Oman Desert Cleaners	A custom robotic cleaning system tailored for desert conditions effectively cleans solar panels in high-dust environments	Oman

recharging. The technologies resemble Ecoppia in having robots that move around panels on their own with brushes and blowers, without using water. Only the solar cells built in them supply power to these watches. With these systems, it is possible to clean at night, since they have sensors that track dirt and share this info with the system to direct the right adjustments to the cleaning program. Interest in new

Table 4 The table below: categorization of automated cleaning systems

Category	Ecoppia E4 [26]	SunBrush (Germany) [30]	Sunbotics [31]	Trina solar robotic system [27]	AI-driven drones [32]	Ultrasonic cleaning robots
Type	Waterless	Semi-wet	Wet	Semi-wet	Waterless (AI-driven)	Dry (ultrasonic)
Technology	Ecoppia E4 (waterless cleaning system)	SolarBrush (rotating brushes)	Hindon Solar's robotic cleaning system	Trina Solar hybrid cleaning system	Skydio Drone (AI-driven flying robot)	Robotex ultrasonic cleaning system
Cost	High	High	High	High	High	High
Water usage	None	Moderate	High	Moderate	None	None
Effectiveness	High	High	High	High	High	High
Maintenance	Moderate	Moderate	Moderate	Moderate	Low	Low
Energy consumption	Low	Moderate	High	Moderate	Low	Low
Regulatory restrictions	None	None	None	None	Regulatory (airspace)	None
Suitability	High	High	High	High	High	Moderate
Power source	Solar/battery-powered	Electric-powered (grid or solar)	Electric-powered (grid or solar)	Electric-powered (grid or solar)	Battery-powered (solar-assisted)	Electric-powered
(Cleaning area)	Large-scale, up to thousands of panels	Large-scale, up to thousands of panels	Large-scale, up to thousands of panels	Large-scale, up to thousands of panels	Medium-to-large solar farms, hard-to-reach areas	Medium-large solar farms, targeted cleaning
Best for	Large commercial solar farms	Large-scale farms	Large-scale solar farms	Large-scale operations	Large solar farms, autonomous areas	Targeted cleaning for hard-to-reach panels

(continued)

Table 4 (continued)

Category	Ecoppia E4 [26]	SunBrush (Germany) [30]	Sunbotics [31]	Trina solar robotic system [27]	AI-driven drones [32]	Ultrasonic cleaning robots
Pros	Reduces labor costs, provides consistent results, and offers waterless options	Efficient for large solar farms, reduces manual labor	Reduces manual labor, efficient in large installations	Flexible, adapts to varying dirt levels, and is efficient	AI-driven, autonomous, waterless	Efficient, waterless, high precision cleaning
Cons	High upfront cost, requires maintenance, can be inefficient in heavy dirt	Still consumes water, not suitable for all panel types	High energy consumption, water usage	Uses water, high complexity, and cost	High operational cost, regulatory challenges	Expensive, limited to specific dirt types

cleaners is growing in Oman, and this has led to the development of local robots such as the RAID Lab/PDO robot and Dymuma's dry-cleaning robot to better address the environmental needs of the country, conserving water and helping oil companies service solar panels (Tables 5 and 6).

Table 5 Table below: categorization of robot cleaning solutions

Category	SolarCleano drone [25]	Joyance drone	Ecoppia E4F [26]	Dust removal air blower
Type	Dry	Wet	Dry	Dry (air blower)
Technology	Autonomous drone cleaning	Water pump	Waterless	Vortex cleaning system (air jets)
Cost	High	High	High	Moderate
Water usage	None	Moderate	None	None
Effectiveness	Medium to high	High	Medium to high	High
Maintenance	Low	Low	Low	Low
Energy consumption	Low	High	Low	Low
Regulatory restrictions	Regulatory restrictions	Regulatory restrictions	None	None

Table 6 Table below: categorization of robot cleaning solutions

Category	SolarCleano drone [25]	Joyance drone	Ecoppia E4F [26]	Dust removal air blower
Suitability for large-scale operations	High	High	High	Moderate
Power source	Battery-powered (solar-assisted)	Battery-powered (solar-assisted)	Solar-powered (autonomous)	Electric-powered
Best for	Large solar farms, hard-to-reach areas	Large installations, hard-to-reach areas	Large-scale, dry regions	Areas with heavy dust accumulation
Pros	Can clean hard-to-reach areas, waterless, and autonomously	Flexible, quick cleaning, and high adaptability	Waterless, reduces environmental impact, and is autonomous	Efficient for dust, low maintenance
Cons	High operational cost, regulatory issues, and limited efficiency on large panels	Regulatory issues, water usage, and high maintenance	Expensive, limited to dry conditions, and needs large-scale installations	Not suitable for heavy grime

1.11 Cleaning Methods Comparison

Ensuring that solar panels are clean is crucial for energy efficiency and can help solar systems last significantly longer. Dust, bird droppings, and other types of environmental dirt reduce panel performance, resulting in losses of up to 30% in high-dust areas. Several methods for cleaning up this type of waste have been developed, each with its strengths and weaknesses. Read the table underneath to help decide on the best way to field train, depending on your specific tasks and overall conditions (Table 7).

1.12 Gaps in Current Technologies

While cleaning technologies have improved, there are still exceptional gaps that challenge existing systems to date, hence the need for constant research and development. Targeting these gaps helps identify areas where this research has room for the field to progress through fresh insights or methodologies.

Sticky Versus Dry Dust Soiling: In many arid climates, dust on panels is not purely loose sand but includes salt and fine particulates that bind when moistened (e.g., overnight dew). For example, Kazem and their team discovered that dew in Oman leads to the creation of sodium, magnesium, and calcium salts. Grow on panels and make it difficult to remove, as it quickly forms a strong and adherent layer. In practice, panels in industrial and urban areas (e.g., Sohar, Liwa) developed chemically adherent dust that required alkaline cleaning (using a sodium solution) to restore output. In contrast, panels found in cleaner desert areas could be cleaned with plain water [33].

High Cost and Scale: Very few robots are affordable for most, as they are pretty expensive. Maintenance costs. Small or distributed solar installations (e.g., on rooftops or small farms) often do not justify the expense of custom robots. Automation frequently means fewer personnel are required—research based on large projects shows Solar farms in the Arabian area have seen that using machines for cleaning instead of people can significantly reduce costs. Plants of the MW scale—the initial outlay and specialized efforts required to care for them (spares, consumables) tend to restrict robots to utility-scale projects. A review published this year observes that robotic cleaning systems generally increase overall O&M cost (due to energy use and mechanical maintenance) even as they reduce water and manual labor [34, 35].

Navigation and Terrain Constraints: Mounted groups that are not complex. Solar sites on uneven or complex terrain (steep slopes, multi-tilt tracking fields, irregular panel geometries) pose navigation challenges that most robots do not know how to deal with yet. Often, operators have to reuse or reorganize cleaning machines. When smaller valleys split the rows of work. Projects on roofs with multiple angles or buildings on a hill are also challenging. Please note that some parts of the material

Table 7 Cleaning methods comparison table

Cleaning method	Description	Advantages	Disadvantages
Manual cleaning	Involves using tools like brushes, sponges, or cloths, often with water or mild detergents, to physically remove dirt and debris from panel surfaces	Low initial cost, simple to implement, no specialized equipment required	Labor-intensive, safety risks, especially on rooftops, potential for inconsistent cleaning quality
Robotic cleaning	Utilizes automated or semi-automated robots equipped with brushes or air jets to clean panels, often without water	Consistent and efficient cleaning reduces labor costs over time and minimizes water usage	High initial investment, maintenance of robotic systems, and specialized infrastructure may be required
Electrostatic cleaning	Utilizes electrostatic forces to repel dust particles from panel surfaces without physical contact or the use of water	Waterless cleaning method, minimal mechanical wear on panels, suitable for remote or water-scarce areas	Emerging technology with limited commercial availability may require specific environmental conditions
Nanoparticle coatings	Applies hydrophobic or hydrophilic coatings to panels to prevent dirt accumulation and facilitate self-cleaning	Reduces the frequency of manual cleaning, enhances panel efficiency by minimizing soiling, and provides long-lasting effects	Additional upfront cost for coating application. Effectiveness may diminish over time and may not eliminate the need for periodic cleaning
Installed hydraulic systems	Incorporates fixed sprinkler or spray systems to periodically wash panels with water	Automated cleaning process, effective in removing loose dirt, can be scheduled during off-peak hours	High water consumption, Risk of mineral deposits if non-deionized water is used, Installation and maintenance costs
Manual dry cleaning	Involves using dry tools, such as microfiber cloths or brushes, to remove dust without the need for water	No water required, suitable for light dust removal, reduces the risk of water-related damage	Less effective on sticky or oily residues, may generate static electricity, not suitable for heavy soiling

have not been cleaned properly. All in all, systems with impressive terrain capability are hard to find [36].

Maintenance of the Robots: Cleaning robots expose moving parts (such as brushes, tracks, pumps, and cameras) to abrasive dust and sometimes water, which can lead to wear and tear. It is essential to replace all your brushes and belts as needed and ensure that your electronics are properly sealed from dust. Field experience (and reviews) indicate that ongoing repairs and downtime can be significant—one study observed that motorized cleaning systems did boost output, but "increased the operational costs, the water usage, and the energy consumption" compared to passive cleaning. As a result, more wear parts are replaced, and the design must be stronger to withstand the increased wear and tear [37].

Smart Scheduling and Data Integration: Most PV cleaning is still performed at fixed intervals (e.g., monthly) or on an ad hoc basis, rather than dynamically. There is growing interest in data-driven cleaning (using dust sensors, weather forecasts, performance monitors), but few commercial systems implement it. For example, the prototype robot in Oman operates in two modes: dry cleaning, when sensors indicate dry soil, and wet cleaning, when sensors detect wet soil. Similarly, research suggests optimizing cleaning frequency through models (one study in Abu Dhabi found the economic optimum cleaning cycle to be ~34 days, longer than the typical 28-day rule of thumb). However, actual cleaning operations in the region rarely use real-time data; automation schedules (if any) are usually pre-set [36].

2 Methodology

2.1 Site Case Study

This case study employs a theoretical design methodology for developing an autonomous solar panel cleaning robot. To anchor the design in a real-world application, a case study of a photovoltaic installation was chosen, providing site-specific parameters that define the system's functional requirements and testing its real-life feasibility. The site is located at the German University of Technology in Oman (GUtech), in Halban, in the coastal region of Al Batinah Governorate, north of the Sultanate of Oman. The facility is situated in an open space near the university campus buildings, the 3D printing house, and workshops. It is only a few hundred meters from a busy expressway, which contributes to air pollution. This facility is a partnership between Shams Global Solutions (SGS) and BP Oman's Social Investment Program (BP Oman) [38].

GUtech's solar training facility features four different solar panel installation systems for various testing and training applications, capable of generating approximately 18 kWp (kilowatt peak) of solar power, as illustrated in Fig. 5. The facility features four different solar systems: two flat-roof solar systems, a pitched-roof solar system, and a ground-mount solar system. The two-flat-roof solar systems are installed on a platform approximately 1.5 m (m) high, as shown in Fig. 6. The system consists of 12 panels, each generating around 3900 kWp. The first one consists of 3 sets of 2 panels, located at the east and west, with an angle of approximately 10° to capture more sunlight. The second is directed south in portrait. In Fig. 7, the pitched-roof solar system has 12 modules facing south in a landscape format, generating the same power as the flat-roof systems.

Fig. 5 Shams solar training facility at GUtech

Fig. 6 The two-flat-roof solar systems on the 1.5 m high platform

Fig. 7 The pitched-roof solar system

The ground-mount solar system is a sub-array installed on the ground, comprising 18 PV modules with a nominal power of 325 Wp each, grouped into nine series and two parallel strings, as shown in Fig. 8. Under Standard Test Conditions (STC), the array's total nominal output is 6.50 kWp; however, in real-world scenarios, at a temperature of 55 °C, it operates at 5.55 kWp. The array's performance at high temperatures is demonstrated by its maximum power point (MPP) values, which are 317 V and 18 amps at 50 °C.

Fig. 8 The ground-mount solar system

All the solar panel systems at GUtech's Shams training facility consist of different components. In this study, the focus is on the ground-mount component system, which consists of:

- 20 solar modules made from multi-crystalline cells 0.15675 × 0.15675 (m), and the dimensions of each panel are: 0.992 m height, 1.960 m width, 0.04 m depth, and it weighs 2.25 kg.
- Huawei inverter—SUN2000-20KTL
  It is a three-phase dynamic string inverter of high efficiency used in medium to large-scale solar power systems. It achieves a maximum efficiency of 98.6%, featuring three MPPTs that support up to six input strings, providing flexibility for adjusting the angle of panels and shading. It has a DC input of up to 22,500 W and an MPPT voltage range of 480–800 V at full load. The output of the inverter is 20,000 W rated AC power with an average of 22,000 VA, while the source of energy can be a 400 V three-phase grid. It can be used at temperatures ranging from −25 to +60 °C. It is a size of 520 × 610 × 255 mm, which is installation-friendly, and a weight of 48 kg. Safety features include Type II burst protection, residual current detection, DC reverse polarity protection, and anti-islanding protection. It has interfaces such as RS-485, USB, and a graphical LCD [39]. Figure 9 shows the Huawei inverter—SUN2000-20KTL on site.
- The weather station sensor
  This sensor makes a crucial contribution to the monitoring and optimization of the solar power plant's performance. It also assists in scheduling maintenance and cleaning exercises. Primarily, it monitors essential environmental parameters, including solar irradiance, humidity, wind speed and direction, and solar module temperature. Figure 10 shows the sensor and its location.
- Thermopile-pyranometers-SMP11
  An apparatus called a thermopile pyranometer measures the sun's irradiance, or intensity, from the hemisphere of a flat surface impacted. The pyranometer structure consists of one or two domes, a black body absorber, and a thermopile (an apparatus that converts heat energy taken in from the sun into electrical energy). Additionally, it features both an RS485 Modbus interface and a current interface

Fig. 9 Huawei Inverter—SUN2000-20KTL

Fig. 10 The weather station sensor

(4–20 mA), allowing for easy connection to any meteocontrol data logger (Fig. 11) [40].

- Silicon-irradiance-sensor

  This sensor, made of a silicon solar cell, can measure radiation since there is a direct proportionality between radiation and short-circuit current. Connecting the

Fig. 11 Thermopile pyranometers SMP11

Fig. 12 Silicon irradiance sensor with additional fixed ambient temperature sensor

temperature sensor to the Silicon irradiance sensor allows it to monitor temperature as well. To measure the sun irradiance that directly hits the solar panels, the sensor is positioned at the same angle close to the panels (Fig. 12) [41].

- DustIQ
  It is an optical soiling monitoring system that records real-time dust buildup on the surface of solar panels and calculates the corresponding soiling ratio (SR), which helps allocate cleaning maintenance schedules more optimally. It is a non-moving and waterless device that has a 4–20 mA analog output and RS-485 Modbus RTU communication. The unit operates on 12–24 V DC, although it consumes only 1.5 W. Compact and robust, it takes 400 × 250 × 85 mm; its weight is around 3.5 kg. It is supplied with a 10-meter cable for easy installation. Designed to be mounted at the same tilt and orientation as PV panels, DustIQ performs well in outdoor environments (IP65 rated) with temperatures ranging from –20 to + 50 °C, requiring only occasional manual cleaning (Fig. 13) [42].

Fig. 13 Dust IQ soiling measurement

2.2 Research Design

The research employed a design-based engineering technique, focusing on the actual application and making repeated adjustments to the design through simulation. The goal for the robot was to design it as a controlled, self-driving solar panel cleaner that is flexible and adaptable, allowing it to increase efficiency at PV plants. To achieve this, the research plan included the following important steps outlined below. Requirements and specification analysis produced a set of precise system requirements for the solar panel cleaning robot.

System Requirements and Specifications Analysis: A set of clear system requirements and specifications was formulated for the solar panel cleaning robot. These requirements, including cleaning performance, safety, and durability, served as a roadmap for the design process, ensuring that the finished system adequately met the specifications (quantitative targets such as cleaning speed and battery life).

Conceptual Design: A brainstorming session was conducted to explore multiple conceptual solutions for developing the system architecture, including selecting mechanisms and materials (for example, considering a rail-guided vs. free-roaming approach and different cleaning mechanisms, such as brushes vs. air blowers). Using a decision matrix that weighed criteria (effectiveness, complexity, cost, etc.).

Detailed Mechanical and Electrical Systems Design: Supported by information from the conceptual design, a 3D model and full electrical components and connections were created, considering the requirements and specifications.

Modeling and Simulation: Using simulation tools to refine design and predict performance (e.g., motion planning, load analyses). Simulations were conducted where possible, for instance, to verify that the chassis could accommodate all components. Additionally, using topology to reduce the weight of the components.

2.3 System Requirements and Specifications

A list of system needs was created to direct the design based on the problem statement and literature review. These requirements and Specifications serve as success criteria of the design. They were developed based on the challenge's requirements and influenced by practical limitations seen in previous solutions (Table 8).

2.4 Design Considerations

Before putting the robot into service, a number of aspects were checked to make sure it could work well in places like hotels. Since it had to work in empty areas, its

Table 8 The table summarizes the key requirements and the target specifications for the robot

Requirement	Target specification
Cleaning effectiveness	≥90% of power output restored (relative to a clean panel baseline) per cleaning cycle; minimal visible dust remains on the surface
Cleaning speed	~2 min per standard 1 m × 1.6 m panel (or ~30 s per square meter); equivalent to ~30–40 panels/hour as a baseline
Autonomy	Fully autonomous traversal of a 10 m long row (typical 20-panel row) and return; no off-board control needed during operation
Power and energy	Onboard battery capable of at least 2 h continuous operation (to clean ~60 panels per charge); rechargeable via integrated solar panel over a day; utilize PV on robot to sustain daily cycles
Water usage	Dry cleaning by default (0 L for typical dust conditions); optional <0.1 L per panel in spot-clean mode for stubborn dirt (wet mode rarely triggered)
Navigation and coverage	Must cover the entire panel surface area, including edges; detect panel edges to stop or turn without falling; tolerance for minor obstacles (e.g., junction box protrusions on panel)
Environmental durability	Operate in ambient temperature 10–50 °C; electronics sealed to IP54 or better (dust-protected, splash-resistant); UV-resistant materials for prolonged sun exposure
Ease of deployment	Weight <15 kg for portability (so one person can lift if needed); able to be placed on a panel row by one person; setup time
Communication (optional)	Preferably wireless communication (Wi-Fi or LoRa) to send status or receive commands; if implemented, range at least 30 m to the nearest base station; fallback to offline operation if no connection
Cost	Prototype BOM costs under USD 1,000; scaled production cost target under USD 2,000 per unit, making it economically feasible compared to labor costs (Assumption: just for guiding design choices of components)

electronics could resist temperatures up to 85 °C and resist dust and its plastic was designed to withstand strong sunlight. So that buildings could withstand big daily temperature changes, joints and connections were made with materials that mixed well.

Being made of glass, solar panels are susceptible to scratching, so a mild cleaning method needed to be applied. Soft brush heads and dust-proof components were added, and these parts were driven either by a rotating motor wheel or by a vacuum made by the electric model. The load was arranged to prevent overloading the panels and the wheels were set on the outer display, so they sit on the panel's support and not on the surface.

Slanted panels (common 20°–30° angles in Oman) affect movement and grip. Wheels were found to be unable to handle heavy loads, so rubber tracks were selected instead, but using them takes more control. There is a trade-off between dust helping grip and dew reducing it which is why track drive was used for improved stability.

Falls off the edge were reduced by using infrared sensors under the panel to find the edges and proximity sensors in front of obstacles. Being power efficient was very important; this meant sizing gearboxes to meet the torque requirements, putting microcontrollers in sleep mode and reducing the amount of time the brushed motors ran. With solar integration, panels can be easily scaled and used for energy regeneration, which supports the management of several robots from one place.

Safeguards were included in case of problems with the battery or the motor. A robot can halt without risk, send messages through wired channels and move to the starting or stopping point. Maintenance was made easier thanks to simple brush attachment for cleaning, easy to fix detachable electronics and the design helps keep the robot clean.

To ensure ethical use, robots replace humans in risky solar cleaning tasks (e.g., rooftop or high-heat environments), while maintaining safety. Wireless connections are isolated, so no one can interrupt or affect the network, so the robot remains safe, avoiding disruption or collisions.

2.5 Mechanical Design

The base design of the robot's mechanics was made using the original concept. The bodies and parts physical connectives. Designers built the robot to be steady and to keep working even when moving across the solar panel while collecting dust (Figs. 14 and 15).

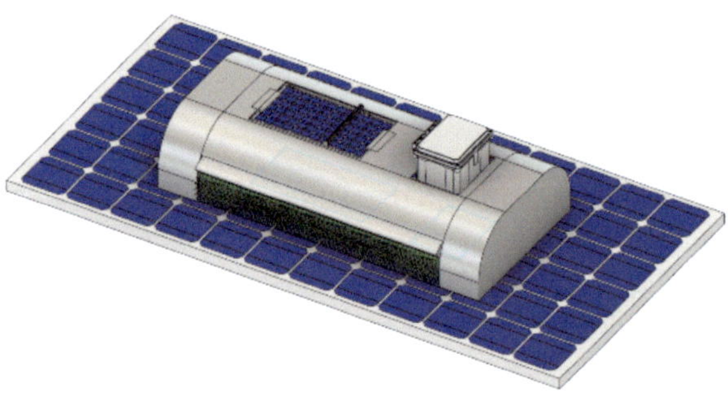

Fig. 14 Overall view of portable cleaning robot

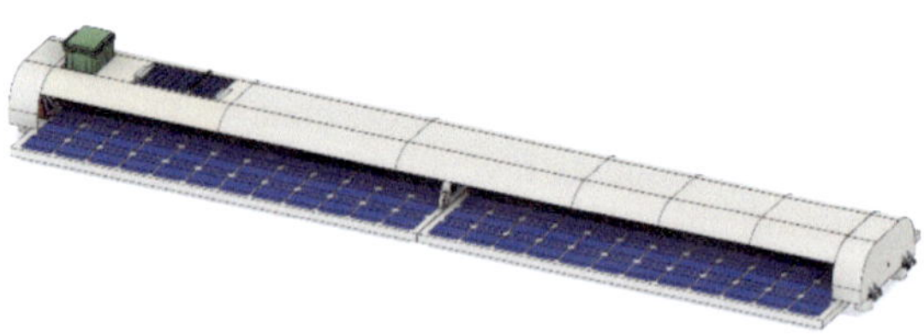

Fig. 15 Overall view of fixed cleaning robot

The robot's body combines a slim cross-shaped frame made of aluminum which is about 50 cm broad, and a long panel slightly extended by travel direction. The height is approximately 25 cm. All the subsystems are linked to the frame, for example, the drive system, cleaning brush assembly, battery and electronics enclosure and sensors. The frame remains rigid as a result of the cross-members which is necessary for the brush to always rest evenly on the surface. The design approach is a compact, bridge-like robot that can straddle a solar panel and travel along its length while cleaning, eliminating the need for a fixed track and can be changed to portable cleaning robot. Here we describe the design process and features:

Chassis and Frame: The frame features an organic shape developed through topology simulation, which reduces the robot's weight while maintaining its high strength. Its dimensions are 50 cm wide, and the length depends on the panel's structure (i.e., whether it has 1 or 2 panel rows). The frame has two side plates that rest near the panel's edges and three crossbeams that support the components. Aluminum is used as a primary material due to its strong yet lightweight properties, its ability to resist corrosion, and its capability to withstand the high temperatures of Oman (Fig. 16).

Mobility (Drive System): The robot features a track drive system mounted on both sides, enabling it to move. Each side has a continuous rubber track looped around front and rear wheels at an angle. The track of the robot runs along the panel's surface or slightly on the frame. This mechanism is designed for the uneven panel steak to cross obstacles. It utilizes sourced rubber tracks that are soft and feature embedded fabric for added strength, approximately 5 cm wide each. The sprockets are 3D-printed in durable nylon and mounted on steel shafts. Two DC gear motors (one for each track) connect the two tracks on one side with a belt and pulley; these are 12 V DC motors with a 150 RPM output speed under load and equipped with planetary gear reduction with an encoder. Each motor has a stall torque of about 20 kg cm,

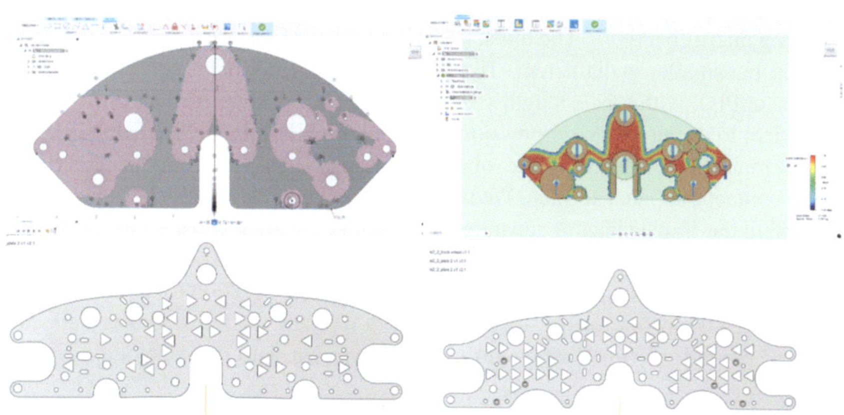

Fig. 16 Some topology simulation results

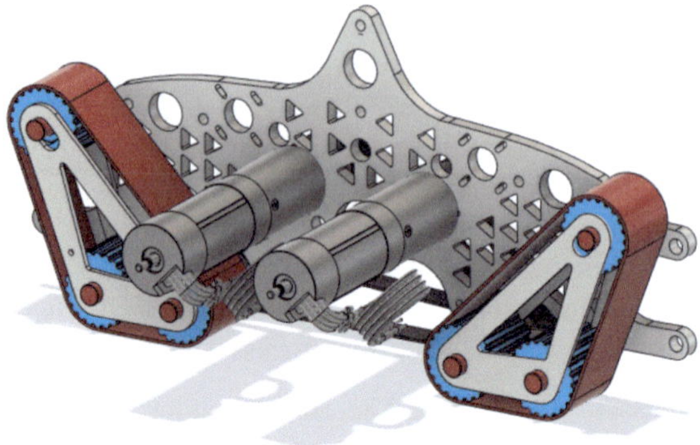

Fig. 17 The mobility system is connected to the side plate

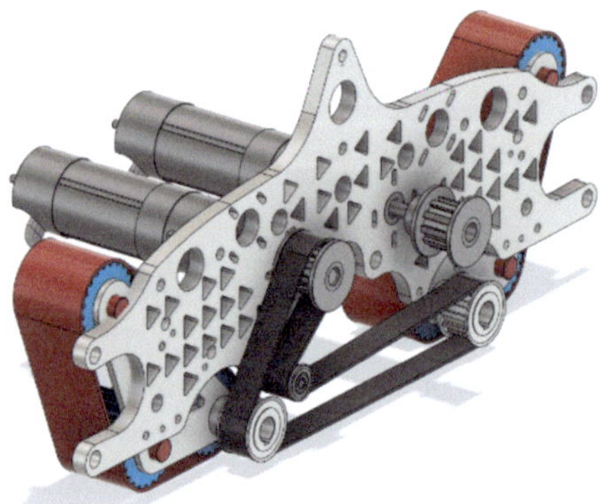

Fig. 18 This figure shows the driving system of the robot

sufficient to move the robot's weight plus overcome the friction of the brush (Figs. 17 and 18).

Cleaning Mechanism: A rotating cylindrical brush spanning the width of the panel or a rotating microfiber cloth. The brush is similar to those found in automatic car-wash systems, but smaller in diameter (~10 cm) and length (1 m). It has soft bristles made of nylon or PET, with bristle length ~5 cm to reach uneven spots on the panel (bristles can deflect to go into small depressions or around panel frame edges slightly). The brush is mounted at the front of the robot, protruding so that when it spins, the bristles lightly contact the panel surface. We mounted it on two bearing blocks, which were attached to the frame sides. A separate DC motor (12 V, high RPM ~3000, with belt drive reduction to ~500 RPM brush speed) powers the brush. The rotation direction can be reversed; we found that spinning in the opposite direction to the direction of travel causes the robot to flick dust away from its path. A thin shield or deflector is placed behind the brush to avoid dust being thrown onto the robot's electronics (Fig. 19).

The center of the robot's body often includes the electronics box and the battery which can be simply attached or removed. There is also a different cleaning system for the built-in solar panel, it has a removeable frame that simply connects to the

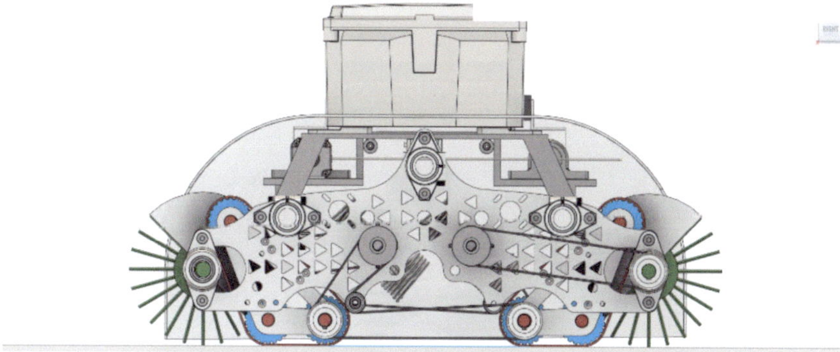

Fig. 19 This figure shows the cleaning and mobility mechanism using belts and pulleys

robot and comes off as well. The robot will have extra support provided by mid-point wheels and a support beam holder which prevents it from bending in the middle. We positioned heavier components (battery, ~2 kg) low and centrally to maintain a low center of gravity. The sensors (IR sensors for edge, etc.) are placed at strategic points: IR sensors at the bottom near the front and back edges of the robot, small caster wheels at each corner to act as supports (these casters roll on the panel surface with very low friction and support if the robot tips) (Fig. 20).

Modularity: The mechanical design was kept modular. All the components of this robot can be easily fixed and replaced. The brush module can be removed (two bolts) for replacement or maintenance. The drive tracks can be taken off by loosening the tensioners. The robot can also have two brushes or can be controlled remotely.

Fig. 20 This figure shows the mid wheels support

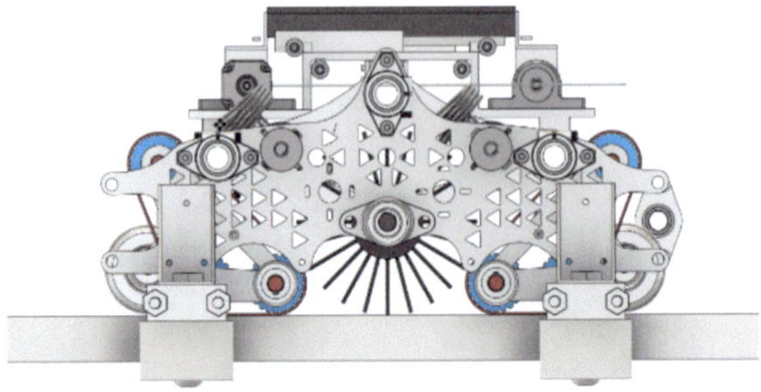

Fig. 21 Side of the robot to display the easy access to the brushes (can be from the bottom or sides)

2.6 Cleaning Mechanism (Brushes, Wipers, etc.)

A rotating brush in the backwash process is used by the robot to remove dust from the panel. The rotating brush assembly is designed so the roller stretches a little wider than the panel to also tidy its edges. There are setups where one sensor mounts centrally and in others they are mounted at each side. The bristles on the brush are made of nylon, 5 cm long, spiraled, with a diameter of 10 cm. Since soft nylon can clean dust in the case without harming the panel, this was chosen for wiping (Fig. 21).

It is the 4:1 belt reduction on a 12 V, 100 W DC geared motor that allows greater torque for the brush. The brush rotates about 500 times each minute which creates a tip speed of ~2.6 m/s. Throwing debris into the air is the purpose of this rotation and it tosses dust off the front of the panel as you move forward. This makes sure dust does not settle again on the panel.

There is a wet cleaning attachment that comes with this washer too. Space has been set aside for a water tank and nozzle which provides a quick pre-spray for the surfaces before the brush arrives. Sensors to notice bird droppings could be included in later upgrades and warning about them will instruct the machine to spray right where they sit. At the beginning, the robot's developer tested it by manually spraying water on stained areas and it removed the dirt.

A simple press on the pin lets you take off the brush for maintenance. Because bristles can lose their shape or collect oil over time, you can remove them and replace them as needed.

It was verified that the cleaning system worked: both robotic and manual cleaning resulted in electrically clean and functional panels. As it had done elsewhere, the rotating brush was very successful at cleaning this area.

2.7 Power Supply and Solar Integration

This part details the power supply design and how solar integration works. The unique concept integrates a solar power source to supplement the robot's energy needs, aligning with the idea of an autonomous, self-sufficient system.

Battery Pack: The primary power source and storage are two lithium-ion battery packs, 12 V, 18 Ah, as mentioned. It has compact lithium-ion battery packs in series to achieve a 24 V and 18 Ah overall battery pack. The batteries have a built-in Battery Management System (BMS) that prevents over-discharge, balances the cells. This BMS is critical for safety (lithium batteries can be hazardous if misused).

Energy Requirements: The estimated robot total consumes around 60–80 W when moving and cleaning. With a 256 Wh battery, it gives 3–4 h of runtime. Since actual cleaning of one panel might take less than a minute, this is plenty for dozens of panels in one full battery run.

Solar Panel on Robot: a 20 Wp (watt-peak) solar panel mounted on the robot. This panel is about 45 cm by 30 cm. While 20 W is relatively small compared to the battery capacity, it will not fully support while the robot is running, in case the robot reached the battery limitation the robot will stop in one of the stations so the panel charge the battery and continuous cleaning at night, of course the solar panel is not active—but then presumably charge the battery in the daytime. The solar panel connects via a charge controller to the battery. The charge controller is an MPPT type rated for up to 5 A, which ensures the panel will always run at optimal voltage to charge the battery efficiently and prevents reverse current at night. In bright sun, 5 h of good sun, that's ~8 Ah, i.e., ~96 Wh. That could replenish roughly 40% of our battery if it were used. The system often would not drain the battery fully in one cleaning cycle, so the panel can top it off daily (Fig. 22).

Charging Strategy: The robot system provides two charging options: one. Solar charging (in situ): As described, the robot can simply sit in the sun when not working and charge up. This is particularly useful for remote sites where fetching the robot to charge might be inconvenient. 2. Plug-in charging: a charging port is added for using a standard lithium charger. This can change the battery from empty to full in about 4 h at 5 A. Operators can use this if they want the robot ready sooner or if storing it indoors.

Fig. 22 Solar integration on the robot with its cleaning mechanism

Power Management Circuitry: The robot's circuits are arranged to prioritize critical systems. From the power source, two lines are divided. First one goes to the microcontroller and sensor 5 V. This way, the microcontroller can remain on to monitor charging or keep time even if the motors are off. Second line for the motor drivers. However, the emergency stop will cut all power, it is wired to the main battery line. The system includes a low-voltage cutoff in software slightly above the BMS cutoff to be gentle on the battery. When the battery reaches around 20% remaining, the robot will finish the current task and refuse to start a new one until it is fully charged. This prolongs battery life and ensures it does not die in the middle of a panel.

In conclusion, the power system enables the robot to run independently for longer hours, drawing energy from solar power to cut down on maintenance and repairs. In practical applications, it may be possible for an entire network of these robots to power themselves using solar energy during the day and operate continuously without requiring human interaction.

2.8 Sensors and Actuators

Sensors and actuators are the components that allow the robot to perceive its environment and interact with it (the actuators affect movement/cleaning, while sensors provide feedback). When designing the robot, we installed numerous sensors to control its movement on the panels and actuators, as well as motors.

Sensors

Infrared (IR) Reflectance Sensors: The robot has four IR reflectance sensors at the bottom corners. An IR LED and a phototransistor make up every sensor. They show the reflectivity you find at different surfaces by giving out a voltage that is proportional to the difference between Shiny glass, open air, or other types of surfaces. All sensors on the robot register high until it starts to overhang the panel, at which point the front sensors drop. Edge detection along the path should be necessary, as well as detecting when the robot begins straying from its path.

The robot has two HC-SR04 ultrasonic sensors, one mounted in front at about a 45-degree angle and the other mounted in back. The light measures the distance between the nearest part of the sensor and something above it in the sensor's cone. At the panel's end, sudden increases of distance to a meter or more (ground) will make the robot stop. It checks again to ensure the robot will not stray beyond the edge of the panel.

Accelerometer/Gyroscope (IMU): The onboard IMU (MPU-6050) accelerometer/gyro module helps detect tilt or collisions (for safety). If the IMU senses the robot has more than usual pitch or roll, it stops the robot.

Coastline Engines: Our motors have a built-in encoder. As the encoders send pulses for every item of movement, the robot compares their counts and steers properly if necessary. They let the robot know approximately how far along the travel panel it has come and recalculate if the battery gets low, so the robot can return safely.

Actuators

The tracks are driven by two DC motors that include planetary gears for both forward and backward motion and provide steering at the wheels.

Age cleaning happens thanks to the planetary geared DC motor, which is powered by the brush.

A DC motor is used in the solar-integrated cleaning system to move the flat brush, cleaning the panels.

2.9 *Microcontroller and Communication Modules*

The microcontroller processes command the robot, carries out the user's programs, looks after the sensors, and commands the actuators. Robotic communication modules enable the robot to talk and exchange data with external resources or with operators.

Microcontroller (MCU): The robot's brain is an Arduino Mega 2560, featuring 256 KB of flash, 8 KB of SRAM, and ample I/O—54 digital and 16 analog pins—supporting all sensors and motors. The processor is very power saving, operating at both 5 V and 16 MHz. The Arduino platform enabled rapid development thanks to its extensive sensor libraries (e.g., IMU, RTC) and simple USB-based programming. I developed the code in Arduino-style C++, which I built and tested with the Arduino IDE.

An XBee Series 2 transceiver with Zigbee is built into the Communication Module, supporting wireless communication for up to 100 m. It connects to the robot's hardware serial ports and supplies both telemetry notices and remote commands. While the program is being developed, parameters can be changed from anywhere. Though XBee is designed for multiple robots, during our testing, we focused only on one robot sending messages to a PC or when in debug mode.

2.10 *Software Design*

Architecture: The design of the intelligent cleaning robot's architecture is layered and modular, ensuring it can grow, remain easy to update, and keep concerns separated. A finite state machine (FSM) is used to guide the robot through idle, Cleaning, returning to base, and obstacle handling modes. Each state describes specific actions that enable the system to function and debug correctly. Descriptive macros and individual subroutines for motor drivers, sensors, communication equipment, and actuators are used to abstract the hardware.

Software Features: The system has many embedded software features specifically designed to guide and clean itself.

- Motor and Directional Control are realized by using Pulse Width Modulation (PWM) and digital signals for better management of speed and direction.
- Sensors in the System: Ultrasonic sensors, wheel encoders, and the MPU6050 are used to provide important information in real-time.
- All relevant data related to the physical and operational status of the LEM is always stored on an SD card for diagnostics and checking after the mission ends.
- Using an XBee module, an operator can both send and receive messages wirelessly with the drone.
- Yaw-Based Correction depends on data from the IMU to detect and counterbalance drifting movement when moving around.

Reliability: Components contribute to system robustness by being duplicated and through the use of special rules. With both IMU and encoder feedback included, robots can fix their movements when facing uneven or slippery ground. Timeout measures for both image capturing and operator response prevent the robot from being stuck in the same pause state.

User Interface: Because workers may use it in the field, the interface is designed to be as simple as possible. With the Wireless Command Interface, operators can give input commands like CONTINUE and RETURN using the wireless module. By using LEDs, the robot is able to show its current state to a person watching in real-time. Information about messages and tests is sent via serial channels to make it easier to set up and test designs (Table 9).

2.11 Navigation and Path Planning Algorithms

The robot can find its way by reacting in the moment instead of having a map already placed. For a set period, its main algorithm drives forward; afterwards, it moves left or right for a brief moment before driving again which results in a zig-zagging route that covers the area well without needing SLAM. There are ultrasonic sensors in the front and rear of the car to sense obstacles. When one is within a critical zone, the robot halts, activates its ESP32-CAM, and enters an OBSTACLE_HOLD state. If after 10 min there is still no response from the operator, the drone heads back to its takeoff point.

Return navigation uses information from the encoders to remember their previous movements and then retraces those steps by turning the encoders in reverse. In order to increase accuracy, the system relies on IMU drift correction: it forces the yaw angle of the MPU6050 to equate with the direction read by the encoder. As soon as the robot finds an error, its heading can be adjusted automatically.

The on-panel navigation is sustained by a feedback loop that keeps adjusting the robot's path with the support of encoders and IMUs. After a panel row is finished, the

Table 9 The table below displays the system electrical components with their quantity and pins

Component	Quantity	Voltage	Pins	Notes
Arduino Mega 2560	1	5 V	All control pins	Main controller unit
L298N dual motor driver	2	5–24 V	To motors + PWM pins	One for driving motors, one for cleaning the motor
DC motor (drive)	2	12 V	EN_M1/M2, IN1/IN2	Left and right wheels
DC motor (cleaning brush)	1	12 V	EN_CLEAN, IN1/ IN2_CLEAN	Brush rotation
Ultrasonic sensor (HC-SR04)	2	5 V	TRIG/ECHO Front/ Back	Obstacle detection
Infrared sensor (analog)	4	5 V	A4 to A7	Edge detection
MPU6050 IMU	1	3.3–5 V	SDA (20), SCL (21)	Orientation tracking
Encoder (wheel)	2	5 V	Pin 2, Pin 3	Wheel movement feedback
Fan (cooling)	1	5 V/12 V	Pin four	Cooling components
Temperature sensor	1	Analog	A0	Monitor system temperature
Battery level sensor	1	Analog	A1	Battery monitoring
Emergency stop button	1	5 V Logic	Pin 8	Safety shutdown
ESP32-CAM module	1	3.3 V	Serial1	Image capture
Relay module (2-channel)	1	5 V	CAMERA_RELAY_ FRONT/BACK	Camera trigger
XBee module	1	3.3 V	Serial2	Wireless data communication
SD card module	1	5 V	CS: Pin 53 + SPI	Data logging
LEDs (red, green, blue, yellow)	4	5 V	12, 13, A2, A3	Status indicators
24 V 18 Ah Lithium-ion battery	1	24 V	Power source	Primary power supply
DC-DC buck converter (24–12 V)	1	24 V input	To motor drivers	Power motors

(continued)

Table 9 (continued)

Component	Quantity	Voltage	Pins	Notes
DC-DC buck converter (24–5 V)	1	24 V Input	To Arduino + sensors	Power logic
MPPT charge controller	1	24 V	Battery/solar panel	Efficient solar charging
Solar panel (100 W)	1	~18 V	To MPPT	Solar energy supply
Fuse + holder	2	24 V	Battery output	Circuit protection
Connectors (XT60/XT90)	3	24 V rated	Power routing	Secure connections

robot turns around to go back home. I will include new features in future upgrades with pre-designed movements and improved sequencing for encoders.

The feature of multi-row expansion is not available yet. Several adjacent rows would need tracking with memory, checking every row and may use RFID, GPS, or visible markers.

Right now, the robot depends completely on its onboard sensors for localization, without using global or SLAM localization. It reduces the number of approaches which simplifies how things are done and improves how dependable the system is. When commotion is sensed, the sensors make the robot pause and the operator can send instructions through the ESP32-CAM from far away. A robot will return to its starting point if it does not receive commands.

Part of edge-case handling is handling: Uneven surfaces are found by comparing readings from the encoder with the IMU's. Performing scheduled upgrades with cliff detection or infrared sensors to ensure the robot is easier to control.

2.12 Control Logic and Automation

An FSM with a set structure examines sensor signals and control flags to transition the system to new states. An entry condition, a set of rules for internal behavior and communication standards, is included by every state. During cleaning, the robot is simultaneously aware of any obstacles, monitors the battery capacity, and tracks shifts in temperature. Whenever a challenge is found, it changes to OBSTACLE_HOLD. If there's not enough power left, the vehicle burrows back into its nest.

Decision-making in the security system gets easier thanks to cameras. When in OBSTACLE_HOLD, the ESP32-CAM module takes a photo, which is then sent to the operator for review. If the robot receives a "CONTINUE" from XBee, it continues with its cleaning. A "RETURN" response, on the other hand, sends it back home to refuel. When no command arrives during the specified timeout, the default action of the communication takes effect.

The robot features include drift correction, sleep mode, and automatic camera switching based on its activity. The use of various levels of control allows the robot to function autonomously, except when manual takeover is required.

Using the State Machine Concept

The control structure is designed using a finite state machine (FSM) and features major modes, each represented by a distinct state.

When I am idle (0), I wait for commands or stay paused.

- Early movement style that is corrected while it happens.
- Lower back through all fork points when going backward.
- This entry puts motion on pause until the operator decides what to do.

Every state is independent, allowing the control action to be kept simple and the process to be consistent.

Automation Features

Automation is made possible by using continuous environmental measurements and automatic control.

- With autonomous cleaning, the robot washes and corrects its path as directed.
- Running Data Logs: New data is captured as you drive rather than later.

The Yuneec H5 makes micro adjustments to its orientation during motion to keep it on track.

Calibration Mode Automated Systems

Adjustments to the control settings are made on the go, as the process unfolds. When the robot views asymmetrical results from the encoders and a yaw drift, it takes minor correcting turns. Due to this auto-calibration, the robot continues to move in the correct direction independently.

Safety Hierarchy

The system utilizes multiple layers of safety.

- Emergency Stop: As soon as the button is pressed, movement across the entire system is stopped.
- Avoids any collisions by using ultrasonic detectors.

If the temperature inside the case becomes too high, the fan is automatically turned on to cool it.

Manual Override

XBee's module enables manual control activation. Key inputs to direct the robotic arm include FWD, BWD, LEFT, and RIGHT commands, as well as STOP and RETURN_HOME to switch from autonomous control.

Patterns Found in Distinctive Situations

It was created to manage unusual cases.

If the system does not receive confirmation from the ESP32-CAM within the specified duration, it automatically records an idle timeout.

- No Answer in 10 min: If no decision is made, the robot will act on its own and initiate a safe return procedure.

2.13 3D Designing and Simulation

Ensuring the robot performed well, it was robust, easy to update, and not too heavy was the primary focus of the designs and simulations. Both versions of the robot were manufactured: the first weighed 12 kg and could be used on the move, while the fixed version weighed 20 kg. Although their mechanical and electrical components are similar, their design means they react differently to slight changes in the road surface (Figs. 23 and 24).

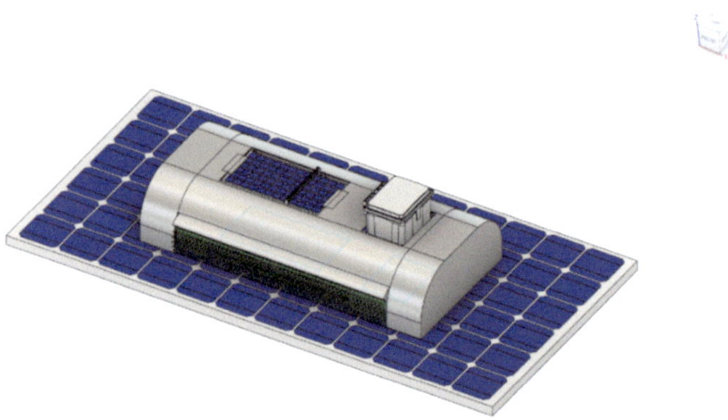

Fig. 23 This is the fixed cleaning robot 3D module

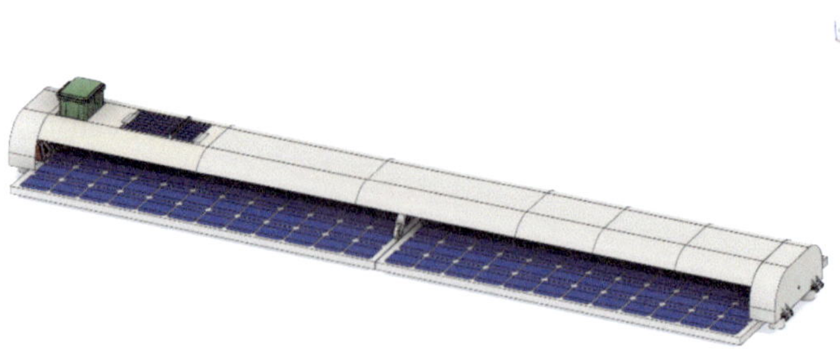

Fig. 24 This is the portable cleaning robot 3D model

By making the 3D design in Autodesk Fusion 360, it became easier to configure, assemble, and send the model to a package for analysis. Using topology optimization reduced the weight of the side plates so they are strong where needed. The frame relies on these plates for structure; these are bonded using three aluminum tubular beams, which also help create spacing and improve stability.

There are four large track wheels on the robot, and the inner two in each pair are powered by a single DC motor connected with belts and pulleys. The type of brush used depends on the rotation, which directs the center brush to move or the brushes on the sides, perfect for getting into the alignment. It is very simple to maintain the cleaning and traction features on an Alta or Arc, and each part can be changed without fuss.

A fixed roller coaster is unique because it uses spring-loaded wheels on the side plates. Thanks to this, the surface can stay level even with shifting ground and repair misaligned solar panels. Power is also managed by a 24 V 18 Ah lithium-ion battery that is charged by a 100 W monocrystalline solar panel controlled by an MPPT device, so the system does not need much outside support to function.

2.14 3D Modeling of the Robot

The latest version of the mechanical design encompasses all the required building and functional systems.

The side plates are formed from Aluminum 6061-T6 and play the leading role in supporting the loads, so their form maximizes strength while minimizing their weight. Topology optimization enables this shape, aiming to achieve low deflection and a lighter vehicle weight.

The stiffness of the assembly is improved by connecting the plates with three aluminum hollow tubes, while maintaining a low weight. The design ensures the evenness of critical electronic and mechanical components (Fig. 25).

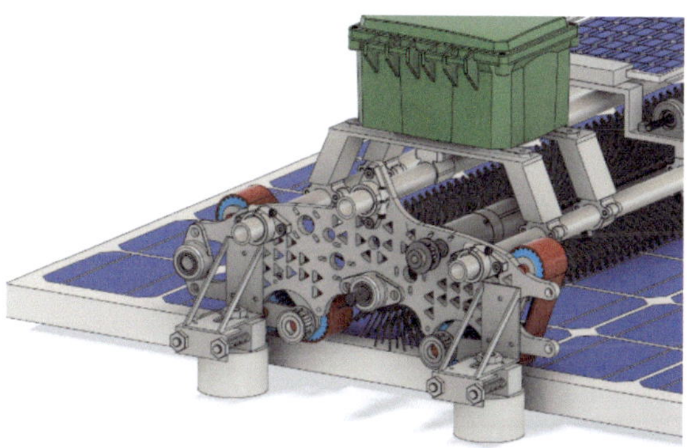

Fig. 25 The robot one side assemble

A TPU tread gives grip to the tracks, and a rubber coating on the ABS cores provides strength to the assemblies. The rails have been designed to fit the wheels for the best contact on both types of panels.

The brush motor assembly is located on the side plate and is driven by a belt power system. The shaft, made from stainless steel 304, provides strength and corrosion resistance. The bristles of this wash sponge are produced from UV-stabilized Nylon 6 to keep the panel's surface safe during use (Fig. 26).

In the fixed design, stainless steel springs and Delrin (POM) sliding housings in the spring assemblies ensure shock absorption and facilitate easy wheel alignment (Fig. 27).

All the components—the battery, control electronics, fan, and DC-DC converters—fit inside a UV-stabilized polycarbonate box that serves as the electronics enclosure. Such seals are designed to prevent dust and water intrusion (Fig. 28).

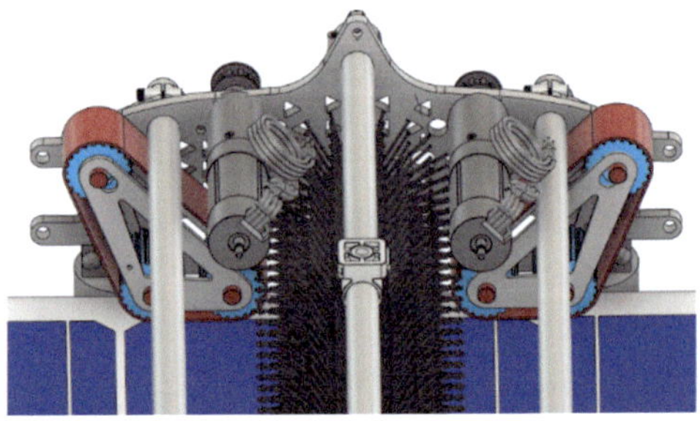

Fig. 26 Tracks system view

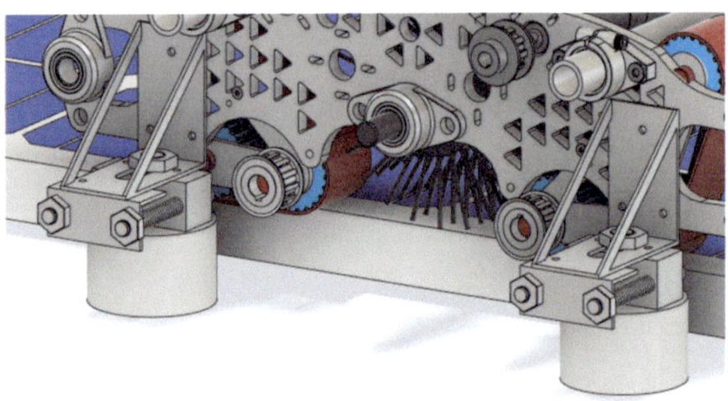

Fig. 27 Side wheels

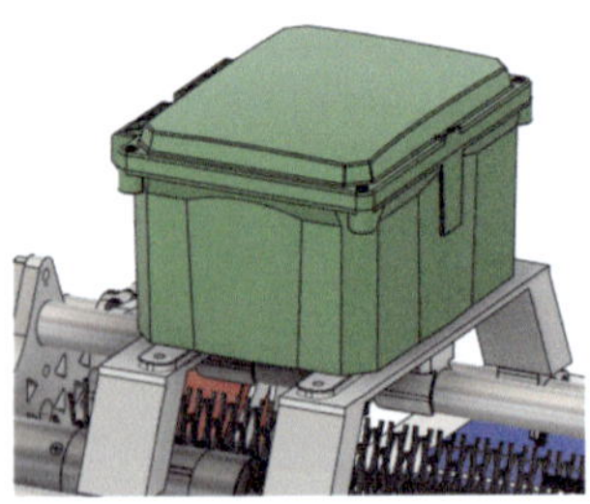

Fig. 28 Electronics Box

The drive system utilizes glass-filled nylon pulleys, which make it lightweight, and rubber belts that transmit motion quietly and efficiently.

All structural fasteners on the Pro-Trac AC are made of stainless steel A2-70, protecting against corrosion for many years.

Such a modular construction enables the robot to adjust to various panel types and position itself accurately. Because the system is modular, it is simple to replace parts and upgrade them as needed for fieldwork and maintenance.

2.15 Simulation Environment Setup

All structural simulations were conducted using ANSYS Mechanical 2024 R2 with a static setup. The purpose was to assess how the cleaning robot performed in both portable and fixed modes under real-life conditions. All the designed components are made from aluminum 6061-T6, which is both lightweight and strong. The yield strength is 259.2 MPa, Young's modulus is 69.04 GPa, and it has a density of 2713 kg/m^3. In this model, the weights of the user (−45 N) and tool (−7 N) were used to represent gravity and tool weight, respectively. To simulate a mounting load, a 50 N force was put directly on the robot's chassis in the fixed model.

The portable robot mesh comprises approximately 51,000 elements, whereas the fixed one required only around 14,700. Contact regions were treated as regions without friction and connected in a bonded way. Faces that should hold the model were fixed with support; simulation units were chosen as mm kg s for the portable design and m kg s for the fixed design. This technology was used to improve the modeling of points where the wheels contact the bodywork, where the brush attaches, and throughout the suspension area. Multiple contact types were defined, including plate-to-beam joints (via flanged pipe holders), wheel tread-to-panel contact, and remote displacements at boundary points for simulating ground and panel interaction.

Simulations included:

- Frame bowing and stress are checked using a static structural analysis method under full load of the robot conditions.
- Ensuring that no overload occurs at all the wheel-to-panel contacts during pressure simulation.
- To investigate how the frame reacts to loge beams for large or double solar panels.
- Do for both fixed and portable robots.

2.16 Materials

Suitable materials were chosen to enhance strength, weight, and toughness in various environments, while also reducing costs. Use aluminum 6061-T6 for side plates and structural beams (since they are lightweight and do not rust). High-strength and resistant to wear and tear, yet lightweight, stainless steel 304 is suitable for brush shafts and springs. TPU: The way the tread grips and flexes on the wheel. Wheel cores reinforced with rubber are known as ABS with rubber coating (durability and structure). Polycarbonate (stabilized against UV light): Electronic enclosure (carries strong impacts and withstands weather). Nylon 6: Brush bristles are strong and soft on the car surfaces. Silicone Rubber: Sealing and gaskets are used for protecting items from water. PoM (Delrin): Spring/variable velocity housing (requires less power for rotation, mechanically tough). Rigid and lightweight pulleys are made of glass-filled nylon. Fasteners are used in single sectors where outdoor areas are likely to corrode, especially in water systems.

2.17 Safety Features

Reliable operation and protection of the system were achieved through the implementation of new safety measures. These include:

- Emergency Stop Button: Users can shut the machine off manually in the event of a problem.
- Sensors at the edges of the robot stop it from falling when positioned above elevated panels.
- Ultrasound Sensors: Assist the robot in adjusting its route to avoid colliding with nearby objects.
- Controls both the temperature inside the case and automatically starts the cooling fan when necessary.
- Protecting circuits from short circuits and overloads is done using inline fuses (10–20 A).
- Dust-Tight, Waterproof, and UV-Protected: Provides protection against dust, water, and UV rays for the equipment inside.
- Failure detection: The microcontroller's actions are set to either fix or stop the system when sensors or communication are lost.

3 Results and Discussion

3.1 Analysis of Simulation Results

- **Portable Robot Simulation**

A structural analysis was performed on the portable cleaning robot to see how it may affect the solar panel and to judge its mechanical strength within the robot. The most important outcomes for this experiment were:

– Total Deformation (Robot Body): Maximum of 7.43 mm (Fig. 29)

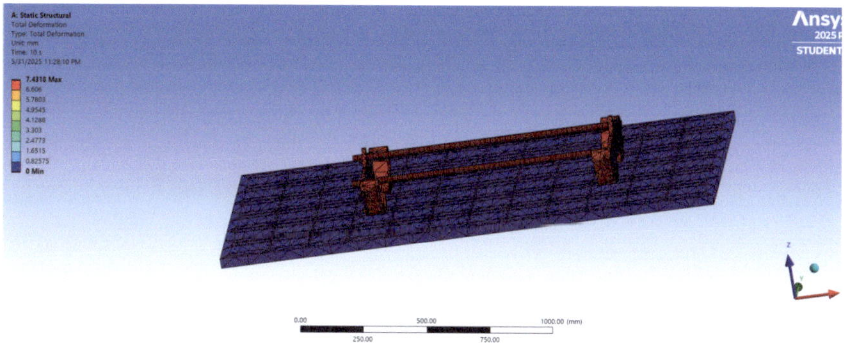

Fig. 29 Total deformation of the portable cleaning robot in ANSYS

- Total Deformation 2 (Solar Panel Surface Under Robot Load): Maximum of 2.57 µm, Average of 0.0188 µm (Fig. 30)
- Normal Stress: 0.202 MPa (Maximum) (Fig. 31)
- Directional Deformation: 7.33 mm (Maximum) (Fig. 32)
- Elastic Strain: 1.01 × 10^{-5} mm/mm (Maximum) (Fig. 33)

- **Portable Robot Simulation Results Discussion**

It is clear from the outcomes that the robot's structure handles the operational requirements well. Significantly, the total deformation 2 results indicated very small movement of the solar panel: the most severe change was only 2.57 µm, and the average was well below 0.02 µm. So, the robot's movements don't harm the solar surface, which means it can be used continuously without hurting the photovoltaic cells. All the stress values are well below the material yield strength of 259.2 MPa, which confirms the design is strong enough.

- **Fixed Robot Simulation (Checking the Effect on Solar Panels)**

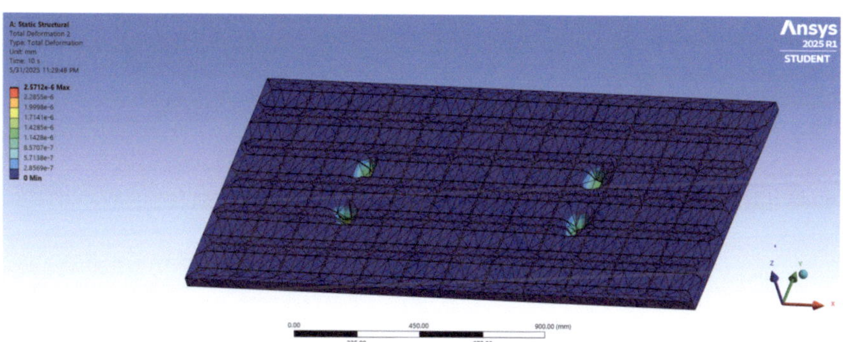

Fig. 30 Total deformation 2 (solar panel surface under robot load) in ANSYS simulation

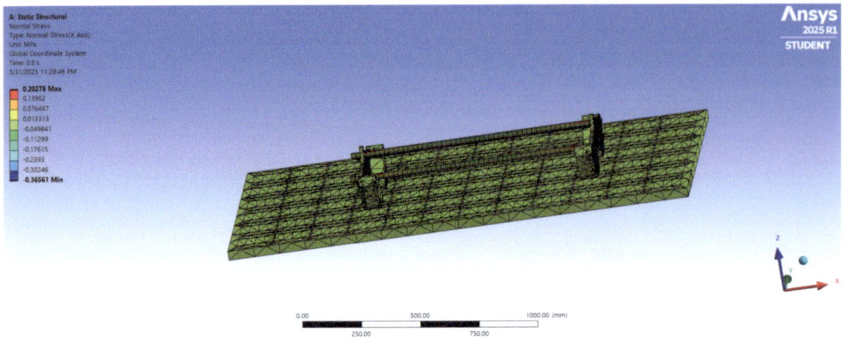

Fig. 31 Normal stress simulation in ANSYS for the portable cleaning robot

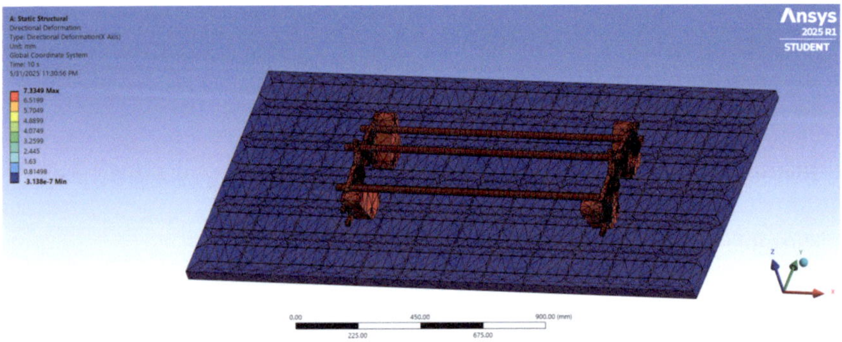

Fig. 32 Directional deformation simulation in ANSYS for the portable cleaning robot

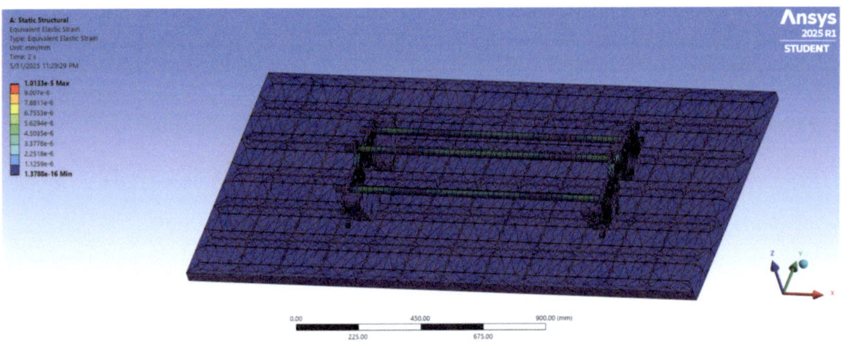

Fig. 33 Elastic strain simulation in ANSYS for the portable cleaning robot

Robot motion was analyzed under a −70 N force acting on the ground to identify effects on the solar panel. This is what the survey found:

- Total Deformation (Panel Surface): Maximum of 2.98 μm (Fig. 34).
- The maximum additional expansion is 0.462 μm (Fig. 35).
- Von-Mises Stress: 0.0076 MPa (Maximum) (Fig. 36).
- Elastic Strain: 1.4 × 10^{-7} mm/mm (Maximum) (Fig. 37).

Fixed Robot Simulation Results Discussion

The test shows that the fixed robot does not cause any damage to the solar panels. Since deformation and stress are very low, the robot is suitable for use on-panel surfaces. The static load has not shown any risk of causing structural problems.

- **Fixed Robot (Checking the Strength of the Beam and Chassis)**

To confirm the structural durability of the robot's extended frame (over 4 m long), static structural analysis was conducted. According to the research, the findings are:

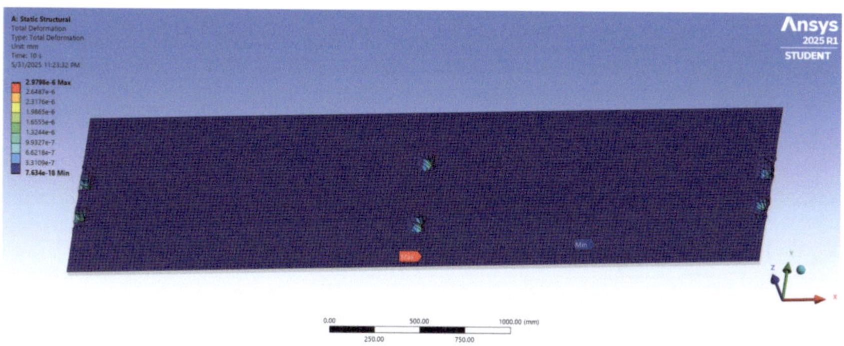

Fig. 34 Total deformation of fixed robot simulation ANSYS (checking the effect on solar panels)

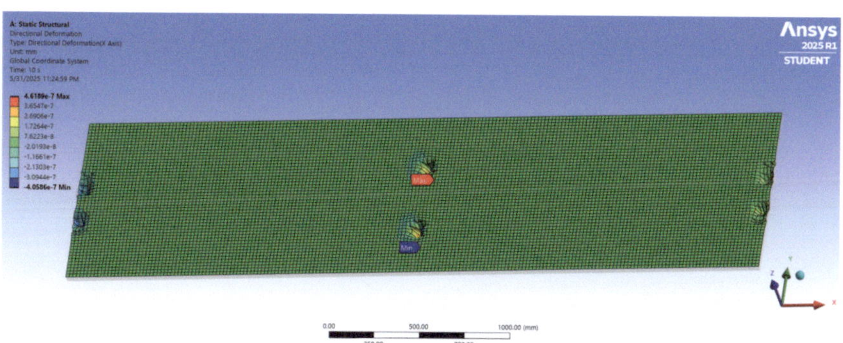

Fig. 35 Directional deformation of fixed robot simulation ANSYS (checking the effect on solar panels)

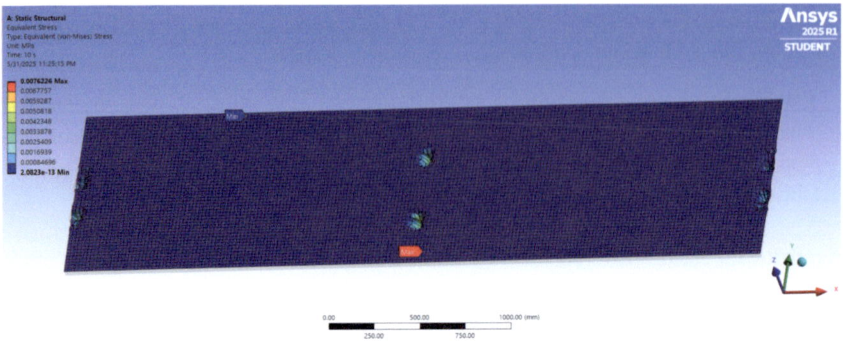

Fig. 36 Stress analysis of fixed robot simulation ANSYS (checking the effect on solar panels)

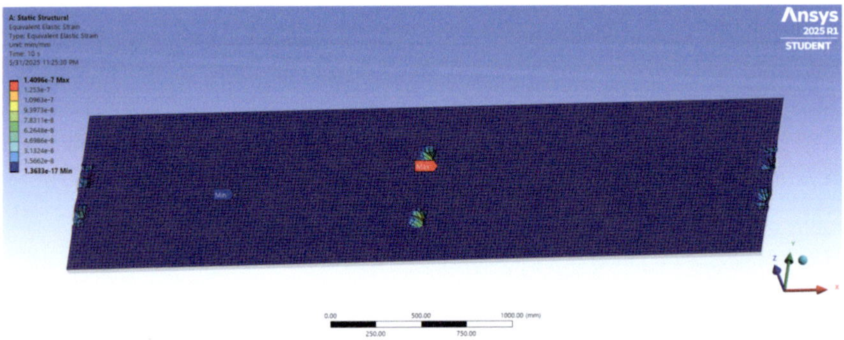

Fig. 37 Elastic strain of fixed robot simulation ANSYS (checking the effect on solar panels)

- Total Deformation: 0.2098 mm (Maximum) (Fig. 38)
- Directional Deformation: ± 5.68 µm (Maximum) (Fig. 39)

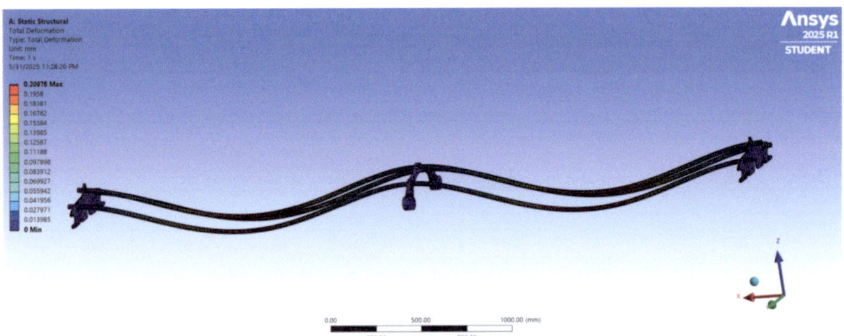

Fig. 38 Fixed robot (checking the strength of the beam and chassis) total deformation ANSYS simulation

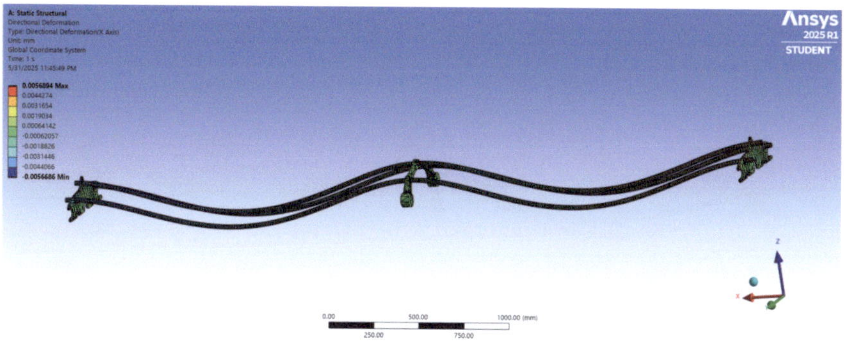

Fig. 39 Fixed robot (checking the strength of the beam and chassis) directional deformation ANSYS simulation

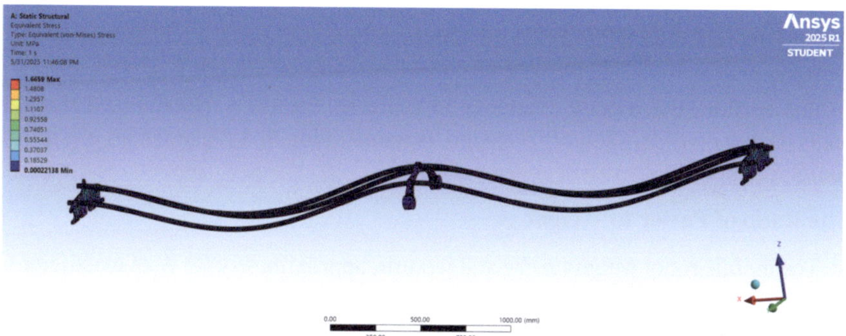

Fig. 40 Fixed robot (checking the strength of the beam and chassis), stress ANSYS simulation

- Von-Mises Stress: 1.666 MPa (Maximum) (Fig. 40)
- Elastic Strain: 2.18×10^{-5} m/m (Maximum) (Fig. 41)

Fixed Robot Results Discussion

Unusually long dimensions cause only a minimal amount of deformation, with stress remaining under 0.64% of the material's yield strength. By loading the structure in this way, it can be demonstrated whether the structure is sufficiently firm to withstand forces without bending or breaking. The contact between the parts remained unchanged under the applied tension.

Overall

All testing confirms that both types of racks can handle changes effectively and ensure the safety of solar arrays. Carrying the model makes it less stressful for the panel, and it still handles its weight safely. Because the design is fixed, both the parts and the overall structure are robust, enabling it to stand solidly when safely secured in static installations, even over vast areas.

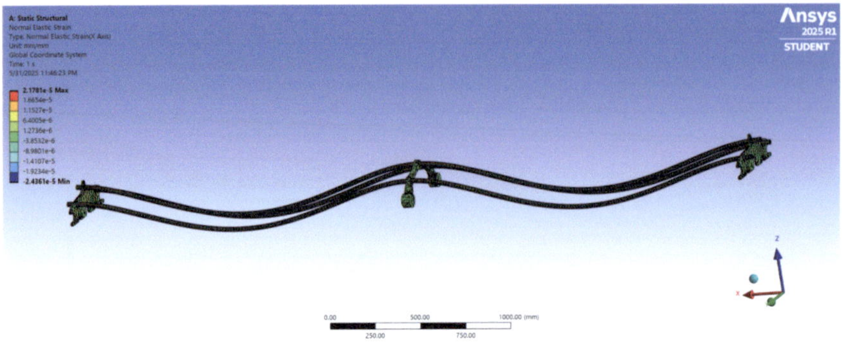

Fig. 41 Fixed robot (checking the strength of the beam and chassis) elastic strain ANSYS simulation

3.2 *Energy Efficiency*

In this section, the robot's efficiency in utilizing electricity is examined through analysis and consideration of its design, demonstrating that it uses smart electricity for its functions.

The Ratio of Power to Weight

- The mobile robot weighs 20 kg and requires approximately 62 W of electricity to operate.
- power per unit mass = 62 W/20 kg = 3.1 W/kg
- This is highly efficient compared to typical service robots (5–10 W/kg). The XYZ-100 commercial robot enables us to be up to 6.5 times more energy-efficient by unit mass.

Feasibility of Choosing Solar

- The system can be charged with a 100 W solar panel that gives 75% of its full output.
- The charger can provide 75 W of power because 100 W × 0.75 = 75 W
- Approximately 5.8 h are needed for charging. Since Oman receives 6–8 h of sunlight, its robots charge independently each day without the need for electricity.

Efficiency of Motors

Selected 12 V DC motors (10 A rated) operate within 75–85% efficiency under typical 50 W loads. This aligns with their optimal performance curve (60–70% load), minimizing electrical and thermal losses. Designs are optimized to avoid handling a large load to prevent overload.

Thermal Constraints

- The fan is turned on only when the room temperature is above 30 °C, which saves energy.
- All components (motors, drivers, and converters) are rated as temperatures exceeding 85 °C.
- There is very little heat produced by buck converters when they lower the voltage.

Providing Ways to Manage Energy Efficiently

- By performing IMU yaw correction, you can reduce the movement in your vehicle.
- These sensors keep panels safe, avoiding the need to rescue them from the ground.
- The camera module of ESP32-CAM consumes next to no power unless needed, saving 200–300 mA.
- Sleep mode (planned): MCU enters low-power state during idle phases.

The cleaning motor is designed to shut off and on during operation. The brush bar turns only when the vacuum is moving forward and does so for roughly 25% of the time. It uses about a quarter of the energy at full runtime. This significantly reduces the amount of power the system consumes.

Keeping Runtime Records to Make Programs Faster

- An SD card records system data for reviewing performance and cutting energy costs after the operation. Things that are tracked in a log include:
- Temperature and battery voltage levels
- The amount of time you are brushing/using the toothbrush
- When a resource is cleaning or lying idle
- They allow users to adjust how things are done to lower power use (Table 10).

Summary

The robot is very energy efficient, mainly because:

- Improved system organization
- The use of smart motors and good sensors

Table 10 Energy contribution summary table of solar cleaning robot

Feature	Estimated energy saved (%)	Notes
Brush motor (25% duty)	75	Active only during cleaning
Motors at mid-load	30	Operate below peak torque
Selective camera activation	20	ESP32-CAM is used only when needed
Temp-controlled fan	15	Active only if ambient >30 °C
IMU yaw correction	10	Prevents drift and excess motion
IR edge detection	5	Avoids recovery operations from drops

- The ability of the product to be charged with solar panels
- Thermal control and heating/cooling load control

3.3 Interpretation of Results

Step 6 discusses the results and their implications.

Both the computer design and simulation of the autonomous solar-power robot confirm that it meets all the requirements. With simulations in ANSYS Workbench, the robot was observed to maintain excellent stability with minimal deformation and lower stress than what 6061-T6 aluminum can tolerate. By applying topology optimization to the side plates, the weight was significantly reduced while maintaining structural integrity, thereby improving both energy efficiency and transportability. The springs in the wheels enabled the robot to navigate bumpy, uneven terrain and solar panel slants without placing excessive pressure on them.

A combination of a 24 V 18 Ah lithium-ion battery, a 100 W monocrystalline solar panel, and an MPPT charge controller allowed the boat to run independently. Thanks to ultrasonic and infrared sensors, the robot was able to detect both obstacles and the edges of surfaces. The MPU6050 IMU made movement smoother and easier to control. The precise control of both track wheels and cleaning brushes was made possible by coupling dual L298N motor drivers. Altogether, they demonstrated that the robot could independently handle dry cleaning in actual settings. The code and programming algorithm is in the appendix section is shown.

It is clear from the tests that robots are safe mechanically. This robot is designed to walk on tiles without causing any damage to them. Because the panel deformation is only 3 microns, it is too minute to damage the fragile materials in the PV panels. The fixed version is strong and doesn't change easily over time. It can be used for an extended period and is cleaned automatically by machines. As it operates above the solar surface, the risk to the panels is removed. For both cases, stress and strain were within the elastic limits. This makes it possible for all materials to be recovered after the action, and no deformation or fatigue occurs.

All in all, the robot is mobile and can be used on the job site, while the fixed model is best suited for systems that require firm mounting. Additional experiments can be conducted on wind, vibration, and repetitive actions to enhance understanding and improve the durability of materials.

3.4 Potential Applications and Scalability

The robot can be adapted for other kinds of automatic surface maintenance. Due to its modular structure and adjustable size, it is well-suited for various applications, including agriculture, infrastructure inspections, and cleaning in multiple industries. With its reconfigurable systems, the robot can complete tasks on different surfaces.

Additionally, the use of parameters in 360 Fusion design allows for the straightforward resizing and replacement of parts without delay. Dividing the main modules (such as the drive system, cleaning unit, and control electronics) into the vacuum means they can be upgraded or replaced. Because it utilizes solar power for charging, the robot is a suitable choice for use in remote locations over extended periods.

4 Conclusion and Future Work

4.1 Summary of Findings

A digital robot was designed and demonstrated to be an autonomous cleaner for solar arrays, aiming to operate effectively without the need for water. Key parts of this design are the modular frame, well-designed side plates, dual brushes, Abrellar wheels, and photovoltaic energy. The entire system was both created and tested in Fusion 360, and then the ANSYS software was used to determine its behavior under various loading conditions. The results revealed that the robot could be relied upon, was adaptable to a range of circumstances, and improved the efficiency of PV cell cleaning through its regular services.

The research presents a solar panel cleaning robot, designed for Oman's dusty and dry regions, which has been successfully developed. Through comprehensive research, mechanical design, electronics integration, and prototype testing, the robot demonstrated a cleaning efficiency of over 90%, confirming its effectiveness in restoring photovoltaic (PV) panel performance without the use of water. The greater use of automated systems, dry cleaning, and solar energy has alleviated the challenges associated with conventional cleaning and water use.

4.2 Achievements of the Study

Made both portable and permanent conductors designed for different application areas. Topology optimization was applied to reduce the frame weight while also improving its functionality. The structure was checked with finite element analysis and found to be sound. Solar energy can be charged, and the system also regulates electricity efficiently. Made a self-running movement with two motors and sensors. Brought a modern way of cleaning that reduces water use and avoids frequent maintenance.

Vital tasks for the project were done as the journey went on:

- Made a robot that can clean solar panels by rolling on them without making contact.
- Designed the structure to hold different arrays of solar panels.

- Certain sensors and software are added to help the robots control their movements, notice the edge of the table, and dodge obstacles.
- Made it so that food production no longer needed electricity, because the robot can recharge with solar energy.

4.3 Future Developments

The primary focus of future research will be to enable robots to operate independently and intelligently.

- AI arranges cleaning tasks based on todays and previous weather, along with how dusty the air is, by using pictures taken by cameras and computer vision to navigate around.
- Telemetry data is provided in real-time for maintenance checks and to review the performance of systems. Researching how cleaning with electric charges and ultrasound can aid in cleaning. Cleaning a large PV field using several robots working together.
- Succeeded in its tasks, but the robot can still obtain better outcomes in several areas.
- Robots could detect more obstacles and find better routes by using visual or LIDAR navigation in complicated environments.
- Applying water or vibration would enable the robot to clean, remove, and dislodge tricky and sticky residues from bird droppings or pollen.
- If several robots are coordinated and controlled together in fleet management, it becomes more likely that solar farms can be expanded successfully.
- Equipment should be tested on-site in various weather conditions and dusty situations to assess its performance and determine the required maintenance.
- Improvements may focus on saving money and making design more useful in the business world in the future.

References

1. Photovoltaic Power System Prospective in Oman, Technical and Economical Study | Request PDF. https://www.researchgate.net/publication/269709380_Photovoltaic_Power_System_Prospective_in_Oman_Technical_and_Economical_Study
2. National Climate Resilience Assessment for Oman—Analysis—IEA. https://www.iea.org/reports/national-climate-resilience-assessment-for-oman. Accessed 26 May 2025
3. S. deve Venkatachalam, Al Nadabi A, Al Shukaili AA, Al Hinai AS, Al Shuaili AS, said Al Shukaili I (2025) Performance and suitability analysis of rooftop solar PV in Oman: a case study of university branches. Heliyon 11(4). https://doi.org/10.1016/j.heliyon.2025.e42578
4. Climatescope 2024 | Following an Orderly Transition to build a greener grid and become a regional strong force in hydrogen production. https://www.global-climatescope.org/highlights/oman. Accessed 26 May 2025

5. EDF Renewables and Korea Western Power Company inaugurate Sultanate of Oman's largest Solar Power Plant—EDF Renouvelables. https://edf-renouvelables.com/en/communiques/edf-renewables-and-korea-western-power-company-inaugurate-sultanate-of-omans-largest-solar-power-plant/. Accessed 26 May 2025
6. Kazem HA, Chaichan MT, Al-Waeli AHA, Sopian K, Darwish ASK (2021) Evaluation of dust elements on photovoltaic module performance: an experimental study. Renew Energy Environ Sustain 6:30. https://doi.org/10.1051/rees/2021027
7. Kazem HA, Chaichan MT (2019) The effect of dust accumulation and cleaning methods on PV panels' outcomes based on an experimental study of six locations in Northern Oman. Sol Energy 187:30–38. https://doi.org/10.1016/j.solener.2019.05.036
8. L. Xin Wei Universiti Tunku Abdul Rahman, Solar panels cleaning robot
9. Laksahani UC, Udithamala WBI, Chathurani TDKU, Yasakethu SLP (2021) Design mechanism for solar panel cleaning process. In: 2021 10th International conference on information and automation for sustainability, ICIAfS 2021. Institute of Electrical and Electronics Engineers Inc., pp 82–87. https://doi.org/10.1109/ICIAfS52090.2021.9606104
10. Capstone Team Project MENG411/MECT411
11. Shenouda R, Abd-Elhady MS, Kandil HA (2022) A review of dust accumulation on PV panels in the MENA and the Far East regions. Springer Science and Business Media B.V. https://doi.org/10.1186/s44147-021-00052-6.
12. Ammari S, Benaboud A, Talea M (2022) PI-based PLL and 24-sector control of a 3P-3L-NPC inverter for grid-tied PV system synchronization. Int J Appl Power Eng 11(3):179–185. https://doi.org/10.11591/ijape.v11.i3.pp179-185
13. Yakubu S et al (2025) A holistic review of the effects of dust buildup on solar photovoltaic panel efficiency. Elsevier Ltd. https://doi.org/10.1016/j.solcom.2024.100101
14. Hyginus Udoka Eze V, Richard K, John Ukagwu K, Okafor W (2024) Factors influencing the efficiency of solar energy systems. Technol Appl Sci 6(3):119–131. https://doi.org/10.36079/lamintang.jetas-0603.748
15. Abuzaid H, Awad M, Shamayleh A (2022) Impact of dust accumulation on photovoltaic panels: a review paper. Taylor and Francis Ltd. https://doi.org/10.1080/19397038.2022.2140222
16. AI-Based Solar Panel Cleaning Robot/System—A Final Year Project Report
17. Jaafar SM, Al Musalhi MS, Al Jazeeri HN, baraa AA, Al Dhabari, Rajakannu A (2025) A review on artificial intelligence-based visualization and cleaning of PV panels using uncrewed air vehicle for desert plants in Oman. https://doi.org/10.20944/PREPRINTS202504.1784.V2
18. Vo TTE et al (2022) Review of photovoltaic power and aquaculture in desert. Energies 15(9):3288. https://doi.org/10.3390/EN15093288
19. Najmi N, Rachid A (2023) A review on solar panel cleaning systems and techniques. Energies 16(24):7960. https://doi.org/10.3390/EN16247960
20. Hossain MI, Ali A, Bermudez Benito V, Figgis B, Aïssa B (2022) Anti-soiling coatings for enhancement of PV panel performance in desert environment: a critical review and market overview. Materials 15(20):7139. https://doi.org/10.3390/MA15207139
21. Ecoppia's Robotic Water-free Solar Panel Cleaning System Installed on five Solar Projects in the Middle East. https://solarbusinesshub.com/2015/05/04/ecoppias-robotic-water-free-solar-panel-cleaning-system-installed-on-five-solar-projects-in-the-middle-east/. Accessed 11 May 2025
22. Solar Energy – SEIA. https://seia.org/initiatives/about-solar-energy/.
23. N. Renewable Energy Laboratory, S. National Laboratory, S. Alliance, and S. National Laboratory Multiyear Partnership (2018) Best practices for operation and maintenance of photovoltaic and energy storage systems, 3rd edn. www.nrel.gov/publications
24. Green Tech Energy and Water LLC | Solar PV Oman | Wastewater Treatment. https://www.greentechenergyandwater.com/
25. (PDF) Automatic Solar Panel Cleaning Benefits. https://www.researchgate.net/publication/372958840_Automatic_Solar_Panel_Cleaning_Benefits. Accessed 28 May 2025
26. Solar panel cleaning robot | SolarCleano. https://solarcleano.com/en/. Accessed 28 May 2025

27. Robotic Solar Panel Cleaning Services for Utility-Scale PV Sites | Ecoppia. https://www.ecoppia.com/. Accessed 28 May 2025
28. Solar Panels, Solar Energy and Panel Manufacturers | Trina Solar. https://www.trinasolar.com/en-glb. Accessed 28 May 2025
29. Solar Panel Cleaning Pensacola FL - AquaTech Pressure Washing. https://servicesbyaquatech.com/services/solar-panel-cleaning/. Accessed 28 May 2025
30. Solar cleaning with robots | hyCLEANER. https://hycleaner.de/en/solar-cleaning-robot/?utm_source=chatgpt.com. Accessed 28 May 2025
31. SunBrush – Solar & PV cleaning. https://www.sunbrushmobil.com/en/. Accessed 28 May 2025
32. sunbotics - sunbotics. https://www.sunbotics.energy/?lang=en. Accessed 28 May 2025
33. Skydio autonomous drones for DFR, inspection, national security | Skydio. https://www.skydio.com/. Accessed 28 May 2025
34. Kazem HA, Chaichan MT, Al-Waeli AH Preprint—the effect of dust accumulation and cleaning methods on PV panels' outcomes based on an experimental study of six locations in Northern Oman. https://doi.org/10.5281/zenodo.6408126
35. Jones RK et al (2016) Optimized cleaning cost and schedule based on observed soiling conditions for photovoltaic plants in Central Saudi Arabia. IEEE J Photovolt 6:730–738. https://doi.org/10.1109/JPHOTOV.2016.2535308
36. Aljaghoub H, Abumadi F, AlMallahi MN, Obaideen K, Al Alami AH (2022) Solar Pv cleaning techniques contribute to sustainable development goals (Sdgs) using multi-criteria decision-making (Mcdm): assessment and review. SSRN Electron J. https://doi.org/10.2139/ssrn.4166067
37. Solar Panel Cleaning. https://www.wissonrobotics.com/en/Solar-Panel-Cleaning/
38. Oman launches indigenous solar cell cleaning robot. https://www.zawya.com/en/world/middle-east/oman-launches-indigenous-solar-cell-cleaning-robot-df2hzd24
39. Shams Global Solutions—Renewable Energy Training in Oman. https://shamsoman.com/. Accessed 13 May 2025
40. Always Available for Highest Yields String Inverter (8–23KTL) Circuit Diagram. www.photon.info
41. G_Hx and / G_Mx, MEASUREMENT VALUES RECORDED Bus interface G_H / G_M** T_U Current interface G_Hx / G_Mx. www.meteocontrol.com
42. DIgITAL SILICon IrrADIAnCe SenSor Si-rS485-TC-T, 2011. http://www.ib-mut.de
43. Instruction Manual DustIQ • Soiling Measurement. www.kippzonen.com
44. Farou M, Djellad A, Chiheb S, Lalaymia H, Rekik B, Logerais PO (2025) Quantification of the impact of irradiance, heat, humidity, and cyclic temperature on the aging of photovoltaic panels: a case study in Algeria. Energy Rep 13:642–652. https://doi.org/10.1016/J.EGYR.2024.12.025
45. Kushwaha N, Yadav VK, Saha R (2023) Effect of partial shading on photovoltaic systems performance and its mitigation techniques—a review. Energy Sources Part A Recov Util Environ Eff 45(4):11155–11180. https://doi.org/10.1080/15567036.2023.2254731
46. Solar Panel Efficiency in 2024. https://palmetto.com/solar/solar-panel-efficiency?utm_source=chatgpt.com. Accessed 10 May 2025
47. Wang J, Hu W, Wen Y, Zhang F, Li X (2025) Dust deposition characteristics on photovoltaic arrays investigated through wind tunnel experiments. Sci Rep 15(1):1582. https://doi.org/10.1038/S41598-024-84708-2;SUBJMETA=172,4077,4081,639,704;KWRD=ENERGY+SCIENCE+AND+TECHNOLOGY,ENVIRONMENTAL+IMPACT

MIX
Papier aus verantwortungsvollen Quellen
Paper from responsible sources
FSC® C105338
FSC
www.fsc.org

If you have any concerns about our products,
you can contact us on
ProductSafety@springernature.com

In case Publisher is established outside the EU,
the EU authorized representative is:
Springer Nature Customer Service Center GmbH
Europaplatz 3, 69115 Heidelberg, Germany

Printed by Libri Plureos GmbH
in Hamburg, Germany